D1527775

Teaching Thermodynamics

Teaching Thermodynamics

Edited by
Jeffery D. Lewins
Fellow of Magdalene College
and Cambridge University
Cambridge, England

Plenum Press • New York and London

Library of Congress Cataloging in Publication Data

Main entry under title:

Teaching thermodynamics.

"Proceedings of a workshop . . . held September 10–21, 1984, in Cambridge, England"
—T.p. verso.
Bibliography: p.
Includes index.
1. Thermodynamics—Study and teaching (Higher)—Congresses. I. Lewins, Jeffery.
QC311.25.T43 1985 536′.7′07 85-25791
ISBN 0-306-42207-7

Proceedings of a workshop on Teaching Thermodynamics,
held September 10–21, 1984, in Cambridge, England

A Division of Plenum Publishing Corporation
233 Spring Street, New York, N.Y. 10013

Printed in the United States of America

PREFACE

It seemed appropriate to arrange a meeting of teachers of thermodynamics in the United Kingdom, a meeting held in the pleasant surroundings of Emmanuel College, Cambridge, in September, 1984. This volume records the ideas put forward by authors, the discussion generated and an account of the action that discussion has initiated. Emphasis was placed on the Teaching of Thermodynamics to degree-level students in their first and second years.

The meeting, a workshop for practitioners in which all were expected to take part, was remarkably well supported. This was notable in the representation of essentially every UK university and polytechnic engaged in teaching engineering thermodynamics and has led to a stimulating spread of ideas. By intention, the emphasis for attendance was put on teachers of engineering concerned with thermodynamics, both mechanical and chemical engineering disciplines. Attendance from others was encouraged but limited as follows: non-engineering academics, 10%, industrialists, 10%. The record of attendance, which will also provide addresses for direct correspondance, will show the broad cover achieved.

I am indeed grateful for the attendance of those outside the engineering departments who in many cases brought a refreshing approach to discussions of the 'how' and 'why' of teaching thermodynamics. It was also notable that many of those speaking from the polytechnics had a more original approach to the *teaching* of thermodynamics than those from conventional universities. The Open University however brought their own special experience to bear.

The record of the Workshop given here is organized in accordance with the six working sessions. Theme and contents of each session are shown in the Table of Contents. The sessions are given in the sequence of the full papers presented to the Workshop, followed by the summaries of short papers discussed in the same session and closed with the record of discussion. I am most grateful to the Session Secretaries who, together with the Session, Chairman, recorded, edited and transcribed the fast moving discussion.

Session One is notable for the theme paper prepared by Professor Bill Woods. In 1980, five years earlier, the Institution of Mechanical Engineers organised a not dissimilar conference in respect of the teaching of Fluid Mechanics and Thermodynamics. That meeting was therefore directed principally to mechanical engineers; the present Workshop sought to integrate views on chemical engineering and engineering science more widely and to omit aspects of fluid mechanics and heat transfer. Nevertheless, it was important to comprehend what the views of teachers were five years earlier and to see what progress has been made. The opening paper served this purpose admirably.

Session Three, held in two parallel sessions, was different from the rest in being devoted to demonstrations of laboratory teaching aids and the usage of microcomputers. Micro- or personal computers were used in two senses: to manipulate data arising from laboratory work and as teaching aids. Several useful programs of the latter sort were made available at the Workshop and can be ordered by the reader direct from the authors reported in Session Three.

The remaining sessions were of the more conventional type and are recorded here in their working order. In one sense the Workshop achieved for its delegates something that at best will only rub off second hand in this more formal record, a sense of *cameraderie* between those who had spent the previous years teaching 'first pass' thermodynamics. Titles in the workshop reflect some of the attitudes of teachers, on the lines 'Why has Thermodynamics become a Difficult Subject?" The act of coming together to evaluate the difficulties had a cathartic effect, removing some of the loneliness that can be felt. It was the evening address by Sir Brian Pippard, FRS, that perhaps brought most relief and solace to the weary; we were comforted to learn that other topics too were difficult to teach; that to the broader view of a physicist, equilibrium systems meant universal helium and our own concerns were somewhat petty metastable and transient considerations!

Why should thermodynamics be difficult to teach? If we compare the concepts involved to those forming the basis of Newtonian mechanics, there are some useful comparisons. The laws of motion have imbued themselves in our civilization for some three hundred years; the laws of thermodynamics for little over one hundred. The difficulties that originally faced the acceptance of Newtonian mechanics are to a large extent those that lie at the heart of thermodynamics. How absurd to say that matter would continue in motion in a straight line when common observation shows mechanisms to run down. Now that our civilization is conditioned to accept Newton, how are we to readapt to Kelvin and Clausius?

The confused history of thermodynamics in the eighteenth and nineteenth centuries has left an immediate legacy of terminology that does not help the student. The teacher, who has studied thermodynamics on at least three levels, has his own difficulties in understanding what the student's difficulties are. The nineteenth century approach to thermodynamics via the process occurring between systems, especially the cyclic process, has left a confusing terminology. If heat is energy in transfer, what is the subject of 'heat transfer'; if heat is that transfer of energy that is not work, have we not *defined* the Second Law that forbids the complete transformation of heat to work; if heat is to do with interactions between systems, why is 'heat capacity' a property and not a process. Terms such as adiabatic (no through path) lack precision; if isothermal means constant *temperature*, why diathermal for *heat*?

An approach favoured by some teachers is to seek rigour. Pippard, in a characteristic remark, warned us that rigour was for mathematicians, not for physicists and certainly not for engineers. Our attempts at rigorous thermodynamics were illusory. What seemed more sensible was to develop the language and laws of thermodynamics in a suitable context. In this regard the work of Haywood and Brazier in developing the 'Law of Stable Equilibrium' as a teaching mechanism seems to me to be of significance. The importance is that one can discuss the nature of systems to include their 'running down' in terms of their properties rather than interactive processes between systems. The material given in Session Four should enable all teachers of thermodynamics to follow up the original Hatsopoulis development in a practical way.

The alternative method of development that was given preference by contributors is perhaps most pungently given by MagLashen to chemical engineers (see bibliography), as an axiomatic development which states the nature of entropy and allows rapid derivation of conventional laws from so succinct a base. There is I think some unease amongst mechanical engineers that the traditional cyclic process route should be a result and not a forerunner of this approach.

In respect of the teaching of foundations, two other areas were explored. Many are concerned with the limitations of conventional thermodynamics to bounds on processes achieved under ideal and vanishingly slow conditions. It must be undesirable to present an optimisation in such impractical terms as the infinitely slow reversible process when real engineering is to do with reconciling capital and running costs. Silver's presentation of irreversible thermodynamics in Session Six then was a welcome attempt to meet this challenge for thermodynamic realism.

The second area concerning foundations would be the place of statistical thermodynamics in an initial course. Our physicist colleagues in many cases have come to grips with the problem, recognising that modern physics is nothing without quantum theory. If indeed quantum theory is accepted, the distinction between heat and work is perhaps made plainer, and more rigorous, than macroscopic processes can ever achieve. Those at the Workshop, with their experience of engineering students were, shall we say, lukewarm to the plea for a statistical starting point. It is puzzling, however, to account for the different capabilities of students of modern physics and modern engineering; one must allow the alternative hypothesis, that the distinction is in the teachers not the taught.

Session Five explored the developing discipline of exergy studies at what would probably be called a 'second pass' level rather than a first teaching level. The terminology 'exergy' is beginning to be accepted in UK after its development on the Continent. Several authors at the workshop attempted an exposition to show how exergy might provide the framework of a practical and useful 'Second Law' analysis, useful in that it would lead to design and synthesis and hence a more valuable role for thermodynamics. We are grateful indeed that the Royal Society should allow us to include Professor Linnhoff's Esso Award Lecture in these proceedings. Perhaps what is needed is a development of nomenclature to distinguish *available* power from power in the same way as exergy distinguishes available energy from energy.

FUTURE WORKING GROUPS

This brings me to discuss the three areas in which The Workshop identified a need for continued study. These three areas were:

(i) The rationalization of nomenclature,
(ii) The evaluation of teaching objectives and methods,
(iii) The development of exergy studies, analysis and synthesis.

Nomenclature

The peculiar terms of our discipline must be seen as one of the difficulties of the subject. Let me put this in the context of the prospect of intelligent or expert systems, computer programs provided with enough information and a set of algorithms that they will guide us through a thermodynamic analysis. What would such a machine make of 'heat', isothermal versus diathermal, or heat capacity? Indeed what do our *teachers* make of heat capacity? One of the most vivid recollections I have of the Workshop is the *identical* arguments put forward by two strong minded teachers, arguments that paralleled each other exactly except that one used the set to promote the term including 'capacity' and the other used identical arguments to exclude 'capacity' from the terminology. (And this twenty years after apparent international agreement on terminology.)

There is a major need to rationalise our terminology. It will be painful for teachers but not I suspect for students. Chemists faced similar problems but have successfully overcome them.

I am glad that one of the initiatives arising from the Workshop is the establishment of Working Group No. 1 on nomenclature and units under the chairmanship of Dr. Y. Mayhew (University of Bristol). In this field, as in the topics of the next two Working Groups, readers are invited to get into contact with the Chairman to contribute to the topic.

Teaching Methods

The second Working Group, under the chairmanship of Dr. W. Kennedy of the Faculty of Technology in the Open

University, has undertaken to review and promote questions of teaching objectives and methods in thermodynamics. Here the new technology associated with computers has importance at many levels. One level might be the availability of computer programs to generate data; another will be the use of computers in a teaching mode if not as an expert system. It was notable that polytechnic teachers had seriously questioned methods of teaching. Proposals for resource-based teaching, relying less on the explicit didatics of the teacher, might indeed have a fundamental role to play.

Exergy Analysis

Finally Dr. T. J. Kotas has undertaken to chair Working Group No. 3 in the prosecution of exergy studies.

Outside the record of the six working sessions, this volume carries the list of those attending with their institutions and a closing bibliography prepared by Professor Button and his colleagues for mechanical engineers with a supplement by Professor Turner for chemical engineers. Quite apart from its value as a record of texts in print, notice might be taken of the suggestion that the act of studying such a bibliography is a learning process, a development in resource based learning.

Copyright difficulties have made it impossible to reproduce the edited text of a reading from the work of Samuel Beckett given after the Workshop dinner by Dr. Nessim Hay and his colleagues. The piece was taken from Lessness and evokes the dead state or the 'entropy death'. Readers who would like to see the original text are referred to the original publication in French as SANS, Les Editons Minuit, Paris, 1969 and the author's translation as LESSNESS, published by Calder and Boyer, London, 1970. A full reading was performed by the BBC in 1971 and a private recording of the presentation to the Workshop was made by Dr. Hay in 1984.

CONCLUSION

I think we may regard the present volume as a valuable source of material relevant to teaching thermodynamics. Many different ideas and some prejudices were ventilated at the meeting; a good teacher should surely be aware of the issues and possibilities open to him. That thermodynamics is an intellectually stimulating discipline is self-evident to those who took part in our discussions. I think we have some hesitation in claiming too important a role for classical thermodynamics in application when the competing claims of rate processes (kinetics) and their connection with capital costs are contrasted with thermodynamics and running costs. Surely the approach should be to take the best of both and to blend them together?

It is intended to review the findings of the Working Groups with expectation of follow-up action that can be reported perhaps to a successor workshop towards the end of this decade. Thermodynamics has its problems and its challenges but it has by no means reached its 'dead state'.

The sentiment for the future of thermodynamics was put superbly by Sir Alan Cottrell in Session Four. Any teacher of our discipline who wants refreshment at the springs of learning would do no better than to taste Sir Alan's refreshing draft.

Teachers of thermodynamics may also want a longer draft from the new philosophical account of non-equilibrium thermodynamics now available in English:

Prigogine, I. and Stengers, I., "Order Out of Chaos: Man's Dialogue with Nature", Heinemann, London, 1984.

Jeffery D. Lewins

ACKNOWLEDGEMENT

We are grateful to the Royal Society for permission to bring the Esso Award lecture, John Wiley for a figure in Session Five and To Ellis Horwood in respect of two papers in Session Five. I have acknowledged the labours of Session Chairman and Secretaries and I would couple with this my thanks to the Steering Committee who guided the Workshop.

Several secretaries assisted me during the preparations for the Workshop but I would particularly thank Mrs. Janis Eagle who typed these proceedings from the submitted papers and draft record of discussion.

CONTENTS

SESSION 1

TEACHING OBJECTIVES

Chairman - Professor W. A. Woods

Secretary - Mr. R. I. Crane

THERMODYNAMICS AND FLUID MECHANICS FOR THE 1980s

(A review of an I. Mech. E. Symposium)

W. A. Woods

Department of Mechanical Engineering
Queen Mary College
University of London

C. A. Bailey

Department of Engineering Science
University of Oxford

The paper outlines the origins and aims of the one-day I. Mech. E. Symposium held in London on 4th December, 1979. The meeting had two main objectives; first, the provision of a forum for the exchange of views on the subject and, secondly, to provide some form of documentation to help those planning and revising syllabuses for use in the 1980s. It is envisaged that this will provide an introduction to the present meeting.

INTRODUCTION

This paper gives an account of a one-day Symposium, sponsored by the Thermodynamics and Fluid Mechanics Group and the Education and Training Group of the Institution of Mechanical Engineers, which was held in London on 4th December, 1979.

Although the Symposium was concerned with both Thermodynamics and Fluid Mechanics, it is hoped that it will provide an introduction to the present workshop on Teaching Thermodynamics. The paper summarises the material presented at the Symposium. The accounts given have been prepared from the speakers' extended synopses which are given in reference (1). The remainder of the paper is based upon observations at the meeting, reports by Session rapporteurs and written contributions; the names of the people concerned are given in Appendix 1.

The structure of the Symposium is followed in this paper, which is:-

1. Fundamentals.
2. Integration of Thermodynamics and Fluid Mechanics.
3. Relative Importance of Fundamentals and Practical Applications.
4. The Value of Laboratories and Computers.
5. Possible New Subject Matter.

In the concluding remarks it is considered that although the majority of engineering students will be required as engineers to deal with a wide range of relatively simple problems in thermodynamics and fluid mechanics, they will need a wider approach including practical design and economic considerations.

A smaller number of students will require new and more advanced courses to take advantage of further developments in the subject.

The integration of EA1 and EA2 engineering courses will no doubt reduce the time for teaching fundamentals, but the EA2 could provide the vehicle by which thermodynamic design, practical design feasibility and economic considerations could be included and put into their proper perspective.

SESSION ONE: FUNDAMENTALS

Basic Thermodynamics (speaker A. P. Hatton)

The speaker in this session stated that definitions and concepts had to be introduced and it took much time to establish a firm understanding of 'heat', 'temperature' and 'irreversible processes'. Some did not appreciate the relevance of what they were doing and this resulted in loss of interest and poor performance. It was suggested that for beginners it was better to stress the practical applications and historical development rather than the philosophical aspects of the subject.

A logical approach such as described by Keenan (2) did not present the subject in the way it developed and it was suggested that the main reason engineers should study Thermodynamics is to determine the amount of mechanical energy that can be obtained from a given quantity of fuel.

The science of Thermodynamics developed from the study of steam engines (3), (4) and coal burning steam engines were in use over a hundred years before the statement of the First Law of Thermodynamics by J. P. Joule in 1845. The basis of the Second Law of Thermodynamics by Sadi Carnot in 1824 (5) was arrived at through his study of steam engines. The speaker suggested that Engineers should appreciate these facts.

The speaker made some comments on the origins of Steam Power and quotes Savery's pump and Newcomen's engine as examples. The speaker had also arranged a display of very good steam engine models to back up his presentation.

Some comments were made on Perpetual Motion machines and that many claims for Patents are still received and some are ingenious (6). The idea of showing a film backwards of irreversible processes brings out visually the degree of irreversibility.

Some comments were made on entropy as a form of currency and an aid to design.

Teaching Thermodynamic Fundamentals
(speaker R. W. Haywood)

The speaker suggested that it was now appropriate to take advantage of recent developments. Thermodynamics was the only important branch of science which was so dependent on so many unproven basic postulates as the zero, first, second and third laws. It was pointed out that there were different ways of expressing these laws; both non-cyclic and cyclic approaches were used.

Thus the question was posed that perhaps they were corollaries of some more basic law.

With the single-axiom approach, Hatsopoulos and Keenan (7) were able to show that the above laws are corollaries of the more basic physical law of Stable Equilibrium (LSE). The speaker explained that he had attempted to produce a more readily understandable book on the topic (8).

The non-cyclic statement of the First Law can be proved as a corollary of the LSE.

The non-cyclic statement of the Second Law declares the impossibility of constructing "a non-cyclic PMM2"; this also can be proved as a corollary of the LSE. Rational definitions of energy and heat follow and the exposition goes on to the cyclic statements of the laws, including the impossibility of a cyclic PMM2.

It was explained that the logical development of the single-axiom approach unifies ideas on concepts which were previously disparate. Starting with non-cyclic processes is helpful in the presentation of concepts and theorems in thermodynamic availability.

The speaker suggested that the theorems of thermodynamic availability have not received any recognition of importance because the first and second laws are normally presented from cyclic statements.

The speaker referred to the efforts of Tribus in promoting the information theory approach to statistical thermodynamics, pioneered by Jaynes and Cox.

The speaker concluded by pointing out that both the above alternative approaches have made little impact in this country but both are worthy of more attention in the 1980s and to help this had produced his book (8).

SESSION TWO; INTEGRATION OF THERMODYNAMICS AND FLUID MECHANICS

The Case for a Degree of Integration
(speaker J. C. Gibbings)

The speaker explained that arrangements within organizations are based somewhat on tradition. Often teaching arrangements are taken for granted. He suggested that in a mechanical engineering course the students are supplied with (a) the basic scientific laws and (b) the application of these laws to (i) Solids and (ii) Fluids.

The topic of electricity is logically separated on the grounds of (a) and then divided on the basis of application to materials and devices. It was also suggested that the division of the subjects of Thermodynamics and Fluid Mechanics is based upon a 'random and irreversible mixture' of (a) and

(b) and the question as to whether this is the best approach does not always arise. The example of what happens to a piece of metal and a piece of rubber when each is stretched - was given as one reason why the question of best arrangement should be addressed.

Separating the subjects dealing with solids, namely, Mechanics of Solids and Mechanics of Machines is more logical since they largely divide into statics and dynamics.

A number of examples were given which supported the case for some merging in the Teaching of Thermodynamics and Fluid Mechanics. They included:-

Generality of the first law of Thermodynamics is not emphasized - e.g.

the way in which it provides the fundamental definition of electric potential and that it requires the assignment of energy to an immaterial force field.

The contravention of the definition of a system, e.g. the analyses with electricity and diffusion.

Teaching dimensional analysis as exclusively Fluid Mechanics, e.g. natural convection and use of more properties as variables than are required to define a state.

Erroneous derivation of reversible work, e.g. assumption of low Reynolds number for pdv.

Poor understanding of the validity pv^{γ} in an irrotational flow, e.g. confusion between irrotational and inviscid conditions.

The invalid statement of the energy relation for a control volume, e.g. the statement that a relation for a fluid machine is the steady-flow energy equation.

The speaker concluded that the case for combining Thermodynamics and Fluid Mechanics in a first year course was overwhelming and he had tried to make this case some years ago (9).

Pitfalls in the Integration of Thermodynamics and Fluid Mechanics
(speaker A. J. Reynolds)

This presentation concerned a substantial course in which the two parts are thoroughly interwoven. The pitfalls were (i) those concerned with the time location in the course, (ii) those intrinsic to the particular course, (iii) students on the course and (iv) interactions with other parts of the curriculum.

In a first year introductory course, there is a risk of giving an 'end on' or stripy presentation. If it is dealyed to the final year, the advantages in understanding associated with an integrated course are lost and teaching time is not saved. Placing an integrated course in the second year could erode the common foundation. There are objections to all possible locations for an integrated course.

The amount of material to be covered is large and an integrated course would have the combined duration of separate courses. Material may be omitted because it did not display thermo/fluid interactions. The omitted topics would have to be located elsewhere. The background of the teacher is also important, the result may be:-

"Thermodynamics with some leakage of fluids or fluid mechanics with some diffusion of thermodynamics".

Perhaps vital elements of component subjects may seem irrelevant for an integrated course, e.g. discussion on irreversibility via cyclic heat engines may seem inappropriate for irreversibility in fluid flow.

It was considered that students have difficulty with understanding Thermodynamics and Fluid Mechanics and it was suggested that the majority of students would have more difficulty with an integrated subject. In addition, there was only a limited range of textbooks attempting to introduce the two subjects in combination.

Links of fluid mechanics to the field of applied mechanics are at least as important as those with thermodynamics. Hydrostatics is part of statics, rotating fluids follow the pattern of rotational dynamics and stress - strain relations for solids and fluids have much in common. Mathematical methods used in fluid mechanics are not closely allied to those of thermodynamics.

One-dimensional gas dynamics provides the most convincing illustration of the interaction between fluid mechanics and thermodynamics - and would form a large fraction of an integrated course. The problem is whether to do more and then find that it may not be needed or to omit further treatment. An integrated course will need more comprehensive revision of both thermodynamics and fluid mechanics.

It was suggested that the combined course is likely to be unstable and destabilize part of the degree program.

SESSION THREE; RELATIVE IMPORTANCE OF FUNDAMENTALS AND PRACTICAL APPLICATIONS

Some Industrial Heat Transfer and Fluid Flow Problems (speaker B. N. Furber)

The speaker first noted with surprise and concern the fact that, although the majority of engineering students would be taking up a professional career in industry, only 7 of the 103 delegates appeared to be from 'industry'.

Then, from his 20 years experience in the Nuclear Power Industry, he presented a few of the problems in thermofluids which had arisen. All reactor systems developed to date involved a heat source, a heat sink and a circulating fluid to transport the heat from the source to the sink. To illustrate the source problems he looked at the difficulty of fuel element heat transfer in gas cooled reactors. For the heat sink, he looked at the banks with optimum heat transfer/ pressure drop performance. For the circulating fluid, he outlined some of the many flow mixing, flow distribution and pressure loss problems. He also outlined the many associated instrumentation problems.

Dr. Furber pointed out that when he trained as an undergraduate the industry he subsequently joined was guessed to be a possibility by only a very select few. Changes now were even more rapid and future progress and requirements are difficult to predict. He, therefore, strongly recommended that student courses should concentrate on the fundamentals of the subject and leave the applications to industry. He also advised that undue reliance on computers should be avoided. No computer can solve a problem the programmer does not understand and there is a tendency to assume that computer output has sanctity which is often quite unjustified. Above all, the engineer is required to be constantly answering the simple question - does it make sense?

Thermo-Fluid Courses - Content and Subsequent Use
(speaker M. G. Cooper)

Although he had been asked to put the academic point of view, the speaker pointed out at the very beginning that he had no disagreement with the views expressed by the previous industrial speaker, who had strongly emphasized the importance of fundamentals. In place of disagreement and fireworks, he planned to lay out some facts and comments in the hope of stimulating discussion on this important topic.

Time on any degree course is limited and, on average, an engineer will each week have 10-12 lectures and 2-3 laboratories. Thus for a 3 year course there will be a total of 600-900 lectures. In a mechanical engineering course about a quarter of these could be thermofluids. In a general engineering course the fraction will be smaller, say one sixth for the student taking a mechanical option and as low as one eighth for the student specialising in some other branch of engineering.

One cannot therefore teach everything and especially for the non-specialist student it is essential to concentrate on the fundamentals. There is no time for dealing in detail with applications, but nevertheless reference should be made to applications in order to illustrate the use of fundamentals.

What are these fundamentals and applications? The following shows the content of a typical course and some (of many) common applications:

Topic	Frequent Applications
Thermo	
1st Law and Steady Flow Energy Equation	vessels turbines
2nd Law and corollaries	cycles, entropy, isentropic efficiency

Properties + Maxwell + mixtures	property tables psychrometry
Reactions 1st law 2nd law	combustion
Heat transfer (conduction, convection and radiation)	many
Fluids	
Hydrostatics	many
Momentum	many
Bernoulli	many
Potential, stream function	cylinder → aerofoil
Fluid machinery	turbomachinery
Compressible flow	shocks, gas tables
Channel flow	channels
Viscous flow	pipe flow
Navier-Stokes equations	boundary layers
Waves	waves
Acoustics	noise suppression

Now the key question - What of this does the student actually use when he becomes a real engineer? For the vast majority of mechanical engineers in industry the answer must be - very little. Probably some incompressible pipe flow using Bernoulli plus friction plus a knowledge of coefficients of discharge and some heat transfer but using mostly standard formulae. In general, he needs to know where to find appropriate formulae and how to recognise them and apply them sensibly. For the small minority who go into high technology industry (e.g. Aero/Turbo or Nuclear) or research, use will be made of advanced aerodynamics and heat flow.

The second category is eye-catching and more interesting for the academics to teach, but we must not allow it to obscure the needs of the vast majority of mechanical engineering students.

There is also the question of balance between the time spent on thermofluids and the other parts of a mechanical engineering course. On that issue we are probably a very biased meeting, but it is an important matter which should be kept under review in the light of present rapidly changing technologies.

SESSION FOUR; THE VALUE OF LABORATORIES AND COMPUTERS

The Impact of the Computer on Teaching of Fluid Mechanics and Heat Transfer (speaker A. D. Gosman)

The speaker proposed that in the 1980s the teaching of fluid mechanics and heat transfer should reflect and exploit the advances of the past decade in three key areas:
a) numerical simulation of thermofluid phenomena, b) computer aided learning and c) low-cost computing equipment.

Increased use of finite difference, finite element and similar techniques to solve the conservation equations of mass, momentum and energy, is being made by universities, government institutions and industry for research and design. It is essential that the teaching curricula should respond to these developments.

If it is accepted that computer-based methods should be introduced for thermofluids teaching and that extra time will not be available, parts of the existing curricula will have to be displaced. This should be done with the aim to equip the student with knowledge of different methods of analysis and the capability to assess the best for a given task. Traditional analytical and empirical methods should be retained for situations where they are superior to computer based ones.

Recent efforts on computer aided learning have focussed on the use of the computer as a simulator of physical proccesses, such as fluid flow and heat transfer. Students are able to carry out computer experiments which are easier and faster to perform, especially when a large number of parameters need investigation.

Simulation methods have been successfully used in the Mechanical Engineering Department at Imperial College for a third year undergraduate course in fluid mechanics and heat transfer and on a second year experiment on transient heat conduction. The simulation was performed in parallel with experiment and both of these generated interest and enthusiasm with students and staff. Further details are reported in refs. (10), (11), (12) and the application to a third year design course in (13).

Implementation of the above ideas requires adequate computing facilities, then interest and enthusiasm of the teaching staff. The first is not difficult to achieve with the low cost of microcomputers. It was suggested that more incentives were needed for staff in these areas.

Program exchange schemes exist and an example is the Engineering Science Program Exchange (ESPE) which is based at Queen Mary College.

The Role of Experimental Work in the Teaching of Fluid Mechanics (speaker, the late Mr. G. Langer).

The speaker suggested that since most fluid flow problems cannot be solved by mathematical analysis alone, laboratory work is essential and the main aims are:-

(1) to convey a physical understanding of the processes of fluid flow,

(2) to demonstrate the interdependence of theoretical analysis and empirical information,

(3) familiarization with experimental methods and measurements,

(4) familiarization with appearance, operation and performance of basic fluid machines,

(5) provide the opportunity for students to learn to write a report on an experiment.

Experiments should be meaningful to students with very little knowledge of the subject. The experiment should enhance physical understanding and the student will need help in knowing what to look for and how to interpret what he or she sees. It is useful to have a number of examples which convey directly the principle under consideration; the hydraulic jump was one such example of a dissipative and irreversible process. Visualization followed by measurement is also very effective.

More advanced experimental work requires more theoretical understanding, perhaps from theoretical models or dimensional analysis, to appreciate the choice of parameters. At the later stage there should be more emphasis on methods of measurement, such as hot wire and Laser Doppler anemonetry, and what principles are involved and the precautions which are needed. An appreciation of the accuracy is important.

The idea of presenting a series of unrelated but simple experiments is useful for the students to explain by recognition of which physical principles are involved. Examples are:-

1. A ball suspended in an inclined air jet (Coanda effect).

2. A plate suspended by the action of a confined radial flow over its upper surface.

3. Super cavitation behind a disc held normal to flow on a water tunnel.

Reference was made to the value of the film loops particularly those produced under the direction of the National Committee for Fluid Mechanics Films (14).

SESSION FIVE; POSSIBLE NEW SUBJECT MATTER

Preliminaries in the Teaching of Irreversible Thermodynamics (Speaker R. S. Silver)

The speaker made a strong plea for the removal in future of the discontinuity between the two conventionally established fields of 'reversible equilibrium thermodynamics' and 'irreversible thermodynamics'. The whole realm of thermodynamics should be a continuum.

He considered the inclusion of irreversibility throughout any thermodynamics course as essential if the students were not to feel unease when dealing with the obvious irreversible non-equilibrium process using mainly reversible equilibrium thermodynamics with a few 'fudge factors.'

The fact that classical thermodynamics had originally evolved from the friction dissipative experiments of Rumford and Joule should not be forgotten.

The discontinuity had unfortunately been widened by placing all modern irreversible thermodynamics on a statistical basis. Statistical analysis is fine for evaluating macroscopic properties in terms of our models, or knowledge of microscopic entities but, as with reversible equilibrium thermodynamics, it should be possible to derive required inter-relations solely by working at the macroscopic level.

The speaker has for some years now been making a contribution towards this objective of a unified presentation on a macroscopic basis (ref. (15), (16), (17), (18)). However until such unification is achieved, the teaching of irreversible thermodynamics to mechanical engineers is probably best done by following the texts of Denbigh (19) and of Benson (20).

The Role of Statistical Thermodynamics in Engineering Applications at High Energy Densities (speaker B. F. Scott)

The speaker began by remarking on the fact that over the last two decades in areas such as power generation and conversion, energy storage, combustion and chemical technology, the applications of arc and energy beams and infra red systems etc., physics based graduates were being increasingly employed in development and design roles instead of mechanical engineers.

There appeared to be a current belief in the power generation industry that physics graduates acquire engineering skills quickly and are then more broadly based and flexible than engineering graduates. That this is so must bring into question, as a contributing factor, current practices in the teaching of thermodynamics and of related topics in fluid mechanics and heat transfer to engineers, for the interface with applied physics lies in these disciplines.

A common feature of the developments referred to is that the physical insight necessary for their practice is more readily achieved by adopting a microscopic rather than a continuum viewpoint of the nature of matter and, more importantly, of energy interactions and associated equilibrium concepts. In physics the foundations laid at the beginning of the century in statistical mechanics, quantum mechanics, the kinetic theory of gases and irreversible thermodynamics were rapidly built upon in an upsurge of interest during the second world war. One result was a steady departure from traditional teaching and a relatively rapid acceptance of statistical descriptions in undergraduate courses in physics.

This was not paralleled by practices in the education of engineers; the view that treats matter as a continuum and the laws of thermodynamics as axioms prevailed, as it still does.

The major task confronting engineering educators in the late 1950s was the establishment of formal teaching in classical thermodynamics (21) to free undergraduate schemes from the "heat engine" straight jacket. Thus, teaching establishments in the United Kingdom, grappling with new concepts, were unreceptive to early American texts in statistical thermodynamics for engineers (e.g. (22)). And, indeed, these texts were greeted with scorn because they proposed that statistical methods should be preceded by a classical treatment and concentrated on the determination of thermodynamic properties with which the student was supposedly familiar. Their purpose was, seemingly, to make complex classical ideas more acceptable by a revelation of their sophistry. This difficulty has been resolved in more recent texts (e.g. (23), (24) and (25)) which propose a parallel exposition of classical and statistical arguments. They rely upon Gibbsian concepts to clarify, for example, work and heat definitions and for descriptions of irreversibility and contain a sufficiency of elementary material in theoretical mechanics and atomic structure. They constitute a realistic basis for a first year course in engineering thermodynamics.

The speaker concluded by reaffirming his opinion that statistical methods can be integrated with classical arguments efficiently and clarify common areas of difficulty in engineering development of thermodynamics. This is the prime benefit at the undergraduate level, although the microscopic viewpoint is a useful aid in subsequent teaching of fluid mechanics and heat transfer. The physical insight gained by more detailed developments of microscopic models is a significant benefit at the postgraduate level and particularly so in applications at high energy densities, where classical description fails.

He believed that engineering graduates equipped with microscopic as well as continuum concepts will find employment in the increasingly critical interface with physical developments forecast in the 1980s.

CONCLUDING REMARKS

In the discussions that followed the presentations various, and sometimes conflicting, views were expressed. It was generally accepted that the majority of engineering students would, later on, have to deal with a wide spectrum of relatively simple problems and the Single-Axiom Approach was too academic for most students.

The need to address overall problems in terms of thermodynamics, practical feasibility and economic considerations was highlighted in several sessions. The meeting attracted a number of written contributions including comments on 'energy economics' by S. S. Wilson. He quoted an extract from Carnot's treatise, which made clear that Carnot himself was well aware of requirements of safety, strength, durability, package space and cost of installation, in relation to the specification of a heat engine.

Having recognised the needs of the majority, new and more advanced courses will still be needed, even if it is for a smaller share of the students, to take advantage of new developments in the subject.

Strong pleas for teaching statistical thermodynamics and for combining non-equilibrium and equilibrium thermodynamics were made in the last session. It is encouraging that this meeting on Teaching Thermodynamics has attracted several papers on new approaches.

The Engineering Council, in taking up Finniston (26) proposals, is to require accredited Engineering courses to have Engineering Applications I (EAI) and Engineering Applications II (EA2) integrated within them (27). This will, no doubt, reduce the time previously available for teaching fundamentals of thermodynamics but the introduction of EA2 could be the vehicle by which the thermodynamic design aspects, the practical design feasibility and economic considerations could be included and put into their proper perspective.

ACKNOWLEDGEMENTS

The authors wish to thank all the members of the Planning Panel and the speakers who contributed to the I. Mech. E. Symposium.

REFERENCES

1. Symposium on "Teaching Thermo-Fluids for the 1980's". Part I: Extended Synopses, Part II, Opening remarks, reports on discussions and written communications, I. Mech. E. 1979.

2. Keenan, J. H., "Thermodynamics", John Wiley and Sons, 1941.

3. Bradley, D., "Thermodynamics - The Daughter of Steam", Engineering Heritage, Vol. 2, I. Mech. E. 1966.

4. "From Watt to Clausius", Heineman, London, 1971.

5. Mendoza, E. (Ed.), "Reflections on the Motive Power of Fire" by Sadi Carnot and papers by E. Clapeyron and R. Clausius, Dover Publications Inc. 1960.

6. Dirks, H., "Perpetuum Mobile" or the search for self motive power during the 17th, 18th and 19th Centuries, E. and F. N. Spon, London 1861 (available only in specialised Libraries).

7. Hatsopoulos, G. and Keenan, J. H., "Principles of General Thermodynamics", John Wiley and Sons, Inc., New York, 1972.

8. Haywood, R. W. "Equilibrium Thermodynamics for Engineers and Scientists", John Wiley and Sons, 1980.

9. Gibbings, J. C., "Thermomechanics", Pergamon Press Ltd., 1970.

10. Gosman, A. D., Launder, B. E., Lockwood, F. C. and Reece, G. J. "A CAL Course in Fluid Mechanics and Heat Transfer", Int. J. Math. Educ. Sci. Technol., 8, 1-16, 1977.

11. Tawney, D. A. (Ed.), "Learning Through Computers", MacMillan, London 1979.

12. Gibson, M. M., Gosman, A. D. and Reece, G. J. "Computer-aided Learning in the Heat Transfer Laboratory - Some Experiences", Computers and Education, 1978.

13. Gosman, A. D., Launder, B. E., Lockwood, F. C., Newton, P. A., Reece, G. J. and Singham, J. R., "Teaching Computer-aided Design of Fluid Flow Engineering Equipment", Proc. CADED Int. Cont. Comput. Aided Design Educ., Teesside Polytechnic, 1977.

14. NCFMF Series of motion pictures depicting fluid-flow phenomena. The films are distributed by the Enyclopaedia Britannica Educational Corp.

15. Silver, R. S., "Introduction to Thermodynamics", Cambridge University Press, 1971.

16. Silver, R. S., "Some Developments in Basic Thermodynamics", Bull. Inst. of Maths and Its Applications, Vol. 7, No. 12, pp.3-7, 1971.

17. Silver, R. S., "Irreversible Thermodynamics of the Steady State with Uniform Disequilibrium," Physics Letters, Vol. 63A, No. 2, pp. 73-75, 1977.

18. Silver, R. S., "Coupling and uncoupling in irreversible thermodynamics", J. Phys. A., Math. Gen. Vol. 12, No. 6, pp. 144-145, 1979.

19. Denbigh, K. G., "Thermodynamics of the Steady State", Lodnon, Methuen, 1950.

20. Benson, R. S., "Advanced Engineering Thermodynamics", London, Pergamon, 1967.

21. Van Wylen, G. J., "Fundamentals of Classical Thermodynamics," John Wiley and Sons, 1959.

22. Lee, J. F., Sears, F. W. and Turcotte, D. L., "Statistical Thermodynamics", Addison- Wesley, 1963.

23. Fay, J. A., "Molecular Thermodynamics", Addison-Wesley, 1965.

24. McQuarrie, D. L., "Statistical Thermodynamics", Harper and Rowe, 1973.

25. Incropera, F. P., "Introduction to Molecular Structure and Thermodynamics", John Wiley and Sons, 1974.

26. 'Engineering our Future', Report of the Committee of Enquiry into the Engineering Profession (The Finniston Report), Her Majesty's Stationary Office, London 1980.

27. 'Enhanced and Extended Undergraduate Engineering Degree Courses', The Engineering Council Discussion Document, August 1983.

APPENDIX I

Symposium Planning Panel

Professor W. A. Woods	Chairman and rapporteur for Session 5
Dr. C. A. Bailey	Rapporteur for Session 1
Mr. G. Bird	Rapporteur for Session 2
Dr. D. C. Chandler	Rapporteur for Session 4
Dr. V. Cole	Rapporteur for Session 3

Speakers at the Symposium

Session 1	Dr. A. P. Hatton Mr. R. W. Haywood
Session 2	Dr. J. C. Gibbings Professor A. J. Reynolds
Session 3	Dr. B. N. Furber Dr. M. Cooper
Session 4	Dr. A. D. Gosman The late Mr. G. Langer
Session 5	Professor R. Silver Professor B. F. Scott

Written contributions

Dr. R. I. Crane
Mr. R. A. Johns
Dr. T. J. Kotas
Dr. J. Patterson
Mr. S. S. Wilson

WHY HAS THERMODYNAMICS BECOME A "DIFFICULT" SUBJECT

B. R. Wakeford

Department of Mechanical Engineering
Heriot-Watt University

INTRODUCTION

Over a period of two or three decades the subject of Applied Thermodynamics has become one presenting considerable difficulty to the average University student, where previously it was generally found to be a reasonably straight forward subject. The author, in common no doubt with many engineers involved in teaching the subject, has been considerably worried by this and has given much thought to the possible reasons for this change.

A number of factors will have contributed to this state of affairs and it is believed that the following are the most significant.

(a) Changes in the cross-section of the population taking engineering degrees.

(b) Changes in attitudes to education in school.

(c) The method of teaching the basic theory of Applied Thermodynamics.

POPULATION ON ENGINEERING DEGREE COURSES

Thirty years ago the most common academic qualification for entrance to the corporate grades of the Professional Institutions was an HNC with endorsements. A first degree was, of course, readily accepted but degree courses were generally intended to meet the needs of those destined for the R and D side of the profession and were concerned with the scientific reasoning required in this area. The HNC courses were more concerned with the knowhow required by the average engineer, including designers.

As a result of the persistent demand for more engineers with academically high qualifications, the number of University places was considerably increased during the 1960s and early 1970s and CNAA degree courses were introduced. The consequent run down of Higher National courses and their eventual loss of status as qualifications acceptable to the Institutions has resulted in a degree in a suitable discipline being virtually the only recognised qualification for admission to corporate membership at the present time.

In the author's experience the objectives and attitudes to teaching of degree courses have not changed significantly. The population now taking degree courses has changed significantly and now includes those who, had they been at their present stage of development twenty or so years ago, would have been engaged on National Certificate courses together with a large number who probably would not have entered the profession at all.

In consequence many students taking degree courses today are not so "academically" inclined as was the norm thirty years ago. They are more concerned to know how to do things rather than with the underlying reasons. This is not to say that the less academically inclined are any less intelligent than their fellows; in some instances the reverse may be true. They do have different attitudes and tend to think differently. Should this different point of view be pandered to in the way that the subject is taught? Since engineering is a world, and as such takes all sorts to make it, maybe it should!

CHANGED ATTITUDES TO EDUCATION

The trend in primary and secondary education has been away from the elite or the vocational and towards the objective that education should be interesting and fun for all. Comprehensive schools have now become the "norm" in the U.K. with non-competitive admission and, in theory at least, with mixed ability classes. Whilst the principles which have directed these trends are laudable the results have posed new problems for the Universities.

In practice few young people find a rigorous, reasoned approach from fundamental principles at all interesting; yet this is what is required of a University student. The topical and somewhat superficial approach to a subject is, for most, as much as they will find of interest. University staff who teach subjects like Applied Thermodynamics cannot help but feel that relevant subjects have been taught superficially or even that large sections of the syllabuses have not been taught at all. How often do staff come across poor English grammar, equally poor mathematical grammar, a lack of practice in forming a logical approach to a problem and a disregard for the need, or inability, to produce a decent diagram? Each of these weaknesses leads to problems in understanding any technical subject; but especially one like Applied Thermodynamics where sound practice in all of the above mentioned areas is essential.

Physics syllabuses have been greatly extended in topical areas over the last three decades. They now contain a considerable proportion of electronics and also nuclear physics. It is apparent that less time is now available for the "heat" part of the syllabus which is not topical. While preparing this paper the author asked a group of his students if they had ever, themselves, carried out an experiment to determine the absolute zero of temperature; about 40% claimed to have done so. In answer to the question as to if they had performed a calorimetric experiment to determine a specific heat of some substance, only 30% claimed to have done so and it was of interest to note that many of the 30% were among those who had done the absolute zero temperature experiment.

The author considers it disgraceful that every student in a Mechanical Engineering Department has not carried out both of these experiments. He also considers it relevant to the difficulties which students encounter in Applied Thermodynamics.

CHANGES IN APPLIED THERMODYNAMICS TEACHING

During the 1950s the subject name was changed from "Heat Engines" to "Applied Thermodynamics" and the teaching method changed; the "Keenan" (1) approach was generally adopted as the basis. On being introduced to the new method, on his return to the academic world after a number of years in industry, the author thought it superb. It had the advantage of getting quickly into the mainstream of the subject allowing more time for applications, an important consideration since many developments in existing and new types of thermodynamic power plant took place around that time.

The degree students of the day seemed to cope with the new method, at least where examination results were concerned. The same cannot be said, on average, for those taking essentially the same courses today. However, since he never taught by the old method the author cannot be certain that students in the 1950s found the new approach as straight-forward as he once found the old one; neither can he be certain that he would himself have found the new method as exciting if first introduced to the subject in this way.

Today more attention is paid to thermodynamic relations and to heat transfer. This has reduced the time devoted to the fundamentals of the subject and to applications in the power plant field.

In the degree courses of which the author has had direct experience the time spent in laboratories, in all subjects, was drastically reduced in the 1960s. At this time it became fashionable for staff, as well as students, to regard laboratory work as a waste of time.

CAN THE DIFFICULTIES BE OVERCOME?

Clearly the population of degree courses is unlikely to change in nature over a short period. Changes in the school systems take several years to show results; in any case it is generally feared, among University staff, that changes currently taking place in the Scottish education system and also those projected for the English system will only make the situation worse.

The only short term solution to the difficulties therefore lies in what is taught and how it is taught in the degree courses. In the remainder of the paper some of the author's views on this are presented.

PRESENTATION OF THE SUBJECT

To the most academically inclined student the method of presentation of a subject is not too important. Indeed (unnecessary) complexities in the arguments used lead to discussion and to a better understanding. With the less academically inclined student this is not so and in the early stages the subject must be presented in a way that builds on knowledge and experience which he has already gained and accepts.

By definition any scientifically based subject is developed through observation of nature, on experiments conducted in order to determine what nature will do in special circumstances and, often, on observation of the performance of manufactured products. This should never be allowed to be forgotten or in any way obscured. Much of thermodynamic teaching should be conducted in the laboratory; the student should himself carry out a number of experiments to convince him of the fundamentals of the subject and then in applications to, for example, types of power plant.

THE FIRST LAW OF THERMODYNAMICS

For a quarter of a century those involved with Applied Thermodynamics in the U.K. have seemed to seek to convince others that the subject is based on laws which are superior to those which govern other sciences (much to the amusement of the practitioners of other disciplines). Two text books which have been extensively used in degree courses over this period (2, 3) are unnecessarily complex regarding the First Law because of this approach.

In fact the First Law is of very little significance, serving only as a short-cut to determining the nett work from a thermodynamic cycle once the heat transfers associated with the cycle have been established. Wrangham (4) seems to have recognised this in granting it a single sentence and then calling upon the general, and far more powerful, Law of Conservation of Energy. The First Law is an obvious corollary of the Law of Energy Conservation and this is recognised by two more recent American texts (5, 6). Keenan (1) himself accepts, with some reserve, that the two laws are equivalent.

Work is, by definition in any context, a form of energy in transition in relation to a system. It is a useful device to avoid having to define what happens to the surroundings as a result of changes taking place within the system. For Joule's experiments relating work and heat to have any significance, it was necessary to measure the quantity of work (energy) entering the system and for a given measured amount of energy entering to have a definite effect on the properties of the system (e.g. the raising of the temperature of a given mass of water in a non-flow system at constant pressure). Now assume that energy is not conserved; on entering the system it is just destroyed and the temperature rise is purely incidental. Then heat entering the system, although it may also produce the temperature rise, might be something quite different; caloric maybe! Indeed if nature had deemed it possible for work to produce a certain form of NUCLEAR reaction, energy would not only have disappeared but a FALL in temperature would have resulted with an immeasurable increase in mass. How would that have been interpreted?

Joule undoubtedly recognised Conservation of Energy since the Law was postulated in mechanics some 40 years before his first papers were published (5). Once it is accepted that the energy entering the system as work is conserved and becomes a part of the energy content, then the whole sequence makes sense. Since a given input of energy causes a definite rise in the energy content of the system and with it a definite change in the measurable properties of the system which define its state, it follows that the energy content is itself a property; the internal energy. If a quantity of heat entering the system now produces the same rise in temperature it follows that the internal energy of the system has risen by the same amount and hence, through conservation of energy, the heat transfer is also energy and is the same quantity as was the work transfer. Heat is seen to be a form of energy in transition, but due to a temperature differential rather than a pressure differential.

Those who deny that they use Conservation of Energy tacitly assume it in stating the First Law and yet again when they purport to show that the summation of heat and work during a process equates to a property (internal energy) change of a system. The trouble is that this complicated approach can be very confusing to the average student of today. He will already be aware of Conservation of Energy in more than one other subject and it will be readily acceptable to him. The First Law readily follows as a corollary.

Having defined heat and work as two different forms of energy in transition and assuming the Law of Conservation of Energy it follows logically that the summed difference between heat and work (using the accepted sign conventions) during a process is equal to the change of energy content of the system. Considering a complete cycle, the First Law is apparent. What could be simpler? There is no need to make it more complex.

THE SYSTEM

The "system" is usually defined as a volume enclosed by a boundary and containing the quantity of substance under investigation. Many teachers then go on to refer to 'closed' and 'open' systems. The term "open system" is contradictory to the above definition of the "system" since at no time does the "open system" contain all of the substance under investigation; indeed in a steady flow situation it is the stuff entering and the stuff leaving which is being investigated.

The author would commend the "system" and "control volume" approach (3) in order to avoid confusion in the minds of students.

THE SECOND LAW

The Plank statement of the Law is the one most often quoted since it is immediately applicable to predictions of power plant performance. It is not, however, the most obvious form of the Law; indeed the author does not find it at all obvious despite his years of experience. On the other hand it has been the student's experience from the instant of his birth that energy flows from a hotter to a colder body and he can accept this as the obvious. The Clausius statement is therefore commended; the Plank form is easily proved as a corollary and the full might of the law can be applied through the consequent corollaries to establish the concepts of entropy and availability.

THE THERMODYNAMIC TEMPERATURE SCALE

The real difficulty here is the necessity of relating numerically the abstract scale to an empirical one in a way which will make sense in a practical situation. The author would advocate that every student should perform the experiment with a simple gas thermometer to establish Charles's Law (7) and to estimate the position of the absolute zero on the empirical scale. The absolute zero is itself abstract yet an experimentally obtained value can be put to it; a lesson to be learned from the experiment which would be of considerable value to the student!

The calibration of the abstract Thermodynamic Temperature Scale should then follow logically and without any misgivings. The use of the tool so formed can then be made apparent.

CALORIMETRY

A number of simple experiments to establish specific heat of liquids, solids and gases and calorific values (or internal energies and enthalpies of reaction) are recommended. Specific heat is established as being heat transfer to unit mass divided by consequent temperature change and the relationships for c_v and c_p in terms of internal energy and enthalpy respectively follow. With the experiments in mind this is easy to understand and provides a satisfactory engineering tool.

CONCLUSION

Changes which have taken place in population of degree courses and school education together with unnecessary complexity being introduced into teaching the fundamentals of the subject have been blamed for Applied Thermodynamics having become a subject presenting difficulty to the average student.

A simplified approach is recommended based upon laboratory experiments to be carried out by the students and with more time being devoted to assimilation of the fundamentals.

Although less time will be left available for in-depth applications it is expected that a better knowledge and understanding of the fundamentals at an earlier period in the course will result in a quicker understanding in the applications. It is believed that a better average standard will be achieved at graduation.

REFERENCES

1. Keenan, J. H. "Thermodynamics", Wiley, 1941, 8-14.

2. Rogers, G. F. C. and Mayhew, Y. R., "Engineering Thermodynamics, Work and Heat Transfer", third edition, Longman, 1980, 14-19.

3. Spalding, D. B. and Cole, E. H., "Engineering Thermodynamics", third edition, Arnold, 1973, 95-107.

4. Wrangham, D. A., "The Theory and Practice of Heat Engines", Cambridge University Press, 1942, 16, 19.

5. Tribus, M., "Thermostatics and Thermodynamics", Van Nostrand, 1961, 136.

6. Faires, V. M. and Simmang, C. M., "Thermodynamics", Macmillan, 1970, 79.

7. Duncan, J. and Starling, S. G., "Text-book of Physics", Macmillan, 1947, 401-410.

TEACHING THERMODYNAMICS TO FIRST YEAR ENGINEERS

C. Gurney

Brooklyn
Habertonford
Devon, U.K.

The paper is based on lectures given to the combined first year engineer class at the University of Hong Kong in the years 1971 - 1973. In a rational presentation of thermodynamics, concepts of heat and temperature are consequent upon, and not preliminary to the Second Law. In the present paper, the Laws of Thermodynamics are presented as the conservation of energy and the non-decrease of entropy of isolated systems.

1. INTRODUCTION

For a discussion of the historical development of thermodynamics, and of its vagaries, the reader is referred to the foreword to the text of Hatsopoulos and Keenan (1). Here we adopt the ideas of Mach (2), that explanation consists in interpreting unfamiliar experience and ideas in terms of familiar experience and ideas; and that the principles of science are an economical description of experience. It follows that explanation is relative to culture. It is currently inappropriate to present thermodynamics initially in terms of excluded perpetual motion, or in terms of heat. Propositions concerning these ideas are best presented as deductions from the theory of energy and entropy.

Increasing restriction of interaction between a system and its surroundings reduces the number of independent elements necessary to specify a particle of the system. Dependence may be described by equations; these may always be presented in the form that some function of the elements is constant. We thus arrive at constancy of mass for closed systems, constancy of momenta for closed and mechanically isolated systems, and constancy of energy, for fully isolated systems, (systems which do not respond to changes in their environment). Aspects of constancy during change are well suited to the human intellect; but any metaphysical context may be eliminated by referring to the above as closure equations.

A weaker form of the dependence of the elements of a system may be presented as a functional inequality; and this form too is well suited to the human intellect, since man has a sense of sooner and later, which he expresses as monotonic time increase. It is thus acceptable to students, to describe possible future states of an isolated system in terms of the monotonic increase of a function of the elements - the entropy. The laws of thermodynamics were thus reduced by Clausius (3) to energy and entropy principles.

Although thermodynamics is of much wider application, it is best introduced to engineers as an extension of rigid body mechanics to deal generally with deformable bodies and chemical processes. A particle is described by position $\underset{\sim}{x}$ and velocity $\underset{\sim}{q}$ (external coordinates) relative to an inertial frame; and internally by specific volume $v = 1/\rho$, by the pressure p, and by the degree of advancement of each chemical process $\xi(1 \geqslant \xi \geqslant 0)$. For simplicity we consider only one chemical process, and we take p as approximately isotropic so that shear stresses are small. Mass concentration of the stoichiometric part is c_s; and of each of the e excess chemical species is $c_x(x = 1,2-)$: both of these are represented by $c_i(i = s,x)$. We note that

$$\sum_1^{e+1} c_i = c_s + \sum_1^{e} c_x = 1 \; ; \quad \frac{\partial c_s}{\partial c_x} = -1 \qquad (1.1)$$

and that $p\ v\ \xi\ c_x$ are independent internal elements or coordinates, measurable by the instruments of mechanics or chemistry. Masses are written as M with suitable suffixes.

2. THE FIRST LAW

The Conservation of Energy

A class of scalar functions of the thermodynamic coordinates are called energy functions. The kinetic energy function is their exemplar, and is defined by,

$$\text{Kin} = \int_0^M \text{kin}\ dM = \int_0^M \tfrac{1}{2}\ \underset{\sim}{q}\cdot\underset{\sim}{q}\ dM \qquad (2.1)$$

Any other functions, which when added to Kin, give a valid relation between the coordinates are also called energy functions. In a constant long range force field of strength k parallel to the z axis, a potential energy function is appropriate. This is defined by,

$$\text{Pot} = \int_0^M \text{pot}\ dM = \int_0^M kz dM \qquad (2.2)$$

From Newtonian Mechanics we know that a description of the motion of a rigid body α in such a force field is

$$[\text{Pot} + \text{Kin} = N = \text{constant}]^{\alpha} \qquad (2.3)$$

We note that Pot and Kin are relative to external axes, and their sum may be called the external energy N n. A body ν so isolated, that only changes in its external coordinates occur, is conveniently called an inertial body. When bodies deform, the numerical value of the sum of their external energies may change. Experiment shows however, that a valid description of the variation of the thermodynamic coordinates of an isolated system of bodies, is obtained by introducing a function U u of the internal coordinates of particles. U is called the internal energy function. The behaviour of two interacting bodies α β in an isolating envelope, can be described by

$$[(U + N) = \int_0^M (u + n)\ dM = \int_0^M e dM = E \text{ constant}]^{\alpha\beta} \qquad (2.4)$$

where E is the total energy (small letters are mass-specific).

The First Law states that changes in the thermodynamic coordinates of a system of interacting bodies can be described in terms of the constancy of the sum of the energy functions of all parts of the system; or more succinctly, but less informatively, "the energy of an isolated system of bodies remains constant". The functional form of the internal energy must be found from experiments. Those suggested here are not necessarily convenient in practice; but practical experiments must be consistent in theory with them.

The internal energy function of $p\ v\ \xi\ c_x$.

In practice the internal energy function of a substance may be very complex and it is not usually presented analytically; instead it is replaced by tables of numerical values at finite increments of the variables. However it is notionally convenient to suggest experiments which give the partial derivatives of u with respect to internal coordinates. To find $\partial u/\partial p$ we conduct experiments in which there is interchange between external and internal energies. Some of Joule's experiments are examples. If p is the only internal coordinate of a fluid α to change, then

$$(\partial u/\partial p)^{\alpha} = \mathrm{Lt}_{\Delta p \to o} - \frac{\Delta n}{\Delta p} \qquad (2.5)$$

Suggestions for finding the other derivatives of u are given by Gurney (4), (5). Knowing the derivatives, we can notionally integrate to find $u(p\ v\ \xi\ c_x)$, or in practice tabulate u over some practical range of the coordinates. For gases at large molar volumes, and over a limited range of $p_x v\ \xi$ experiment shows that the molar specific internal energy u^x is given by (with a b c constant)

$$u^x - u^x_o = pv^x(a\xi + b) - p_o v^x_o(a\xi_o + b) + u^x_o c(\xi - \xi_o) \qquad (2.6)$$

The Effect of Size of Measuring Instruments

Measuring instruments do some sort of averaging over areas, volumes, and times; but if they are small enough, they will record some of the kin and pot associated with Brownian motion, which experiment shows to be a function of the same coordinates as u. On the other hand if measuring instruments are too coarse they will underestimate kin and pot. In practice, the size of each instrument should be chosen to suit (2.4) in which u represents all the energy which is a function of internal coordinates.

3. THE SECOND LAW

Introduction

We may imagine many impossible mechanical and chemical processes, which are nevertheless consistent with the First Law; examples are given by Gurney (4) (5) with emphasis on their careful specification. Typically, the external energy of all interacting bodies, when a piece of chalk is dropped, decreases, and their internal energy increases. A burnt match does not reconstitute itself, so that $d\xi \not> 0$.

Experience suggests that all such inequalities for an isolated system of bodies, $\alpha\beta$, can be codified by

$$\left(\frac{dS}{dt}\right)^{\alpha\beta} = \frac{dS^{\alpha}}{dt} + \frac{dS^{\beta}}{dt} = \left[\int_{0}^{V} \frac{ds}{dt} \frac{dV}{v}\right]^{\alpha} + \left[\int_{0}^{V} \frac{ds}{dt} \frac{dV}{v}\right]^{\beta} \geq 0 \qquad (3.1)$$

where s is a function of the same elements as u, and t is time. The function S was invented by Clausius, who named it entropy. Its meaning lies in its use. Relation (3.1) represents the monotonic overall aging of isolated systems-wearing, decaying, corroding, burning, heating and cooling etc. By overall, we mean that renovation of a part of α of an isolated system, is accompanied by aging of other parts β. Imagined processes for which $dS^{\alpha\beta} = 0$, typically ideal mechanical processes, and some processes which occur as a sequence of equilibrium states, are called reversible.

In words the Second Law States, "Changes in the internal coordinates of a system of bodies, are such that the sum of the entropy functions of all interacting bodies is stationary or increases". Expression (3.1) is a minimal statement, and may be made more precise in special cases. (Prigogine (6)).

The first task associated with the Second Law is the discovery of the function $s(p\ v\ \xi\ c_x)$ for substances; or in practice the generation of suitable tables. Since (3.1) expresses an inequality, the method of accomplishing this task is obscure. However, for an isolated system, in however chaotic an initial state, experience suggests that a time invariant equilibrium is eventually reached, so that near equilibrium S is stationary. It is found experimentally that the $s(p\ v\ \xi\ c_x)$ found from stationary conditions is applicable with useful accuracy over a wide range of practical conditions for fluids.

Equilibrium of a System of Bodies - the Entropy Function.

We consider a system of deformable bodies, initially in a chaotic state, and bounded by a rigid isolating envelope of volume V, which may include some empty space. Then the energy of the system relative to an inertial frame is

$$E = \int_0^V e \frac{dV}{v} = \int_0^V \frac{1}{v} \{u + \tfrac{1}{2} \underset{\sim}{q} . \underset{\sim}{q} + j(x_i)\} dV \tag{3.2}$$

where $j(x_i)$ is the energy of configuration of long range forces, typically the mutual gravity of the system. The linear momentum is represented by the vector

$$\underset{\sim}{L} = \int_0^V \frac{\underset{\sim}{\ell}}{v} dV = \int_0^V \underset{\sim}{q} \frac{dV}{v} \tag{3.3}$$

The angular momentum is represented by the vector

$$\underset{\sim}{\Theta} = \int_0^V \frac{\underset{\sim}{\theta}}{v} dV = \int_0^V (\underset{\sim}{r} x \underset{\sim}{q}) \frac{dV}{v} = \int_0^V \varepsilon_{ijk}\, r_j\, q_k \frac{dV}{v} \underset{\sim}{I} \tag{3.4}$$

where $\underset{\sim}{I}$ is an orthogonal triad of unit vectors, and ε_{ijk} is the alternating tensor.

The mass of each chemical element (j) may be expressed as

$$M_j = \int_0^V (k_j c_s + \Sigma_1^e\, k_{jx} c_x) \frac{dV}{v} = \int_0^V \Sigma_1^{e+1}\, k_i c_i \frac{dV}{v} \tag{3.5}$$

where the c's are described in section 1, and where k_{jx} are constants, depending on the relative molar mass of each element; k_j depends in addition, on the stoichiometric coefficients of the chemical process.

From the initial chaotic state, with all variables randomly distributed throughout the volume, we consider the approach to equilibrium. The entropy S will approach a maximum, subject to constancy of the energy E, the linear momentum $\underset{\sim}{L}$, the angular momentum $\underset{\sim}{\Theta}$, and the mass of each chemical element M_j. The conditional maximum of S is found by introducing Lagrangian multipliers, which allow equations containing additive terms of otherwise physically incompatible dimensions to be meaningful. We thus consider the integral

$$W' = \int_0^V w' dV = \int_0^V \frac{1}{v}(s + \alpha' e + \underset{\sim}{\beta}' . \underset{\sim}{\ell} + \underset{\sim}{\gamma}' . \underset{\sim}{\theta} + \Sigma_1^{e+1}\, \varepsilon_i' c_i) dV \tag{3.6}$$

where $\alpha' - \gamma'$ are Lagrangian multipliers, and ε'_i are compounds of Lagrangian multipliers and of the coefficnets k of (3.5). Since s is a scalar, we note that the multipliers $\underset{\sim}{\beta}'$ and $\underset{\sim}{\gamma}'$ are vectors, so as to generate scalar products with the vectors $\underset{\sim}{\ell}$ and $\underset{\sim}{\theta}$. The system includes deformable bodies, whose internal coordinates are for the moment left general.

The stationary value of W' is found by equating to zero derivatives of w' with respect to the independent variables. It should be noted that the physical dimensions of the bracketed terms of (3.6) are those of specific entropy, and are therefore as yet unfamiliar. It is thought that the reader may be more comfortable with an expression whose physical dimensions are those of specific energy. This may be achieved, by dividing (3.6) by α' and by resymbolising the new fractional multipliers.

Thus we shall obtain identical results by considering stationary values of

$$W = \int_o^V w dV = \int_o^V \frac{1}{v} (\alpha s + e + \underset{\sim}{\beta} . \underset{\sim}{\ell} + \underset{\sim}{\gamma} . \underset{\sim}{\theta} + \Sigma_1^{e+1} \varepsilon_i c_i) dV \qquad (3.7)$$

When we are not interested in the effects of different chemical substances, we may replace the last term of the above by the single multiplier ε, which will thus ensure that the total mass of the system is constant. Comparing (3.6) and (3.7) we see that the process of finding a stationary condition for the entropy at given energy etc. is identical with the process of finding a stationary condition for the energy at given entropy etc. This idea is due to Gibbs (7). We shall work from (3.7) but the reader is at liberty to use (3.6) if he prefers.

(a) Variation of w with q

We deal first with mechanical processes and consider the stationary variation of w with respect to q_i: thus

$$\frac{\partial w}{\partial q_i} = \frac{1}{v} \frac{\partial}{\partial q_i} (e + \underset{\sim}{\beta} . \underset{\sim}{\ell} + \underset{\sim}{\gamma} . \underset{\sim}{\theta}) = 0 \qquad (3.8)$$

where the derivatives of s and c with respect to q are zero. Thus referring to (3.2, 3.3, 3.4), we find,

$$q_i + \beta_i - \varepsilon_{ijk} r_j \gamma_k = 0; \ (ijk - 123) \qquad (3.9)$$

Calculation easily shows that the time rate of change of distance between two points x'_i and x_i is zero, so that (3.9) represents a rigid body movement, which we know by Chasles's Theorem to be that of a nut on a screw.

The equilibrium axial and angular velocities are easily shown to be respectively $-\beta$ and $-\gamma$. We shall call this rigid body motion of deformable bodies, kinetic equilibrium. In passing we note that the motion of the solar system is not that of a rigid body, but there is some historical evidence and every expectation that it is evolving toward rigid body motion. An agent of this evolution is irreversible tidal action.

The equilibrium kinetic energy is

$$\int_0^V \text{kin}\,\frac{dV}{v} = \int_0^V \left(\tfrac{1}{2}\beta^2 + \tfrac{1}{2}\gamma^2 r^2\right)\frac{dV}{v} \tag{3.10}$$

where r is the radial distance to a particle, measured from the axis of the screw.

The kinetic contribution to wv is (from 3.2 - 3.7)

$$\left(\tfrac{1}{2}\beta^2 + \tfrac{1}{2}\gamma^2 r^2\right) - \beta^2 - \gamma^2 r^2 = -\,\text{kin} \tag{3.11}$$

We now consider the variation of W with internal coordinates, considered for simplicity to be $p\ v\ \xi\ c_x$.

(b) Variation of w with p.

Since only s and u vary with p we have,

$$\frac{\partial w}{\partial p} = \frac{1}{v}\left(\alpha\,\frac{\partial s}{\partial p} + \frac{\partial u}{\partial p}\right) = 0 \tag{3.12}$$

Thus

$$\frac{\dfrac{\partial u}{\partial p}}{\dfrac{\partial s}{\partial p}} = \frac{\partial u}{\partial s} \equiv T = -\,\alpha \tag{3.13}$$

where T is defined by (3.13). We name it the thermodynamic or Kelvin temperature; at equilibrium we see that this function is uniform throughout the volume. Later it will be easy to see that an ideal gas thermometer may record the Kelvin temperature. Equality of temperature is called thermal equilibrium, and (3.13) leads to a statement of the Zeroth Law: "Two bodies in thermal equilibrium with a third body, are in thermal equilibrium with each other".

We note that the deduction of rigid body movement at equilibrium, and the deduction of uniformity of temperature, are independent; so that we could have thermal equilibrium, before rigid body movement is attained, and vice versa.

Equation (3.13) suggests that it may be more convenient, (although not necessary), to eliminate p as independent variable, and to use s instead. We therefore continue with $s\ v\ q\ \xi\ c$ as independent variables, and differentiating w with respect to s we again find expression (3.13).

Variation with the other coordinates is treated in Gurney (5), where it is shown that equilibrium of a simple fluid capable of undergoing a chemical process is characterised by:

Kinetic equilibrium; rigid body motion or motionless.

Thermal equilibrium; $\partial u/\partial s = T$ and is spatially uniform.

Chemical equilibrium; $\partial u/\partial \xi = -a$ (affinity) $= 0$.

Diffusion equilibrium; (μ_i - kin + pot) is spatially uniform, where μ_i is intrinsic or chemical potential of substances s or x; or of each chemical species.

Equilibrium with respect to density; $\partial u/\partial v = w$. Connection with Newton's Second Law, and subject to the first two or three of the above:

$$(\partial u/\partial v)_{s\xi c_x} = -p; (\partial u/\partial v)_{sc_x} = -p \qquad (3.14)$$

These equilibria are all independent, and states of partial equilibria may exist, satisfying some but not all of the above conditions.

By differentiation of $u(s\ v\ \xi\ c_x)$ we can calculate $T\ p\ a$ and μ_x. Alternatively, if we know these derivatives, we can find $s(u\ v\ \xi\ c_x)$ by integrating, subject to thermal and kinetic equilibrium,

$$ds = \frac{du + pdv + ad\xi - \Sigma_1^{e+1}\ \mu_i dc_i}{T} \tag{3.15}$$

and hence knowing $u(p\ v\ \xi\ c_x)$ we can find $s(p\ v\ \xi\ c_x)$, or in practice we can tabulate s at increments of the variables. The discovery of s and of other functions is treated further in Gurney (4) (5).

Some Inequalities near Equilibrium

So far we have considered only the consequences of the entropy function's being stationary at equilibrium. We now consider the consequencies of its being maximum, so that the hyper-surface $w'(s\ v\ q_i\ \xi\ c_x)$ is convex.

(a) To prove $T > 0$

Differentiating w' of 3.6 twice with respect to q_i we find near equilibrium

$$\frac{\partial^2 w'}{\partial q_i{}^2} = \frac{\alpha'}{v}\frac{\partial^2 e}{\partial q_i{}^2} = \frac{\alpha'}{v} = -\frac{1}{Tv} \tag{3.16}$$

Hence for w' convex

$$-\frac{1}{Tv} < 0\ ;\quad T > 0 \tag{3.17}$$

(b) To prove $\left(\frac{\partial s}{\partial T}\right)_{v\xi c_x} > 0$

Referring to (3.6)

$$\frac{\partial^2 w'}{\partial s^2} = \frac{\alpha'}{v}\frac{\partial^2 e}{\partial s^2} = \frac{\alpha'}{v}\frac{\partial^2 u}{\partial s^2} = \frac{-1}{TV}\frac{\partial T}{\partial s} \tag{3.18}$$

Hence for w' convex

$$-\frac{1}{Tv}\frac{\partial T}{\partial s} < 0\ ;\quad \frac{\partial T}{\partial s} > 0\ ;\quad \frac{\partial s}{\partial T} > 0 \tag{3.19}$$

Other well known inequalities similarly proved are

$$T\left(\frac{\partial s}{\partial T}\right)_{p\xi c_x} > T\left(\frac{\partial s}{\partial T}\right)_{v\xi c_x} > 0 \; ; \; \left(\frac{\partial a}{\partial \xi}\right)_{vsc_x} < 0 \qquad (3.20)$$

$$\left(\frac{\partial p}{\partial v}\right)_{s\xi c_x} < \left(\frac{\partial p}{\partial v}\right)_{T\xi c_x} < 0$$

There are very many other inequalities similarly to be discovered.

(a) Another form of the Second Law.

We see from (3.16) and (3.17) that α' is < 0. It follows that w of (3.7) is of opposite sign to w' of (3.6). Hence since W' is maximum at equilibrium, W is minimum at equilibrium. The form of W therefore minimises E at constant entropy etc. We therefore state that the energy of a system with constant total volume, entropy and momenta, and with no diffusion of matter across its envelope, decreases or is constant. The above is an alternative form of the Second Law. To maintain the entropy constant, in a system in which entropy generating processes occur, it is necessary to provide cooling (see Section Four). The system is then no longer completely isolated.

4. HEAT AND WORK

General

The inconvenience of considering only isolated systems lead to ideas of heat and work which we define as follows. Heat Q and work W are interactions across closed interfaces between bodies; they involve energy transfer, but not interfacial transfer of matter. Heat is such that the direction of energy interchange is in opposition to the temperature gradient, while the direction of work interchange is independent of the temperature gradient.

In general, increments of heat and work are inexact differentials, and their sign is not implicit in the change in magnitude of a function of thermodynamic coordinates, but must be separately defined. If the energy of a body increases due to interactions, we write them as Q_{in} and W_{in}; and if it decreases, we write them as Q_{out} and W_{out}. The definition of an interaction between bodies α and β requires

$$dQin^{\alpha} + dQin^{\beta} = 0;\ dQin^{\alpha} = -\ dQ^{\beta}in = dQ^{\beta}out \qquad (4.1)$$

and similarly for W. Then

$$[(dQ + dW)_{in} = dU + dN]^{\alpha};\ [(dQ + dW)_{in} = (dU + dN)]^{\beta} \qquad (4.2)$$

where for a chemical system in a gravitational field U is $U(s\ v\ \xi\ M_i)$ and N is $N(q\ z\ M)$.

Heat Work and Entropy for Bodies in General

Since we have specified that u and s are functions of the same coordinates, we may write using (4.2) and with $r = v\ \xi\ c_x$, or other relevant coordinates:

$$Tds = dwin + dqin - dn - \int_{o}^{1} (\Sigma \partial u/\partial r\ dr)dM \qquad (4.3)$$

We can now show that heat interactions can occur without corresponding changes in any other term to the right of (4.3). Consider two interacting particles α and β, and of masses ΔM^{α} and ΔM^{β}, and so small that their coordinates are effectively uniform. Then,

$$\left[(\Delta Mds)\text{diathermal} = \frac{dQin}{T}\right]^{\alpha};\ \left[(\Delta Mds)\text{diathermal} = \frac{dQin}{T}\right]^{\beta} \qquad (4.4)$$

$$\left[(\Delta Mds)^{\alpha} + (\Delta Mds)^{\beta}\right]\text{diathermal} = dQ_{in}^{\alpha}\ (\frac{1}{T^{\alpha}} - \frac{1}{T^{\beta}}) \qquad (4.5)$$

Our definition of heat ensures that this entropy change is positive. The solely diathermal interaction is thus possible (dia = across). For closed bodies let us write,

$$[ds = ds\ (\text{diathermal}) + ds(\text{adiathermal})]^{\alpha} \qquad (4.6)$$

$$\text{where, } Tds\ (\text{adiathermal}) = dw_{in} - dn - \int_{o}^{1}(\Sigma\ \partial u/\partial r\ dr)dM \qquad (4.7)$$

Then, for adiathermal interactions between two otherwise isolated particles, the Second Law requires

$$\left[(\Delta Mds)^{\alpha} + (\Delta Mds)^{\beta}\right]\ \text{adiathermal} \geqslant 0 \qquad (4.8)$$

Substituting in (4.7) gives

$$\left[\Delta M \frac{(\sim)}{T}\right]^{\alpha} + \left[\Delta M \frac{(\sim)}{T}\right]^{\beta} \gtrless 0 \qquad (4.9)$$

Since by definition, adiathermal interactions are independent of temperature, the two terms of (4.9) may be unrelated, and in general,

$$ds^{\alpha}(\text{adiathermal}) \gtrless 0; \; ds^{\beta}(\text{adiathermal}) \gtrless 0 \qquad (4.10)$$

It is instructive to write ds(adiathermal) as ds^{+} (pronounced ds plus). Then

$$\left[ds = \frac{dqin}{T} + ds^{+}\right]^{\alpha}; \quad \left[ds = \frac{dqin}{T} + ds^{+}\right]^{\beta} \qquad (4.11)$$

where $ds^{+} \geqslant 0$; or $ds^{\alpha} \geqslant (\frac{dqin}{T})^{\alpha}$; $ds^{\beta} \geqslant (\frac{dqin}{T})^{\beta}$ (4.12)

Actions which contribute to ds^{+} are: reducing temperature differences, approach to kinetic equilibrium, surface frictional effects, non-equilibrium chemical processes, and internal diffusion effects. For a bulk body α

$$\left[dS = dS\ (\text{diathermal}) + dS^{+}\right]^{\alpha} \qquad (4.13)$$

and for a body at uniform temperature,

$$\left[dS = \frac{dQin}{T} + dS^{+}\right]^{\alpha} ; \; dS^{\alpha} \geqslant \frac{dQin}{T^{\alpha}} \qquad (4.14)$$

Expression (4.14) is a sharper statement of the Second Law than (3.1) and is called the Clausius inequality. By placing notional envelopes of bulk bodies away from surfaces where there is friction, or sudden temperature change, or other entropy generating process, we may adopt Prigogine's (6) notation, and write,

$$\left[dS = d_eS + d_iS \; ; \; d_iS \geqslant 0\right]^{\alpha} \qquad (4.15)$$

where d_eS is the flow of entropy from outside, and d_iS is the entropy generated within α. Bodies for which dQin = Tds, and for which dWin = dN are respectively called heat and work reservoirs.

5. CLOSURE

The Hong Kong course continued with the treatment of exergy, and of heat engines, refrigerators and heat pumps. Treatment of internal combustion engines was in terms of ideal gas reactions, including evaluation of affinities. The final topic was heat transfer. The treatment outlined eliminates some topics of current text-books, including the Zeroth Law, Carathéodory treatment of the Second Law, and air standard cycles.

This course of forty lectures was fast moving and logical, and it was enjoyed by student and teacher. Examination results were good. A parallel course in properties of materials dealt with atomic modelling, and statistical treatment, as well as thermodynamic properties arising from phase equilibria.

REFERENCES

1. Hatsopoulos, G. N. and Keenan, J. H., "General Thermodynamics", Wiley, 1965.

2. Mach, E. "Conservation of Energy," Open Court, 1911.

3. Clausius, R. "Mechanical Theory of Heat", Macmillan, 1879.

4. Gurney, C. "Pure and Applied Chemistry", 22, 519-525. 1970.

5. Gurney, C. A fuller version of this paper 1984 (available on request).

6. Prigogine, I. "Thermodynamics of Irreversible Process", 3rd ed. Wiley, 1967.

7. Gibbs, J. W. "Collected Works 1", 56, Longmans Green, 1928.

BETTER THERMODYNAMICS TEACHING NEEDS MORE THERMODYNAMICS RESEARCH

J.R. Himsworth

Dept. of Chemical Engineering, U.M.I.S.T.
Manchester

INTRODUCTION: WHY THERMODYNAMICS IS DIFFICULT TO TEACH

There is a limit to the number of possible explanations of why engineering students should find thermodynamics difficult to understand.

The source of the difficulty might be thought to reside in the students themselves or in their teachers. Doubtless this is true in individual cases but the problem is so widespread, and of such long standing, that any general explanation along these lines must be unconvincing.

The only other possibility is that there is something about the subject itself which makes it difficult to understand. Three primary sources of difficulty are suggested below.

THE SUBJECT MATTER IS ABSTRACT

Thermodynamics is, for the most part, concerned with concepts; that is, with abstract and intangible quantities. Although mathematicians and theoretical physicists are trained for, and at ease with, matters of this sort the same cannot be said for most engineers.

THE INTELLECTUAL APPROACH IS ARCHAIC

Anybody today who reads Carnot's seminal paper for the first time will be astonished by how many thermodynamic concepts stem from this single source (1). Still more will they be astonished by the way in which Carnot set the tone - the intellectual 'style' as it were - which has been followed by most thermodynamicists since his time.

The mode of reasoning that Carnot employed is rather reminiscent of Euclidean geometry. We must suppose that this style of thought was perfectly familiar and natural to Carnot and his contemporaries. It is certainly neither familiar nor natural to students, or their teachers, over 150 years later.

IT IS TOO NARROWLY CONCEIVED FOR ENGINEERING PURPOSES

The thermodynamics that is taught to engineering students starts off by excluding the effects of motion, gravity, electric and magnetic fields, and all irreversible effects. But practicing engineers are concerned with all of the diverse phenomena that occur within their machines, with the overall performance of these machines and, ultimately, with the performance of complete processes comprising a number of machines acting in concert. Thus, the subject that is called thermodynamics is only a part of a wider body of knowledge that is concerned with the transformations that occur within systems and the interactions that take place between them. This wider body of knowledge encompasses not only the subject matter of equilibrium and irreversible thermodynamics, but also that of transport phenomena, mechanics, strength of materials ... etc.

CONCLUSION: THERMODYNAMIC CONCEPTS REQUIRE RE-APPRAISAL

The foregoing considerations suggest that the difficulty that students experience with thermodynamics is due, at least in part, to the fact that the subject is incomplete and, further, that it employs concepts and an intellectual approach which are distinctly different from those that are employed in related subjects. If this diagnosis is correct then what is required is that we should seek out a general purpose set of concepts and axioms that can be applied in a consistent fashion to all those problems where energy is a relevant consideration.

It has been said that a subject needs to be re-examined when it reaches the stage where it is generally agreed that the subject is understood in all its essentials and, thus, that any further progress is likely to be of marginal significance. On this basis alone it would seem that thermodynamics is overdue for a fundamental re-appraisal.

REFERENCES

1. Carnot, S., 'Reflections on the Motive Power of Fire', E. Mendoza (Ed), Dover, 1960.

THERMODYNAMICS - A DIRECT APPROACH

J. W. Rose

Department of Mechanical Engineering
Queen Mary College
University of London

The main advantage of a clear and concise presentation, along similar lines to the more advanced treatments of Tisza (1) and Callen (2), is that students are brought very quickly to a point where they have an overall understanding of the basics of macroscopic equilibrium thermodynamics. They are then in a position to study applications, advanced topics, alternative approaches or history. The presentation, outlined very briefly below, has been used by the author for around 15 years.

Energy E and entropy S are introduced together at the outset as extensive properties such that, for *isolated* systems

$$dE = 0 \text{ and } dS \geq 0$$

These basic postulates can be made plausible by reference, for example, to the 'spontaneous slowing-down and warming-up of a spinning flywheel'. 'Possible states' of isolated and non-isolated systems (i.e. sub-systems of isolated systems) are thought of as states through which a system passes while attaining equilibrium. The stable equilibrium state of an isolated system is the possible state with the highest entropy.

Simple systems are those which are defined (i.e. the possible states are fixed) when E, together with a fixed number of other extensive, conserved properties X_1, X_2, etc. are specified. For a simple, closed fluid system there is only one X property, the volume V. When E, X_1, X_2 etc. are fixed, the system is isolated. Hence for *equilibrium states*

$$S = S(E, X_1, X_2 \ldots)$$

The terms 'reversible' and 'irreversible' are used only in the context of isolated systems. For non-isolated systems the terms 'non-dissipative' and 'dissipative' are preferred. When an isolated system undergoes a reversible process both it, and all sub-systems of it, undergo 'non-dissipative' processes. A system undergoing a non-dissipative process passes through a continuous sequence of equilibrium states (quasi-equilibrium process).

The 'energy minimum principle' dictates that the equilibrium state of a simple system, among possible states having the same values of S, X_1, X_2, etc., is that with the lowest energy. An ideal frictionless mass-spring oscillator would execute reversible (constant entropy) processes indefinitely. When coupled to a viscous damper, the energy of the mass-spring sub-system would pass spontaneously to the damper. The entropy of the combination and hence that of the damper must increase, i.e. $\partial E/\partial S$ for the damper is positive. This serves both as a plausibility argument for the energy minimum principle and to demonstrate that temperature ($T = \partial E/\partial S$) is essentially positive.

If the boundary separating two systems is permeable to energy it is necessary that, at equilibrium, the systems have the same value of T. Hence T is a function of empirical temperature θ. Practical approximations to adiabatic boundaries may then be found i.e. boundaries such that $\theta_A \neq \theta_B$ (and hence $T_A \neq T_B$) at equilibrium between systems A and B.

The additional postulate that when nechanical work W is done on a system, it experiences an increase in E equal to W, together with the existance of adiabatic boundaries, enables measurement of E (apart from an arbitrary datum value). When applied to the special case of a closed simple fluid system undergoing a quasi-equilibrium constant entropy process, the work postulate serves to identify normal stress or hydrostatic pressure with thermodynamic pressure $P = -\partial E/\partial V$.

The facts that W and E are measurable, that $\theta = \theta(T)$ and with adiabatic boundaries, enables T_1/T_2 to be measured for empirical temperatures θ_1 and θ_2. Thus, for simple fluids, e, v, P and T can be measured. Difference in specific entropy may then be found from

$$s_2 - s_1 = \int_{e_1}^{e_2} (1/T)\, de + \int_{v_1}^{v_2} (P/T)\, dv$$

where the integrals are for specified paths through tabulated states. The fact that the entropy difference is independent of the path demonstrates that entropy is a property.

The approach may be extended to open systems when it is readily demonstrated that for equilibrium between two homogeneous phases of a single-constituent system the molar and specific Gibbs functions are equal for the two phases implying $\psi(P, T) = 0$ and $dP/dT = s_{fg}/v_{fg}$.

REFERENCES

1. Tisza, L., "Generalised Thermodynamics", M.I.T. Press (1966).

2. Callen, H. B., "Thermodynamics", John Wiley (1960).

TEACHING THE SECOND LAW:

EXPERIENCE FROM THE OPEN UNIVERSITY TECHNOLOGY FACULTY

Rosalind Armson

The Open University
Milton Keynes

Confusion about the concepts arising from the second law is a well known phenomenon. A conscientious attempt has been made in the Open University course 'Thermofluid Mechanics and Energy' to face up to this by a new treatment of the teaching of both First and Second Laws. This treatment has involved several factors. Firstly, there is an explicit attempt to dismantle students' pre-existing misconceptions about energy. Energy is treated as the capability of effecting change.

Secondly, to make explicit the relationship between energy and energy transfer, the terms 'heat' and 'heating', 'work and 'working' are deliberately and consistently avoided. Instead the formulae 'energy transfer by heating' and 'energy transfer by working' are used for both the substance and the activity.

It is many people's experience that while entropy and the second law are difficult to understand, they are not difficult to use. This observation led to the third feature of the O.U. Course. No attention at all is given to formal statements of the second law. Students at first and second year level need a clarity of thought in _manipulating_ the second law concepts and it was our judgement that confidence in doing this is not enhanced by proofs of the theoretical equivalence of the statements or an emphasis on imagining things (e.g. infinite numbers of quasi-static engines operating between negligibly different temperature reservoirs) which strain credibility to breaking and despair. Instead, rather more attention than usual is given to the concepts of reversibility and irreversibility. This gives a confidence about the domain in which the second law operates. A major part of the students' time is given to examples and calculations (some worked, many more self assessment) to reinforce the grasp of the practical outcomes of the Second Law. Available and unavailable energy give rise naturally to the concept of entropy and there is no appeal to notions of disorder.

In retrospect, this approach has been a success, albeit qualified. Insofar as one can tell from students' submitted assignments and examination answers, there is comparatively little confusion about the Second Law, the direction of energy transfers or the resulting changes in entropy and energy availability. Students with no previous background find the topics stimulating and interesting and problem-free. Where problems have occurred, they appear to arise where students have previous experience of thermodynamics teaching. Our sign conventions and our treatment of available energy do not map conveniently with those other teachers and texts.

Responses from teachers and students at other institutions are favourable and we know of a number of Universities using our teaching materials.

TEACHING ENGINEERING APPLICATIONS

R. W. Szymanski

Faculty of Engineering, Science and Mathematics
Middlesex Poltechnic
London

Teaching applications means no more than teaching how to apply general concepts and principles to specific and particular 'hardware'. It goes without saying, therefore, that we must first be well versed with fundamental principles, i.e., systems, their properties - clear definitions, dimensions, units - Laws of Thermodynamics and their application to the systems in their solid, liquid, vapour and gaseous states, basic concepts on Heat Transfer Mechanisms. The teaching of these fundamentals, from the very beginning of any course, can be interwoven with illustrations of their meanings and use in the real everyday environment. Examples can be drawn from our immediate surroundings - an overhead projector, a chair, a lamp, the human body - are these closed or open systems? Of course illustrations of other machinery, turbines, compressors, pumps, heat exchangers, combustion chambers, etc., are used. Properties like pressure and temperature and parameters, fluid velocity and flow rate, torque, rotational speed and power, can all be introduced with demonstrations of 'hardware' and/or actual use and calibration of equipment in laboratories. In these lecture-demonstration-application situations we can introduce information on construction, principle of operation, response, commercial availability, cost, installation aspects, lifespan, maintenance and other factors like corrosion and erosion problems. At every stage and as frequently as is possible, these introductions can be backed up by simple numerical examples and computer-aided studies. Initially a student may not fully appreciate all these aspects, but he will grow in the atmosphere of reality and at every stage of analytical work he will appreciate the necessity and value of assumptions made connecting his simple calculations with the real world. A similar approach can be made to the 1st Law applications. Numerical examples concerning simple problems on closed and open systems can be backed up by laboratory evidence when operating such systems. The necessity and appreciation of assumptions becomes then apparent to the student. Given the cooperation of electronics and material science specialists, with established links between these areas and thermodynamics and fluid mechanics, coordinated teaching can be set up, even at this early stage of the Course. In the second, third and possibly fourth year or stages of study - we must be open to the many possibilities which may come our way with the new B.Eng. and M.Eng courses - we can construct a wide range of flexible modular blocks of studies in thermodynamic and allied subjects interwoven with other course subjects. The overall aim should be the appreciation by students of the totality of the final product or system. Some examples are given at the end of this contribution. In all such modules in the early stages, aspects of economics, production, environmental considerations, maintenance, lifespan and other such problems are introduced qualitatively to make students constantly aware of overall requirements of the plant under design study. More hard hitting and quantitative economic analysis would come in the advanced modules. The problem of linking various subjects does rest on the good will of those teaching them and on the work co-ordination. All one can do is to suggest a possible manner of forming a matrix of themes which would indicate the link. Good knowledge of details of material covered in various subjects is essential so that the same or very similar examples of hardware is analysed in respective areas such as thermodynamics, fluid mechanics, design, materials, stress analysis, mechanics of machines, electrical, electronics and control production and management, economics, social and environmental. Examples of modular blocks should help to illustrate this problem.

<u>Some Examples of Modular Blocks</u>

I. Mixtures of Gases and vapours.

1.	Basic conceptual work, backed up by number of numerical examples built on realistic systems.)	Mainly thermo-fluids work with qualitative link made with other subjects.

2. Realistic examples of applications e.g., air-conditioning systems, condensers, cooling towers, exhaust systems, evaporators.	Thermo-fluids linked with design, material science, stress analysis, production and management, and through instrumentation and services with electronics and electrical systems.

3. Advanced work on total systems as above leading to basic specification, sizing, costing, marketing in face of existing products.

Level 3 type of modules will form a link between undergraduate work (say fourth year or stage of M.Eng course) and/or postgraduate work with openings for re-training of personnel from industry or close work with industry or specialist consultants.

<u>Some Other Examples of Group of Modules</u>

II. Rotodynamic Machines

III. Combustion Systems

<u>Note</u>: Modular blocks are not uniform in size (teaching hours) and involvement with other subject areas.

DISCUSSION IN SESSION 1 "TEACHING OBJECTIVES"

R. I. Crane

Department of Mechanical Engineering
Imperial College of Science and Technology
London

On the Paper by W. A. Woods and C. A. Bailey, "Thermodynamics and Fluid Mechanics for the 1980s"

This paper being a factual review of an Institution of Mechanical Engineers Symposium held in 1979, Professor W. A. Woods (as Session Chairman) declined to accept discussion until the end of the session.

On the Paper by J. R. Himsworth, "Better Thermodynamics Teaching Needs More Thermodynamics Research" (printed as an extended abstract)

Professor R. S. Silver referred to Dr. Himsworth's remarks on the archaic mode of reasoning employed by Carnot; he considered that we had become used to taking short cuts in reasoning and did not recognise when we were being logical or, much worse, being illogical. With the advent of machine logic, he envisaged students in years to come not knowing the principles of logic at all, and believed it would be extremely serious if we were to lose logic from the subject. Dr. Himsworth agreed on the necessity of logic; he reiterated that people of Carnot's time were familiar with that strict style of reasoning, the problem being that present-day students were not.

Dr. Y. R. Mayhew commented on the abstractions involved in the subject, quoting an example the difficulties in understanding the Kelvin-Planck statement of the second law. The Clausius statement should, logically, be equally difficult to accept; how did we know what was meant by one body being hotter than another? From whichever point we were to begin the study of thermodynamics, we would stumble across ideas which initially could not be defined.

On the paper by B. R. Wakeford,
"Why has Thermodynamics become a "Difficult" Subject?"

Mr. Wakeford's belief that textbooks of the last quarter-century were unnecessarily complex in their approach to the first law, and accorded it too much importance, was supported by Professor J. Patterson. However, Professor Patterson preferred to teach the equation of state, in its most general form, as the first law, claiming it to be more fundamental; the definition of entropy was taught as the second law, in the form

$$dS = dQ/T + dQ_R/T$$

where Q_R represents internal "heat" generation. Mr. Wakeford accepted that state equations were not far removed from the principle of conservation of energy, but felt that differentials were best left out of an introduction to the second law; he would prefer to reserve them for a later stage in the course.

Dr. J. W. Rose suggested that the word "become" in the paper's title was inappropriate; thermodynamics is either not difficult or has always been difficult. Being baffled as a student but nevertheless finding it possible to get good examination marks was a familiar experience to many of the participants.

Dr. P. Mukundan spoke of problems in teaching students with some industrial experience; they needed to be taught in a phenomenological manner, starting with conservation of energy, keeping mathematics to a minimum but ensuring that the subject's jargon was understood. To make students think, rather than rely heavily on formulae, he found it useful to give the occasional wrong answer to a problem, and also asked students to consider how Carnot arrived at his conclusions. Mr. Wakeford agreed that methods of teaching should differ according to the students' backgrounds.

Dr. Lewins added that some teaching problems were nevertheless self-inflicted, by the teachers' continual desire to improve their approach. As an example, he noted that if special attention were to be given to segregating heat and work, students might consider it "obvious" that one could not be completely converted to the other; in other words, the second law would have been "defined".

Commenting on jargon, Dr. Mayhew remarked that we meet energy in real life, so statements such as "E is constant for an isolated system" are merely playing with words. Mr. Wakeford concluded by noting that energy, in all its forms, must be measured by what it *does*; we could not say what it *is*, and this lack of understanding was what made conservation of energy a law.

On the Paper by C. Gurney,
"Teaching Thermodynamics to First Year Engineers"

The abstract, mathematically-based approach described in Professor Gurney's paper prompted Professor Patterson to note that the conclusions reached in that course were inevitably the same as those in the usual type of course, and that no time was available for such detailed derivations. Professor Gurney agreed that unless the students were mathematicians, it did not matter that they might not understand the analysis, only that they should be aware of the results (tabulated, in any case) and when designing equipment, for example, they should not forget to consider the way in which mass, momentum, energy, entropy enter the problem. Concepts such as entropy could only be properly appreciated by using them; he likened the process of grasping the meaning of entropy to learning the meaning of hunger. Professor Patterson returned to the mathematics of Professor Gurney's derivations, saying that there were verbal equivalents of all his symbolic expressions, having real physical meanings; a descriptive interpretation of the analysis saved a great deal of time.

Recalling his experience of students from Hong Kong, Dr. R. I. Crane felt that they would have little difficulty with Professor Gurney's treatment but wondered how the average British student would react. Mr. D. H. Bacon, supporting Professor Patterson in saying that one should not worry too much about the derivations, also regarded Hong Kong as untypical and queried the stage in the degree course at which this analysis was introduced. There was a feeling among the participants that some of the economical mathematics used at the beginning of Professor Gurney's analyses to represent everyday experience might not immediately be recognized as such. Professor Gurney agreed that his students had been exceptionally gifted.

Further comments on the theme of answering the questions "What are entropy, energy, etc.?" included that of Dr. J. W. Rose, who felt that students wanted explanations in terms of the concepts of geometry and mechanics, with which they were familiar. When asked such questions, he reminded students that they did not know what are mass, time, etc., but the meanings were contained in the prescriptions for their measurement.

Dr. T. H. Frost stressed that learning through usage must be accompanied by demonstrations that thermodynamics works as a complete structure, accurately predicting physical phenomena. Examples quoted by Dr. Frost were predictions of the propagation velocity of weak pressure waves, of choking in nozzles, of temperature changes in throttling and of the relationship between saturation pressure and temperature.

On the Paper by J. W. Rose,
"Elementary Thermodynamics: a Direct Approach"
(printed as an extended abstract)

Dealing with requests to clarify some aspects of his approach to teaching the basics of equilibrium thermodynamics, Dr. Rose stated that motion, gravity, etc. were excluded when defining energy E, also that he did not mention heat.

Mr. R. W. Haywood asked Dr. Rose why he used the terms "reversible" and "irreversible" only in the context of isolated systems and preferred "non-dissipative" and "dissipative" for non-isolated systems; use of the term "open system" in place of "control volume" was also criticized. Dr. Rose replied that any process could be reversed if one allowed an external agency; "irreversible" did not seem a very appropriate word. He added that he did not object to using "control volume" but was thinking more in terms of diffusion etc.; he might prefer "control region" in a flow context.

On the paper by R. Armson, "Teaching the Second Law: Experience from the Open University Technology Faculty"
(printed as an extended abstract)

The treatment of energy as the capability of affecting change was disliked by Professor Gurney, who regarded energy as one of the two scalar invariants of an isolated system (the other being mass). Miss Armson said that in the context of effecting change, energy transfer was meant, rather than energy. Professor Gurney repeated that an understanding of energy and entropy came from using these concepts to descirbe experience.

Mr. Bacon saw a danger in using long pieces of terminology like "energy transfer by working/heating", in that students would abbreviate and then confuse them in conversation. However, these expressions were applauded by Dr. Rose, who added that if they must be shortened, it was better to say, "work (or working) done" and "heat (or heating) done" than to use the words "supplied" and "rejected" since the latter tended to imply that work and heat might be conserved quantities or properties. Students were not so dim, according to Dr. Mayhew, that any harm was done by teaching them to accept definitions and thereafter using abbreviated forms.

Commenting on the stress on manipulating second law concepts and the deliberate lack of attention to formal proofs etc. in the Open University course, Dr. Mayhew approved of the technique of introducing a difficult concept by means of examples, but felt that by continuing to rely on examples of use this course never distilled the principles of thermodynamics. Miss Armson replied that the phenomenological outcomes of the laws were nevertheless listed in her course, and denied that the fundamental concepts remained unsolidified.

GENERAL DISCUSSION ON TEACHING OBJECTIVES

Most of the time remaining at the end of this Session was taken up with discussion stimulated by the remarks of Dr. B. N. Furber, who claimed to represent the academics' customers in industry. Since the very existence of engineering departments supported the case for specialization in universities and polytechnics, Dr. Furber suggested that such departments should be more sensitive to the requirements of industry. From his personal experience in industry and from conversations with recent recruits to his company, it was clear that the subject of heat transfer made a much larger contribution to their work than that of thermodynamics. Quoting a well-known textbook whose content was about four-fifths thermodynamics and one-fifth heat transfer, he did not accept that this was the right balance and made a plea for more detailed treatments of heat transfer in degree courses.

To back up this point, he invited participants to compare the number of international conferences on heat transfer with the number on thermodynamic topics.

It was noted that work involving analysis of cycles etc. was frequently reported at conferences on themes such as power generation or gas turbines rather than meetings with a discipline-orientated title; the imbalance was perhaps not so great as had been implied. However, there was no disagreement on the importance of heat transfer, only on the notion of associating it closely with thermodynamics. Mr. Haywood pointed out that his own book contained no heat transfer material whatsoever; the decision to exclude it had been deliberate, since heat transfer was better treated in a book of its own. To have included heat transfer would in any case have made the book so unwieldy in size as to be unsaleable. Dr. Mayhew defended the heat transfer content of his book, remarking that it reflected the balance between topics as seen in the early 1950s. The courses taught nowadays at Bristol University contained a much higher proportion of heat transfer than suggested by his book, including much material not contained in the book.

Acknowledging the importance of heat transfer and the need for more of it to be taught, Dr. Rose declared that the additional material must not be at the expense of thermodynamics. He expressed the generally accepted view that heat transfer was much more akin to fluid mechanics (except for topics such as radiation which still had no strong link with thermodynamics). This point was taken up by Professor Silver, saying that heat transfer should be taught separately, its rational basis being in fluid mechanics. Professor Silver continued by stating that too much reliance was still placed on correlations; more work was needed on the underlying theoretical basis of heat transfer, and this in turn would stimulate interest in teaching the subject.

A general comment was made by Professor B. L. Button, in the light of the many different approaches to teaching the fundamentals of thermodynamics which had been aired during the Session. He asked whether anyone would care to distil the merits of the various approaches and apply them to the same set of problems, so that students could choose the approach best suited to them as individuals. Not surprisingly, no volunteers were forthcoming; a legitimate excuse could have been that by the time a student was ready to solve a wide range of practical problems, all approaches should have converged to the same set of working rules.

Summing up the Session, Professor Woods began by admitting that it had made him feel inhibited once again about using words such as "system". He remarked on how thermodynamics teachers became very constrained by terminology and recalled that he had once campaigned strongly to stamp out the term "system"! However, it was quite unrealistic to expect engineers to stop using such barbaric terms as "heat release" (a favourite in the I.C. engine community). Finally, Professor Woods restated his own view of the three objectives of teaching: that students should know, understand, and be able to use the principles of the subject. Understanding everything was not a realistic goal, but it was important that teachers should successfully impart an understanding of _some_ of it. Ability to use had to be at the top of the pinnacle of objectives.

SESSION 2

INNOVATIVE METHODS OF TEACHING

Chairman - Sir Brian Pippard, F.R.S.

Secretary - Mr. D. Walton.

MATERIALS THERMODYNAMICS FOR ENGINEERS

D. R. H. Jones

Department of Engineering
University of Cambridge

We begin by outlining the engineering materials course at Cambridge. We show that much of this course overlaps with thermodynamics. Three case studies are presented - on creep, diffusion, and solidification - which illustrate how thermodynamics, coupled with simple kinetic theory, can greatly aid a student's understanding of the underlying physical processes.

INTRODUCTION

Because of the importance of materials in all aspects of engineering design, engineering undergraduates at Cambridge study materials in both the first and second years of their course. In order to show the range of topics covered in the materials courses (1,2) we have summarized them below, underlining the subjects that overlap with thermodynamics.

First-Year Engineering Materials:
Classes of Property (27 Lectures).

Price and Availability.
The Elastic Moduli: data for stiffness, atomic bonding, packing of atoms in solids, designing stiff materials, case studies in moduli-limited design.

Yield and Tensile Strength, Hardness and Ductility: testing and data, dislocations and yielding in crystals, strengthening methods, continuum plasticity, case studies in yield-limited design.

Fast Fracture, Toughness and Fatigue: fast fracture and toughness, micromechanisms of fast fracture, fatigue failure, case studies in fast fracture and fatigue.

Creep Deformation and Fracture: creep and creep fracture, kinetics of diffusion, creep mechanisms, designing creep-resistant materials, the turbine blade - a case study in creep-limited design.

Oxidation and Corrosion: oxidation of materials, case studies in dry oxidation, wet corrosion of materials, case studies in wet corrosion.

Friction, Abrasion and Wear: friction and wear, case studies in friction and wear.

Final Case Study: materials and energy in car design.

Second-Year Engineering Materials:
Classes of Materials (27 Lectures).

Metals: generic types, structures of metals, phase diagrams, case studies (segregation in castings, zone refining electronic materials, bubble-free ice), phase transformations, case studies (single-crystal silicon films, recrystallization, artificial rain making), aluminium alloys (precipitation and age hardening), steels (heat treatment, alloy steels), case studies in steels (detective work in a boiler explosion, welding steels safely), production, forming and joining of metals, design with metals.

Ceramics: generic types, relation between structure and properties, production, forming and joining, design with brittle solids, cement and concrete, case studies (the strength of sea ice, glass windows for vacuum chambers).

Polymers: generic types, property-structure relationships, production, forming and joining, design with polymers, case studies (plastic for pop bottles, redesigning the bicycle).

Composites: wood, fibre composites, foams, case studies (materials for violin bodies, packaging and executive jets).

The areas which involve thermodynamics may be collected together under two main headings:

(a) Equilibrium Properties

Phase equilibrium: phase diagrams, activities of components in solution.

Entropy: entropy of gases, liquids and solids viewed in terms of increasing order, statistical entropy of polymer chains (entropy elasticity).

(b) Rate Processes

Rates of chemical reactions: oxidation, corrosion, setting and hardening of cement, polymerization.

Rates of phase transformation: crystallization, allotropic changes, precipitation reactions.

Rates of defect movement: migration of surfaces and interfaces (diffusion creep, recrystallization, grain and phase coarsening, sintering), thermal movement of dislocations (dislocation creep, recovery).

Rates of atomic or molecular movement: diffusion and viscous flow.

It is obviously not possible in a single paper to cover all of these areas in detail. We have therefore chosen to look at a few topics in some depth in order to show how the intellectually curious student can enliven his materials with some thermodynamics, and his thermodynamics with some materials.

CREEP

When materials are loaded mechanically, and the temperature is high enough, they get progressively longer with time by creep, so that when the load is finally removed they become permanently stretched. This process is, of course, of immense importance - the creep of turbine blades limits the temperature at which gas turbines can operate and, at the other end of the temperature scale, glaciers flow by the creeping of ice.

Creep is a typical example of a rate process. And, like all rate processes, it needs both a driving force and a mechanism for it to take place. This concept can be illustrated pretty easily by simple situations like the car parked on a slope: a driving force (provided by a gradient in potential energy) acts so as to push the car down the hill, but movement can only occur if a suitable mechanism (such as a slipping handbrake!) is able to operate.

The driving force for creep is almost as obvious. If we look at the schematic creep process shown in Fig. 1, it

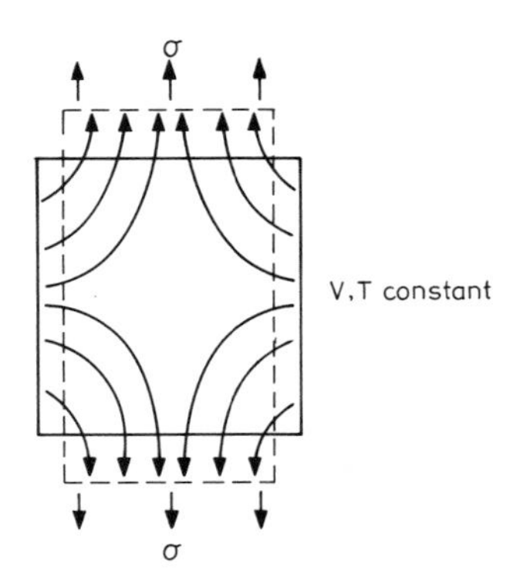

Fig. 1 Schematic illustration of creep in a uniform specimen of material subject to a tensile stress σ at constant temperature. Since creep involves the transfer of atoms from the sides of the specimen to the specimen ends there is no change of volume associated with the process.

is pretty clear that the creeping out of the specimen allows the environment to do mechanical work on the material. It is also obvious that the mechanism involves the transfer of atoms or molecules of the material from the sides of the specimen to the ends. But in many other rate processes, it is not nearly as easy to identify the driving force and we might as well do things properly from the start by borrowing the Gibbs free energy

$$G \equiv H - TS \tag{1}$$

from our friends in thermodynamics. We can then calculate the amount $\Delta\mu$ by which the free energy of an atom decreases as it goes from the sides to the ends of the specimen. (We point out, beforehand, that μ is the average free energy, G/N_o). Thus

$$dG = Vdp - SdT, \tag{2}$$

giving $\Delta G = V\Delta p$ at constant V and T, and producing

$$\Delta\mu = \Omega\sigma, \tag{3}$$

where Ω is the atomic volume and σ is the stress under which the material is creeping.

How does an atom know that it should move from the sides to the ends? We can present this problem in two ways - a continuum approach or a statistical one. Students are already familiar with a continuum approach to the movement of electrons through conductors in an electrical potential gradient as Ohm's Law, $\underline{J}_e = \sigma_e \underline{E}_e$, or

$$J_e = - \sigma_e(dV_e/dx). \tag{4}$$

This is merely a special case of the general equation

$$J = - Bc(d\mu/dx), \tag{5}$$

where J is the number of particles travelling through unit cross section per second, B is their mobility, c is their concentration, and $d\mu/dx$ is the free energy gradient of the particle.

In our creep process

$$(d\mu/dx) = \Omega\sigma/d, \tag{6}$$

where d is a specimen dimension. Thus the number of atoms flowing down a specimen of unit cross sectional area per second is given by

$$J = - Bc(\Omega\sigma/d). \tag{7}$$

For a pure material $c = 1/\Omega$. The creep speed $v = J\Omega$. The creep strain rate $\dot{\varepsilon} = v/d$. Substituting these results into equation (7) gives

$$\dot{\varepsilon} = B\Omega\sigma/d^2. \tag{8}$$

This result agrees with experimental results for diffusion creep which show the creep rate to be linear in B and σ and inversely dependent on the square of the specimen size (or the grain size in a polycrystalline sample). Home and dry, and a nice little application of thermodynamics!

Most material properties are governed by processes which operate at the level of single atoms, or small assemblies of atoms, and no self-respecting student of materials can claim to have understood the subject without getting down to this level of scale. In any case, visualizing what atoms may be up to can be entertaining as well as informative. We approach the statistics of atom movement using the familiar kinetic treatment adopted by Eyring (3). Although Eyring's theory is based on questionable assumptions, and has been superseded by recent developments, it is attractive from a teaching standpoint because of its easily understood intuitive approach.

Our approach to kinetic theory is then to say simply that the atoms in a solid have a mean thermal energy of order kT; that they are in a state of thermal agitation; that collisions occur frequently so that there is always a chance of a particular atom having an energy greater than the mean; and that the probability of an atom having an energy greater than the mean by amount q is $\exp(-q/kT)$ at any instant. For all their naivety these ideas make obvious intuitive sense - one will hope that students have come across Brownian motion and the Boltzmann distribution before they get to university!

Let us now apply this kinetic theory to the movement of atoms from the sides to the ends of our creeping material. We imagine atoms hopping - or <u>diffusing</u> - down the free energy gradient from one atomic site to the next. In order to hop from site to site the atom has to "squeeze" through the gap left by its neighbours and this requires it to overcome an energy barrier. This is shown schematically in Fig. 2. Here, the number of atoms jumping from sites in plane A to sites in plane B per second is

$$n_{AB} = n_A \frac{\nu}{6} \exp-(\Delta\mu^* - \tfrac{1}{2}\delta\mu)/kT. \qquad (9)$$

n_A is the number of atoms in plane A able to take part in diffusion and ν is their atomic vibration frequency. Thus each atom makes ν attempts to overcome an energy barrier every second; of these only 1/6 are aimed in the forwards direction from A to B (attempts can also be made in the backwards direction, to left and right, or up and down); and <u>of these</u> only a proportion $\exp-(\Delta\mu^* - \frac{1}{2}\delta\mu)/kT$ actually make the jump. Similarly

$$n_{BA} = n_B \frac{\nu}{6} \exp-(\Delta\mu^* + \tfrac{1}{2}\delta\mu)/kT. \qquad (10)$$

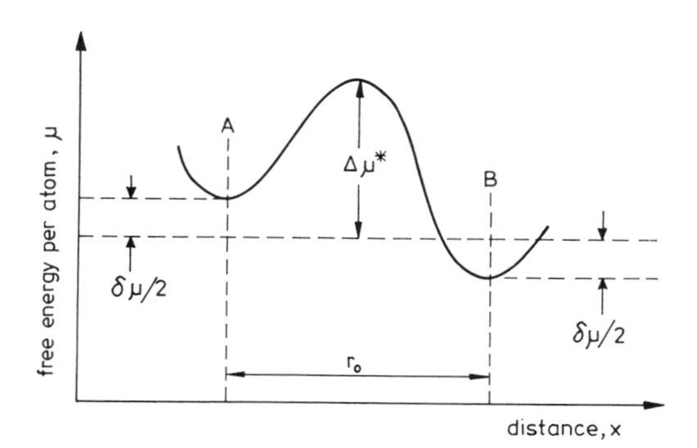

Fig. 2 A and B are two adjacent planes, a distance r_o apart, in a crystalline solid. There is a gradient of free energy in the solid such that atoms at A have a free energy that is $\delta\mu$ higher than atoms at B. The curve shows, schematically, the free energy of an atom as it moves from A to B via an intermediate energy barrier.

The net rate of jumping from A to B, $n_{AB} - n_{BA}$, is therefore

$$n_{net} = \frac{\nu}{6} e^{-\Delta\mu^*/kT}\{n_A e^{\delta\mu/2kT} - n_B e^{-\delta\mu/2kT}\}. \tag{11}$$

This expression can be "brought down to earth" quite easily. First, for materials processes other than chemical reactions, $\delta\mu/2kT$ is small and we can expand exponentials to first-order terms in $\delta\mu/2kT$. Then, if we take a pure material $n_A = n_B = n$. For unit cross section $n_{net} = J$ and $c = n/r_o$. Finally $\delta\mu/r_o = d\mu/dx$. We then end up with

$$J = -\frac{\nu}{6}(r_o^2/kT)e^{-\Delta\mu^*/kT} c(d\mu/dx). \tag{12}$$

This equation, derived statistically, is identical with the continuum result, equation (5), with

$$B = \frac{\nu}{6}(r_o^2/kT)\exp{-\Delta\mu^*/kT}. \tag{13}$$

Equation (13) is usually written in the form

$$D = BkT = \frac{\nu}{6} r_o^2 \exp{-\Delta\mu^*/kT}, \tag{14}$$

where D is the <u>diffusion coefficient</u>.

The reward for the student who has worked through these equations is to see that kinetic theory gives exactly the same exponential temperature dependence that one gets from experimental measurements of diffusion coefficients. Indeed, thermal activation shows up directly in all sorts of situations: Araldite glue sets much faster in a warm place; cement goes off horrifyingly quickly on a hot day; wooden window frames rot faster on the sun-facing side; and one can even estimate the energy barriers crossed in breeding bacteria from the cold storage lives given on the back of a Marks and Spencer frozen flan!

DIFFUSION IN CONCENTRATION GRADIENTS

Many processes in materials use the fact that a concentration gradient can provide the driving force for atomic diffusion. This can be useful, as when crankshaft journals are hardened by diffusing carbon or nitrogen into them; or when different metals are stuck together by diffusion bonding. But diffusion can also be a menace, as when hydrogen diffuses into steel welds from damp welding rods to cause "cold cracking"; or when interdiffusion occurs at electrical connections between copper and aluminium, depositing poorly conducting intermetallic compounds which cause burn-out.

It is well known experimentally that, in a single-phase solid solution, the solute will diffuse down a concentration gradient according to Fick's first law

$$J = -D(dc/dx). \tag{15}$$

Let us first of all try the continuum approach. We start, as before, with the basic equation

$$J = -Bc(d\mu/dx). \tag{16}$$

Every physical chemistry student should know that

$$\mu = \mu_o + kT \ln c \tag{17}$$

for an ideal solution. Few engineering students will know this, but they can easily follow the proof, which involves only $dG = Vdp - SdT$, $pV = RT$, and Raoult's law. Equation (17) allows us to calculate

$$(d\mu/dx) = (kT/c)(dc/dx) \tag{18}$$

so that, by using $D = BkT$ as well, we can get equation (16) to drop out as Fick's law.

Now for the kinetic theory approach. We again look at a pair of adjacent atomic planes (see Fig. 2) but this time we assume that we have n_A solute atoms in plane A and n_B solute atoms in plane B. Then the net rate of jumping of solute atoms from A to B can be found from equation (11). $\delta\mu$ can be calculated from equation (17) to give

$$\delta\mu = kT \ln(c_A/c_B) = kT \ln(n_A/n_B). \tag{19}$$

Substituting this result into equation (11) and expanding exponentials to first-order terms in $\delta\mu/2kT$ gives

$$n_{net} = \frac{\nu}{6}\, 2(n_A - n_B)\exp-\Delta\mu^*/kT. \tag{20}$$

Writing $n_{net} = J$ and $c = n/r_o$ as before gives us

$$J = \frac{\nu}{6} 2(c_A - c_B) r_o \exp-\Delta\mu^*/kT, \tag{21}$$

or

$$J = -\frac{\nu}{6}\, r_o^2 2(dc/dx) \exp-\Delta\mu^*/kT. \tag{22}$$

Now, if we compare this last equation with Fick's law, we can see that

$$D = \frac{\nu}{6}\, 2r_o^2 \exp-\Delta\mu^*/kT, \tag{23}$$

which is <u>twice</u> what it should be (see equation 14). This discrepancy is our just punishment for having been cavalier enough to calculate free energy differentials between adjacent atomic positions from macroscopic thermodynamic gradients. There is no doubt, of course, that there is an <u>overall</u> free energy gradient driving atomic diffusion. Because we are dealing with an ideal solution this gradient arises entirely from changes in entropy. But it is questionable to talk about entropy changes of single atoms when they jump from site to site.

In most cases this problem is brushed under the carpet by considering atoms as jumping over "energy" barriers, suggesting "potential energy" barriers, so that entropy barriers are implicitly excluded. Thus, in the ideal-solution case - where there <u>are</u> no potential energy differences -

diffusion can simply be treated as a population effect. In other words, because there are more solute atoms in plane A than in plane B, more A atoms cross the barrier per second and net diffusion occurs from A to B. This produces the "right" value for D, as can easily be checked by putting $\delta\mu = 0$ in equation (11). This approach, however, involves the loss of an important piece of physics. For an atomic analogue of entropy undoubtedly has a powerful effect on how easily atoms can squeeze between their neighbours when jumping from site to site.

SOLIDIFICATION

Solidification is important to materials people. Most crystalline solids are produced by solidification, and the way in which solidification occurs can very much affect the properties of the final solid. And solidification can be just as important at an everyday level - the chaos caused by a harsh winter is, after all, only due to a liquid-solid transformation.

To study solidification we start, as we have done all along, by examining the driving force for the process. From equation (2) we can write

$$(dG/dT)_p = -S. \tag{24}$$

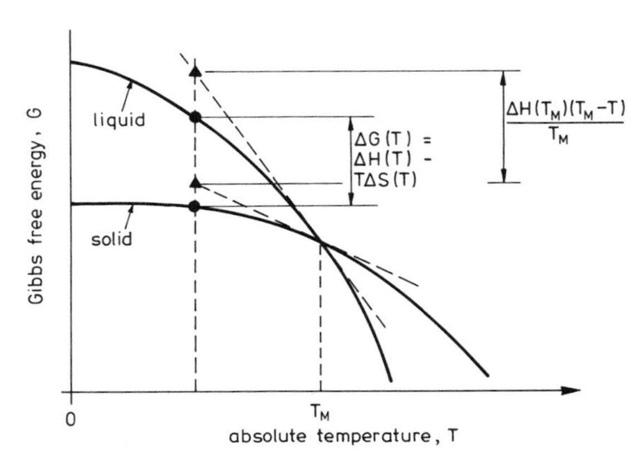

Fig. 3 Schematic free-energy curves for two phases (liquid and crystalline solid) which coexist in equilibrium at T_M.

This equation allows us to construct schematic free-energy curves for the liquid and the solid phases (Fig. 3) by introducing the concept of disorder to give us a qualitative measure of entropy. Thus, at any temperature, the slope of the G curve for the liquid is greater than that for the solid (liquid more disordered than crystalline solid); both slopes must become more pronounced at higher temperatures (increased thermal agitation); if the crystalline solid is defect-free the slope of its G curve must be essentially zero at 0 K (Third Law); and of course the curves must cross at the melting point T_M. It is then easy to show that, close to T_M, the driving force for solidification is given quite well by

$$\Delta G(T) = \Delta H(T_M) - T\Delta S(T_M) = \Delta H(T_M) - T\Delta H(T_M)/T_M,$$

or

$$\Delta G(T) = \Delta H(T_M)(T_M - T)/T_M. \tag{25}$$

This result provides insight in several ways. Thus at $T = T_M$, $\Delta G = 0$ which is the condition for equilibrium; as the undercooling $T_M - T$ increases, ΔG increases roughly in proportion; and finally, ΔG can be related to an experimentally measurable quantity, the latent heat of solidification $\Delta H(T_M)$.

The solidification mechanism involves the jumping of atoms from sites in the liquid to adjacent sites on the surface of the growing solid. The free energy gradient is therefore given by

$$\frac{d\mu}{dx} = \frac{\Delta H(T_M)(T_M - T)/T_M}{N_o r_o} \tag{26}$$

and the continuum approach, using $J = -(D/kT_M)c(d\mu/dx)$, $v = J\Omega$ and $c = 1/\Omega$ gives us

$$v = \frac{D\Delta H(T_M)(T_M - T)}{r_o R T^2_M} \tag{27}$$

for the speed of the solid-liquid interface.

Starting from equation (12), the kinetic approach leads to

$$v = \frac{\nu}{6} r^2_o e^{-\Delta\mu^*/kT} \frac{\Delta H(T_M)(T_M - T)}{r_o R\, T^2_M} \tag{28}$$

which (see equation 14) is identical with equation (27). Since $\Delta\mu^* = \Delta h^* - T\Delta s^*$, equation (28) can be written as

$$v = \frac{\nu r_o}{6} e^{-\Delta h^*/kT} e^{\Delta s^*/k} \frac{\Delta H(T_M)(T_M - T)}{RT_M^2} . \tag{29}$$

We now point out some of the consequences of this highly useful result, which should convince the interested student of the power of simple rate theory. Now, a major feature of equation (29) is the way in which v increases with undercooling (T_M-T). As the undercooling is increased from zero, the growth speed increases in proportion because the driving force term dominates. However, as T_M-T increases further, the exponential thermal activation term begins to take over, ultimately leading to a net decrease in v. This behaviour is plotted schematically in Fig. 4 in terms of the time it would take a sample of liquid to become completely solid at various temperatures.

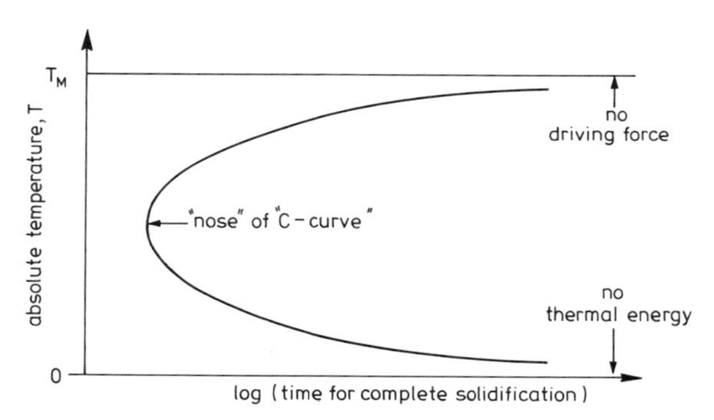

Fig. 4 Schematic plot of the time taken for the complete solidification of a sample of liquid which has been cooled below the melting point T_M and then seeded with small solid crystals to initiate solidification.

It is not easy to see this effect in liquid metals. Solidification in metals is easy, and is really dominated by the transport of latent heat away from the growth front. However, new processes have been developed which can quench liquid metals past the "nose" of the "C-curve", avoiding solidification altogether, and producing a whole range of metastable amorphous metals with exciting new properties.

C-curves can, however, be demonstrated to students by using organic compounds such as salol. Salol melts to a transparent liquid at the conveniently low temperature of 41°C and a petri dish will cool nicely on the overhead projector to show the outline of growing crystals on the screen. With care, classic C-curve behaviour can be demonstrated and, indeed, if the dish is removed to the refrigerator overnight there will be no visible growth at all! Of course, the curious will wonder why it is that an organic compound like salol solidifies so much more slowly than a metal. The entropies of fusion of typical metals are close to 1.0 R whereas that of salol is 7.5 R. Physically, "spherical" metal atoms have rotational freedom in the solid state but the non-spherical molecule of salol has to be oriented in specific directions. There is therefore a much larger ordering required when salol solidifies and the activation entropy term $e^{\Delta s^*/k}$ in equation (29) can be expected to be about 400 times less for salol than for a metal.

C-curve behaviour does, of course, apply to diffusive transformations in the solid state as well. And probably its most important application is in the hardening of carbon steels by quenching. Here, the C-curve applies to the $\gamma\alpha$ phase transformation. Quenching steels past the nose of the C-curve leads to the formation of the spectacularly hard, but metastable, phase called martensite on which so much of engineering design depends.

CONCLUSION

Thermodynamics is, at any rate to the author, a subject where an understanding of the basic equations comes only after they have been applied to many different situations at both the macroscopic and the mechanistic level. It is hoped that the illustrations that we have given above will show that materials can provide a wealth of illustrative material for thermo-dynamicists; and that thermodynamics is a great help in understanding materials.

REFERENCES

1. M. F. Ashby and D. R. H. Jones, "Engineering Materials", Pergamon, Oxford (1980).

2. M. F. Ashby and D. R. H. Jones, "Engineering Materials II". To be published by Pergamon.

3. J. W. Christian, "The Theory of Transformations in Metals and Alloys, Part I, Equilibrium and General Kinetic Theory", 2nd ed. Pergamon, Oxford (1975).

THERMODYNAMICS IN A BROAD BASED COURSE

C. R. Stone

Engineering and Management Systems
Brunel University
Uxbridge, Middlesex

SYNOPSIS

The Special Engineering Programme (SEP) is one of the original eight enhanced engineering courses introduced by the UGC. The course incorporates Mechanical, Electrical and Production Engineering with Business Studies. SEP is taught at Brunel in a specially created department - Engineering and Management Systems (E. and M.S.); and like all Brunel undergraduate courses it is a 4 year thin sandwich programme. Entry to the course is limited to about 30 highly able and motivated students each year. The scale and philosophy are such that much use is made of student active learning and a variety of small group techniques; project work is significant in all four years.

Whilst the thermodynamics syllabus is conventional, the integration of the material into the course is less conventional; applications are emphasised throughout the programme. In year 1 a Thermofluid Dynamics course provides a background that can be utilised in project work based on Artefacts. In Year 2, a Power Systems course combines the thermodynamics of prime movers with electrical power engineering. The course is continuously assessed using a variety of techniques, including: tests, essays and precis. In Year 3 a Mechanical Technology course combines fluid mechanics and thermodynamics, with heat transfer providing a useful linking element. For the associated labwork students devise their own experiments based on selected test rigs. An important part of Year 4 is project work; in addition students take a selection of final year options chosen from the whole of the Technology Faculty.

Throughout the course the emphasis is on application, and this is supported by close industrial links - for example all 3rd and 4th Year projects should be industry based. The department also runs an M. Eng. course for SEP graduates; the main component of this is an industrial project. In conclusion, the aim of the programme is to produce versatile, creative, articulate engineers, with the ability to tackle wide ranging projects in a commercial context.

INTRODUCTION

The Special Engineering Programme (SEP) started in 1978 and produced its first graduates in 1982, and is one of the original eight enhanced courses set up by the UGC. Engineering should attract a proper share of the most able students, and it was with this in mind that Government set up these 'Dainton' courses. These courses all included Business and Management Studies, and one of the aims was to produce 'high flying' engineers for manufacturing industry. The purpose of this paper is not to discuss the effectiveness of these measures, but to describe how thermodynamics is shoe-horned into one such course. Prior to this a brief description of SEP would be useful, to show how thermodynamics fits into the programme; a fuller description is presented by Clark, et al. (1984).

Entry to the course is limited to about 30 highly able and motivated students each year.

The selection procedure includes:

i) satisfactory performance in two interviews and two written tests.

ii) high 'A' level attainment (usually better than 2 'A's and a 'B').

iii) sponsorship from a collaborating company.

Like all Brunel undergraduate courses SEP is a 4 year thin sandwich programme, and the structure is shown in Fig. 1.

The M. Eng. occupies the 5th year and is industry based. The M. Eng. comprises of two components: Directed Objective Training and Project Work. Thus the M. Eng. complements the undergraduate programme and completes the training requirements of the professional institutions (IEE, I. Mech.E, I. Prod. E).

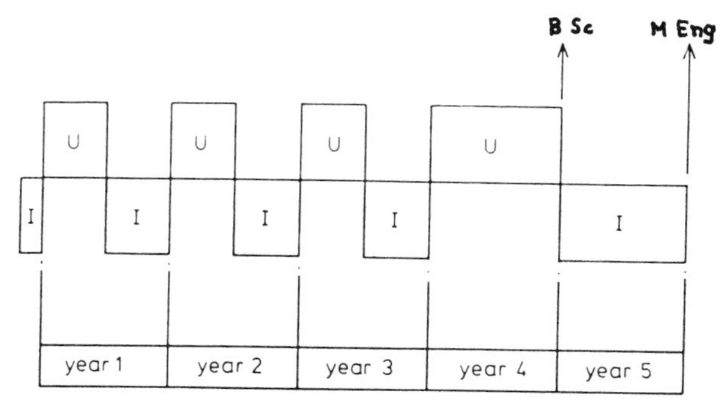

Fig. 1. Structure of SEP.

Close liaison is maintained with the 40 or so collaborating companies to provide a high level of integration between the University and Industrial elements of the programme. These companies have agreed to follow a particular training pattern that supports the University based element. Students undertake University originated assignments during the industrial placement, since this can provide a more appropriate environment for learning about certain topics e.g. aspects of Production, Management and Business. Each student is visited a minimum of two times during each industrial placement.

Finally, the scale and philosophy of SEP are such that much use can be made of student active learning, a variety of small group techniques, and project work. These include:

A Tasks Course which is described by Clark and Ainscough (1984); the Artefact Study, reported on by Wild (1984); and CAD modules described by Medland (1982).

YEAR 1

In Year 1 thermodynamics is taught in a Thermofluid Dynamics module that is part of an Engineering Systems and Principles course, Fig. 2.

The thermodynamics course is based on lectures (20 hours) with tutorial classes to cover problem sheets. The course provides an introduction to Thermodynamics by covering:

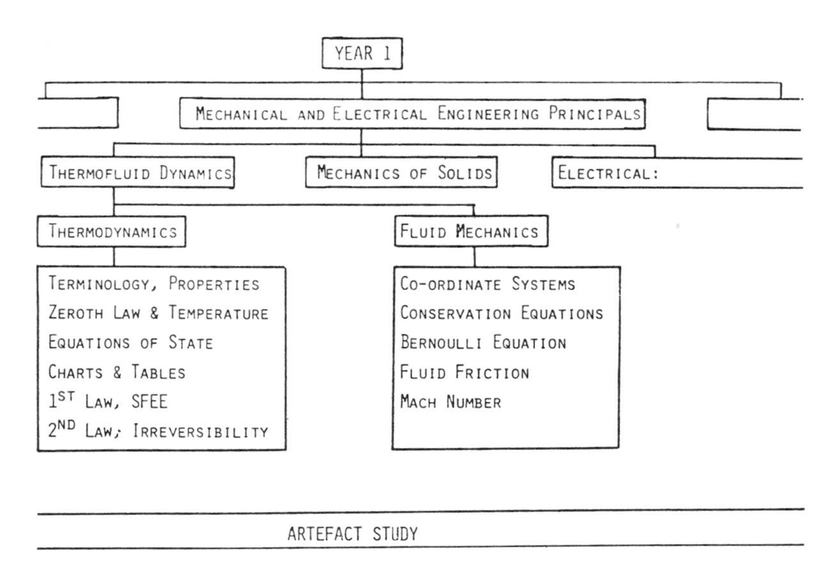

Fig. 2. Structure of SEP Year 1.

Basic Thermodynamics:- terminology; thermodynamic properties, systems and processes; equations of state and relations between properties; phase change; and thermodynamic tables and charts.

Energy Conservation:- 1st Law for closed systems and internal energy; steady flow energy equation and enthalpy; applications to simple processes.

Irreversibility:- 2nd Law and entropy; energy degradation; Clausius inequality; absolute temperature scales; thermodynamic efficiency.

Thermodynamics is assessed as part of a 3 hour Mechanical Engineering exam, that also covers fluid mechanics, structures and mechanics. There is no formal labwork in Year 1, instead 4 hours is timetabled each week for the Artefact Study.

The artefact is an item used or made by a student's collaborating company, which the student uses for a range of open-ended investigations. The artefact can also help the student to identify him/herself with his or her sponsoring company. In addition, the artefact provides a link between the first university and industry periods, since the manufacturing and business elements of the artefact study are completed during the industrial training period.

The artefact study is run as a clinic, with members of staff acting as consultants to discuss with students the investigations they might conduct on their artefacts. Engineers from the students' sponsoring companies are also invited to attend, to provide further insight into the artefact. In the area of thermofluid dynamics most experiments involve no more than flow, pressure and temperature measurements. However, tests have included the full thermodynamic performance of an air-compressor, and the evaluation of an oxygen concentrator.

During the artefact study students are encouraged to write relevant computer programs. This may be solely for the data processing, but for thermofluids, students have also evaluated series solutions for heat transfer problems.

This approach to labwork provides a bridge between the closed form of 'A' level type experiments and the uncertainty of project work.

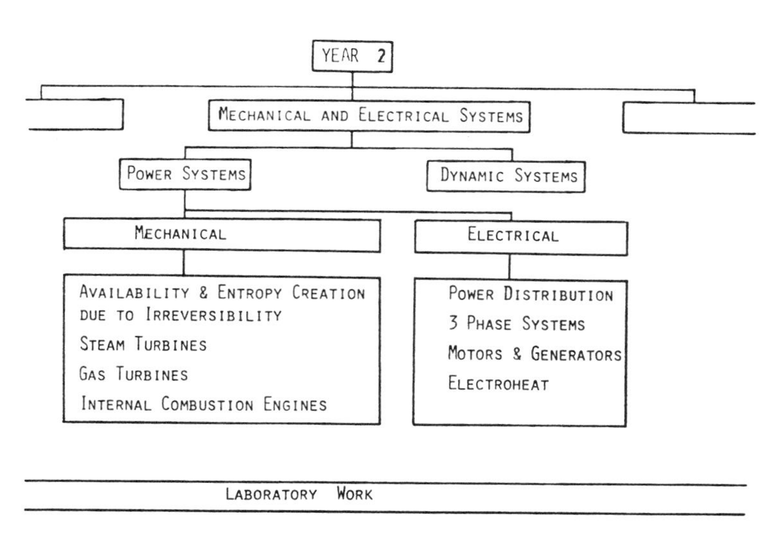

Fig. 3. Structure of SEP Year 2.

The Year 2 Power Systems Course has 20 hours of lectures divided between Thermodynamics and Electrical Power Engineering, Fig. 3.

The Thermodynamics component concentrates on prime movers (internal combustion engines, gas and steam turbines), whilst the electrical component focuses on power generation, transmission, and usage. Prior to discussing prime movers students are introduced to availability; this provides an excuse to briefly revise the 1st and 2nd Laws, and to introduce the concept of entropy creation due to irreversibility.

Availability is applied to steam power plant, where the concepts of mean temperature of heat reception and heat rejection are also introduced. This provides a convenient means of assessing qualitatively the impact of more complex cycles. This approach is continued in the discussion of gas turbines. Students are also reminded to include all the elements in a system, - Fig. 4. shows the comparison between a liquid fuelled and electric vehicle, - this then leads to economic as well as thermodynamic considerations.

Vehicle Efficiency Comparison			
Electric	%	Liquid Fuel	%
Power Station	30	Distribution & Refining	90
Distribution	90	Part Load (Engine)	20
Battery	75	Vehicle	95
Traction	90		
OVERALL	18	OVERALL	17

Fig. 4. Comparison of Electrical and Conventional Vehicles.

The associated tutorials involve student group presentations based on problem sheets, and this leads naturally to peer group assistance and involvement. The Power Systems Labwork is conventional, with groups of students conducting experiments on Gas Turbines, Spark Ignition and Compression Ignition engines.

Since the Power Systems course is continuously assessed a variety of assessment techniques can be incorporated, including:

i) solutions to specific problem sheet questions
ii) a short answer, end of term test
iii) essay assignments.

The essay assignments are included since it is essential for engineers to be able to write good reports that are both literate and technically sound. The first essay assignment involves writing a precis of a selected paper. Students are given a choice of review type papers that are relevant to the lecture course, e.g. Steam Turbines, Traupel (1979); Combined Heat and Power, Clifford et al (1980).

The second essay requires students to research their chosen topic and to be discerning in their choice of sources. The topics include:

Alternative Energy Sources.
Nuclear Power.
Recent Developments in Gas Turbines.
Modern Steam Power Plant Practice

The Year 2 course obviously builds on the Year 1 Thermofluids course, and it is also possible to use the students Year 1 Artefacts to illustrate aspects of the course, e.g. RB211 blades, gas compressors, pistons and piston rings.

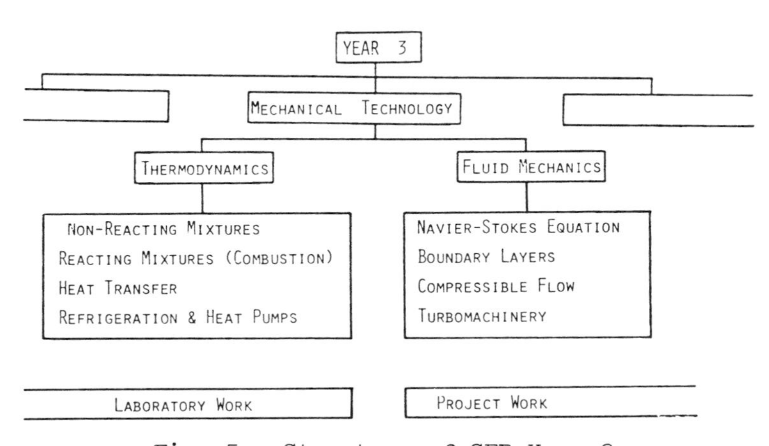

Fig. 5. Structure of SEP Year 3.

In Year 3 the Mechanical Technology course (40 hours of lectures) consists of Fluid Mechanics and Thermodynamics with convective heat transfer and the diffusion equations providing a useful link between the two elements. Dimensional analysis is used in both conduction and convection and this provides a further link with fluid mechanics, Fig. 5.

There is no time until Year 3 to discuss refrigeration and heat pump cycles. In this context it provides useful revision of cycle analysis and the 1st and 2nd Laws of Thermodynamics. In particular, the discussion of heat pumps can lead to a reminder about energy sources and resources.

However, the main theme in Year 3 is the examination of the processes that enable cycles, devices and open circuit plant to operate. Heat Transfer is self-evidently one important theme and another is Reacting and Non-Reacting mixtures.

Non-Reacting mixtures introduces the kilomole and leads through the Gibbs-Dalton Laws to the properties of mixtures. The most important application is in psychrometry where the illustrations are based on air conditioning equipment and cooling towers. Reacting mixtures are illustrated by combustion, where the 1st Law is applied to both closed and open systems. Dissociation is introduced through Le Chatelier's Principle to save a rigorous derivation of equilibrium constants.

For the associated tutorials the class is divided into two groups, and again presentations are given by students to their peers on specified problems. The course is assessed by a 3 hour Mechanical Technology paper that is divided equally between Fluid Mechanics and Thermodynamics.

The labs involve students devising their own experiments on specified rigs, namely: Orsat Gas Analysis of the exhaust from a Ricardo E6, a Refrigeration rig, and a forced convection Cooling Tower. Wherever appropriate the students are encouraged to write or use computer programs to process the experimental data. In the case of the cooling tower the students are provided with correlations for the water saturation temperature and pressure, and for the partial pressure of the water vapour as a function of wet and drybulb hygrometer readings.

YEAR 4

In their final year SEP students draw from a range of final year options that are mostly taken from departments within the Faculty of Technology. For example, the Mechanical Engineering Department runs a Thermal Power option (60 hours of lectures) that covers gas turbines, internal combustion engines and steam turbines in somewhat greater depth. In the near future it is likely also to cover energy management and conservation - there are many more users of gas turbines than designers.

SEP students also undertake a major project that is normally originated from within their collaborating company. Some recent projects include: the modelling and optimisation of advanced turbofan/prop engines, the microprocessor control of multiple air compressor installations, and a self optimising ignition controller for a spark ignition engine. The multi-disciplinary nature of these projects is well matched by the students' abilities and the staff interests.

CONCLUSIONS

The constraints of time eliminate the possibility of repeating material, yet the 6 months industrial placement provides students with ample opportunity to forget material that is not in constant use. Consequently, each Thermodynamics unit has to be as self-contained as possible. Nonetheless it is also important to illustrate as many links as possible with other subjects.

Applications can be quickly illustrated by using 35 mm transparencies, and there seems to be some advantage in re-using the same transparency to illustrate different points in subsequent lectures and courses.

REFERENCES

1. Clark, C., Medland, A. J., Rakowski, R. T., and Wild, R., "A New Enhanced Engineering Programme for Manufacturing Industries", World Conference on Education in Applied Engineering and Engineering Technology, Köln, 1984.

2. Clark, C., and Ainscough, J. A., "An Engineering Design Tasks Course for Small Groups, with Continuous Assessment",

World Conference on Education in Applied Engineering and Engineering Technology, Köln, 1984.

3. Clifford, D., Coates, R. and Park, A., "Repowering of Slough Estates for Optimum Energy Conversion", Proc. I. Mech. E., 1980, Vol. 194, No.34, pp.279-289.

4. Traupel, W., "Steam Turbines, Yesterday, Today and Tomorrow", Proc. I. Mech. E., 1979, Vol. 193, No. 38, pp.391-400.

5. Wild, R., "The SEP Artefact Study", SEP Monograph 84/1, Brunel University, 1984.

MAXIMIZING THE TEACHING OF EXPERIMENTAL THERMODYNAMICS

P. W. Foss and D. R. Croft

Dept. of Mechanical and Production Engineering
Sheffield City Polytechnic

SYNOPSIS

Current and well tried techniques in imparting maximum information on experimental thermodynamics to the student are discussed, ranging from the basic laboratory sheet to work books, tape-slide and video presentation. However, in order to maximise the student's involvement and awareness of topical Industrial applications of the hardware that he is using, it is suggested that a combined video-computer package be used.

This is intended to supplement the role of the lecturer by leading the student through the experiment, extending his analysis to include topics not examined experimentally.

The possible implementation of an interactive computer-video system is described.

INTRODUCTION

Experimental work can play a vitally important and stimulating role in the students' understanding of Engineering Thermodynamics. Besides strengthening ideas, assumptions and theoretical models discussed in lectures, the "hands-on" approach gives the student a feel for the hardware and implementation; how well it runs, vibrates and sounds like, how it smells, the accuracy with which experimental results may be made, and agreement or otherwise with theoretical predictions.

The student must be able to relate to the equipment; not be frightened of it and confident of being able to obtain the results required. This requires the equipment to be designed ergonomically by experienced Thermodynamicists! Also, the students must see the need for doing an experiment, rather than blind acceptance of the results of other people.

Engineers in the outside world do experiments in order to establish empirical coefficients, and to confirm concepts and assumptions that may not be in the reference books. Whilst of necessity the set laboratory experiments will confirm well known ideas, the student must be given some feeling for this aspect of experimental work. Unfortunately, experimental subjects are often poorly taught and the student may be given the impression that the academic staff are not interested in practical work, and that he is merely expected to spend time in the laboratory to satisfy the requirements of the Examination Board. This state of affairs is often brought about by the timetabling officer's view that the lecturer may be more profitably employed in the lecture theatre and that enthusiastic, but relatively inexperienced research assistants can conduct the laboratory session. The number of students involved, and the placing of experimental equipment in different laboraties mean that, if a lecturer is present, he will have to share his time between several groups of students in different parts of the building. This may sound unrealistic but does often happen in practice.

Maximum benefit to the student may be achieved by integrating the actual experiment with a thorough description of the hardware; its operation, limitations, safety aspects, and possible extensions. Ideally this should be done by the lecturer, supplemented by all relevant teaching aids such as written material and tape-slide or video presentations. However, care must be taken not to use technology for its own sake. For example, the advent of the magic lantern gave the less conformist churches the ability to project verses of Victorian hymns on to a screen, accompanied by suitably chosen pictorial ideas, but one wonders whether this produced better results than the written words backed up with spoken thoughts!

Finally, the student must be encouraged to extract the maximum amount of information from his measured results, and to supply answers to questions designed to broaden his thinking. To give an example, if the experiment is to measure the output power from an engine, the student should go on to extract the various engine efficiencies including Volumetric efficiency. Although primarily serving to confirm whether his results are satisfactory, an insight into the behaviour of the engine test unit can be gained. Unfortunately, because of the original standard of teaching at College, Industrial employees often do not perform experiments correctly. A well known supplier of educational equipment sold a small engine test unit for many years which worked well and produced acceptable results. General development, use of modern materials and the pressures of customers and sales engineers to update the test bed resulted in a better looking unit with electronic readout which, it was noted, produced a higher output power. This was considered to be a good selling point by the salesmen and it was not until much later that somebody decided to calculate the engine volumetric efficiency. When this turned out to be 150% for a normally aspirated two stroke engine, it became obvious that the air box and induction system was too small and the system had to be re-engineered. Had the designer taken the trouble to evaluate the complete set of performance data, this would not have slipped through to the production stage.

SURVEY OF STUDENT OPINION OF EXPERIMENTAL WORK

Students were questioned to discover their views on effectiveness of current experimental work and teaching methods. Typical results are shown in Table 1, which contrasts the opinions of first year, first term full time Degree students direct from school, with those of much older part-time Degree students aged 25+ having considerable practical Industrial experience.

Laboratory Sheets

Traditionally, the student is given a laboratory sheet outlining the experimental equipment, procedures to follow, analysis required, and some questions to answer relating to the results achieved. Generally, students found the set experiments relevant to the course although additional to the lecture material. Whilst students from Industrial backgrounds found the experiments interesting, those direct from school tended to regard them as a hurdle to pass.

Students regarded the laboratory sheets as only partially helpful - they were not always designed with the actual hardware in mind and tended to be somewhat academic.

Table 1

EXPERIMENTAL THERMODYNAMICS QUESTIONNAIRE

Table 1. EXPERIMENTAL THERMODYNAMICS QUESTIONNAIRE		ANALYSIS OF RESULTS (% YES, % NO)	
		BSc (PART TIME)	BSc (FULL TIME)
1. Your course	BSc / BSc (PART TIME) / EBS / HND / TEC / OTHER		
2. Year of your course	1 / 2 / 3 / 4 / 5	YEAR 3	YEAR 1
3. Was the experiment (a) Relevant to course	YES / NO	94, 6	100, 0
(b) Additional to lecture material	YES / NO	76, 24	65, 35
(c) Interesting to carry out	YES / NO	88, 12	43, 56
(d) Informative	YES / NO		
4. Was a laboratory sheet provided?	YES / NO	82, 18	96, 4
5. Was the laboratory sheet helpful?	YES / PARTIALLY / NO	35, 59, 6	9, 74, 17
6. Did you learn anything from the experiment? (a) About the theoretical aspects	YES / NO	76, 14	67, 33
(b) About the hardware	YES / NO	65, 35	61, 39
7. Who looked after you?	LECTURER / RESEARCH ASSISTANT / DON'T KNOW / TECHNICIAN / NO-ONE	LECTURER AND TECHNICIAN	RESEARCH ASSISTANT AND TECHNICIAN
8. Roughly how many groups were there in the laboratory at the time?	1 / 2 / 3 / 4 / 5 / 6 / 7 / 8	2	3
9. Did somebody explain (a) Theoretical aspects?	YES / NO	94, 6	78, 22
(b) Hardware and operation?	YES / NO	76, 24	74, 26
(c) What to look for in results/ conclusions)?	YES / NO	94, 6	52, 48
(d) Safety Aspects?	YES / NO	23, 77	17, 83
10. Did you need computing facilities?	YES / NO	6, 94	4, 96
11. Did you have access to computing facilities?	YES / NO	65, 35	56, 44
12. Could the experiment have been (a) More interesting?	YES / NO	65, 35	91, 9
(b) More relevant?	YES / NO	35, 65	52, 48
13. If YES, suggest how this could have been done?		See text	
14. What questions would you have wanted the answers to, if different from mine!?		Nil return!	
15. What is your subjective impression of the way in which experimental thermodynamics is taught? (a) Teaching Methods	GOOD / REASONABLE / POOR	29, 65, 6	18, 65, 17
(b) Experiment Design	GOOD / REASONABLE / POOR	11, 76, 13	18, 65, 17

Whether a lecturer or research assistant conducted the experiment, the theoretical aspects and hardware used were explained, but it was disturbing to find that safety aspects were not covered very well.

Computation

Experimental analyses were relatively simple so that calculators were adequate, but despite micro-computers or main frame terminals being situated in all laboratories, students did not seem to realise that they were available!

Interest Level of Experiment

Students generally felt that experiments could be made more interesting, the following points being made:-

(a) Ensure that experiment "works" and that the "right" results are obtained.

(b) Groups too large, too little time allocated, just result collecting.

(c) Give student more control of what to investigate.

(d) Give more detail about hardware.

(e) Label the equipment more reasonably so as to identify data more easily.

(f) Make moving parts visible, or use sectioned models.

(g) Use more modern equipment.

MAXIMISING TRANSMITTED INFORMATION TO STUDENT

There are three principal ways in which information can be imparted to the student.

(a) Lecturer and simple laboratory sheet.

(b) Lecturer and more involved text.

(c) Lecturer and tape-slide or video programmes.

The first method tends to be used since it is the most cost-effective approach, but is limited by the availability of the lecturer.

Provision of Work Sheets or Books

The second technique involves the provision of work sheets or books which may be related to a single experiment, or to a range of experiments. They may be written with a specific item or range of equipment in mind made by a single manufacturer, or be generally applicable for use with equipment supplied by several different manufacturers. Clearly, a work book is most valuable if it relates to a specific item of equipment.

Most work books have their origins in the mid 1960's when programmed learning texts were fashionable, and the student progressed at his own rate through both academic and practical studies. At one time the Royal Air Force College Cranwell considered a scheme whereby Engineering Officers direct from various University disciplines could progress through a conversion course to emerge as Air Force Specialists.

The concept failed because of logistic and administrative problems, but one of the authors was responsible for designing a package of experimental work in Aerodynamics[2], which set a precedent for the other Engineering Squadrons.

A book was devised containing three Sections. The first described the equipment: wind tunnels, models, and instrumentation. The second Section explained the theoretical basis for the experiments and the calculation of results, whilst the third Section led the student step by step through the procedure, data to be recorded, calculations to be made, and conclusions to be drawn.

Questions were included to relate the experiment to full scale values. For example, after an experiment on a model aircraft of a type in service the student was given typical full scale values of the speed range, service ceiling, rates of climb and stick forces, asked to scale his values and compare with full size results.

When completed and marked, the book was valued by the student as a useful and interesting set of notes.

Tecquipment Ltd.[3] produced a range of experimental sheets, manuals and workbooks for its products along similar lines.

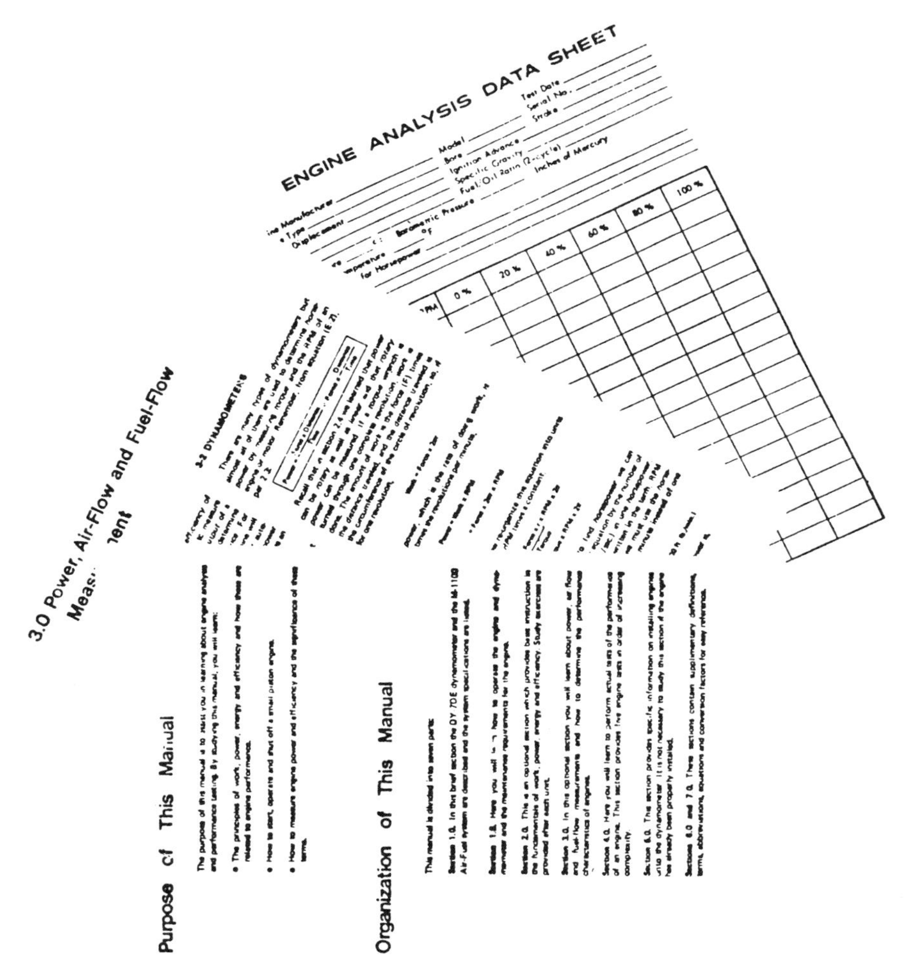

Figure 1 Typical Workbook Presentation

"Go-Power Systems"[4] also supply a manual to assist the student in learning to run, and take performance measurements from a small piston engine. Figure 1 gives a copy of the purpose and organisation of the manual, and shows the layout adopted. It is orientated towards the Technician Engineer, but a similar book for Degree students would be invaluable.

Lomax and Saul[5] wrote a book on experimental work in Hydraulics which again leads the student along well-defined paths of experimentation, showing typical results to be achieved, and providing the theoretical bases. Such a book in Thermodynamics would be valuable.

Not quite in the same category of being written with the student in mind, but nevertheless an excellent introduction to experimental work which is useful to have in the laboratory for students to read is Engineering Experimentation by Tuve and Domholdt[6].

Tape-Slide Presentations

At one time tape-slide presentations were considered to be an excellent way of introducing the student to the layout and operation of the test rig and the measurements to be taken.

Tecquipment included both a general introduction to the topic and examples of larger scale equipment in the real environment.

The programmes were designed for groups of 2 or 3 students to watch. Although paced automatically, the student could stop the programme at any stage or replay parts that had not been fully understood.

Slides have the disadvantage of being static pictures, although Philips experimented with a system that could show slides in rapid succession to provide an illusion of motion, as well as having stationary pictures.

Such a programme still requires written back up material for the student to retain and work from when he is away from the laboratory.

Tape-slide programmes are now little used and Tecquipment have ceased supplying them.

Video Presentations

Clearly the role of the tape-slide programme can be taken over by a video presentation, with the added realism of motion and sound. The student can be given a far better impression of the equipment in its industrial environment.

Computer Involvement

For the majority of calculations encountered in the traditional Thermodynamics experiments there is no need to have access to a computer. Indeed, there is an argument in favour of calculation by hand so that the student gains an appreciation of the magnitude of the parameters involved at each step. However, all experiments at Sheffield City Polytechnic have micro-computer programs and hardware available in the laboratory if required, but students do not seem to wish to use them.

The computer, however, can play an important role in verifying complex mathematical models, and demonstrating facets which time or equipment does not permit.

For example, a standard experiment is to study the performance of an internal combustion engine, taking measurements of torque, air and fuel flows, and various temperatures and flow rates. The laboratory engine is limited in size, but the student could be asked to make predictions for the performance of larger units. "Economy of Scale" would then become apparent.

Computer models are available, such as Benson and Whitehouse[7] and Sorensen[8] which enable a detailed analysis to be made of a particular engine unit. The student is able to confirm the mathematical model for the engine tested. Subsequently, besides being asked to provide an estimate for the performance of a very much larger unit, the student can assess the effect of modifying his engine in various set ways in order to improve the performance, or to predict the performance of the engine when operating on alternative fuels.

Simulating Industrial Useage

The student is at once immersed in the real world. Leeds Infirmary, for example, have a power station that uses multi-fuel diesel engines which can burn either diesel oil or gas. The gas may be natural or manufactured biologically. The station must produce both electricity and hot water as economically as possible, both the load and cost of the fuel varying daily. From a theoretical model of the plant, and given typical operating costs, the student may use the results gained from his actual experiment to suggest how best to run the real power station.

Ideally, the computer program must be in an inter-active style so that it both prompts and teaches the student. The references quoted do not use this format, and should therefore be re-written.

Similarly, having completed experiments on the laboratory petrol and diesel engines, gas and steam turbines, and being given typical operating costs for real versions of each, the student can be led to a decision on the best power plant for a particular power station. Such material can be backed up by video case studies, such as the choice of heavy oil engines for the Hereford power station.

THE ULTIMATE LEARNING SYSTEM, COMBINING VIDEO AND MICROCOMPUTER

The maximum transmitted information and student involvement can be obtained by integrating video and microprocessor.

The VTR is controlled by the microprocessor and computational facilities are available. The program may be stored on the beginning of the video tape. It is technically possible to build such a system around a relatively cheap micro-computer having a user port to accept count impulses from the VTR so that the tape position can be controlled by the computer program. The VTR should have an electronic counter rather than a mechanical device to obviate the need to build in a system for obtaining impulses. At current prices a minimum cost system could be assembled for less than £350. A typical DIY system is described in reference 9.

The more expensive items of laboratory equipment could be supplied by the manufacturers complete with such a system already installed, rather as both Plint and Tecquipment can supply engine test beds with computer control. Indeed, computer control of the test bed can be built into the teaching computer program.

The computer program should again be written in an interactive style, and lead the student through the experiment and subsequent expansion of results to Industrial practice.

The flow chart given in Figure 2 shows how such a system could be implemented.

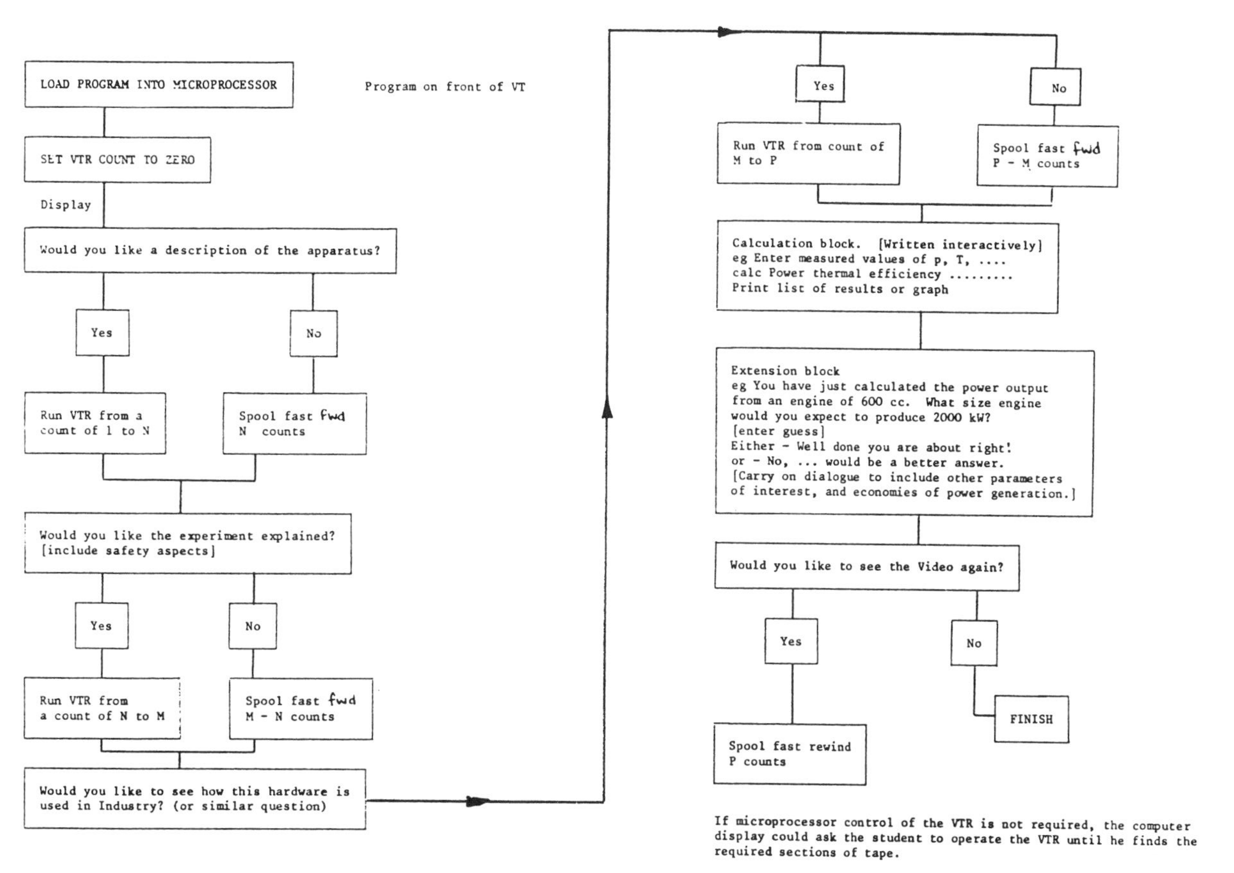

Figure 2 Combined Video-Microcomputer Presentation

REFERENCES

1. Foss, P. W., "The design and Development of a Multi-fuel, Variable Compression Engine Test Bed Intended for Teaching and Research", Second Polytechnic Conference in Thermodynamics, Leicester 1983.

2. Foss, P. W., "Experimental Aerodynamics", Internal Publication Royal Air Force College, Cranwell, 1967.

3. Several authors, "Various Technical Manuals and Literature", Tecquipment Ltd., Nottingham, 1976-1980.

4. Anon. "Principles of Engine Analysis", Go Power Systems Inc., Palo Alto, California.

5. Lomax, W. R., Saul, A. J., "Laboratory Work in Hydraulics", Granada Publishing, 1979.

6. Tuve, G. L., Domholdt, L. C., "Engineering Experimentation", McGraw Hill, 1966.

7. Benson, R. S., Whitehouse, N. D., "Internal Combustion Engines", Vols. 1 and 2, Pergamon Press, 1979.

8. Sorensen, S. C., "Simple Computer Simulations for Internal Combustion Engine Instruction", Int. Journal Mech. Eng. Education, Vol. 9, No. 3.

9. Hallgren, R. C., "Interface projects for the Apple II". pp.170-140. Prentice-Hall, Inc., 1982.

ENGINEERING LABORATORY TEACHING - A CASE FOR CO-ORDINATION

M. R. Heikal and T. A. Cowell

Mechanical and Production Engineering
Brighton Polytechnic

ABSTRACT

This paper presents the results to date of an ongoing exercise to identify and implement an appropriate philosophy for laboratory teaching within a Mechanical Engineering Department. The overall objectives of the course are stated and the measures taken to achieve them described. The key features of the programme are:-

i) the appointment of a course coordinator to oversee the whole of the three year programme.

ii) gradual progression during the three years from controlled assignments, through open-ended experiments to project based work.

iii) the integration of computerised data acquisition and computer aided learning into the overall scheme.

INTRODUCTION

Through the years the usefulness and objectives of laboratory teaching programmes have been the subject of numerous discussions and studies (1-13). The results of these studies have varied greatly in their definitions of the objectives of laboratory teaching and the ways in which they are best achieved. At one extreme are those which conclude that laboratory work should be abandoned, so little is it worth. Other studies have defined whole lists of objectives which practical work could/should achieve.

The advance of modern instrumentation and automatic data acquisition systems followed rapidly by the development of Computer Aided Learning (CAL) techniques have not helped to simplify views on laboratory teaching. It is argued that the rapidly developing engineering world requires today's engineering graduates to be well equipped with all the tools for engineering problem solving. However there are now several schools of thought on the ways in which laboratory teaching can be used in achieving this goal.

One school of thought (14) warns against the dangers of using 'automated' laboratory experiments, stating that they cannot enhance the experimentation skills of the students. Another school (15-20) gives support to the use of modern instrumentation and data acquisition techniques; saying that students need exposure to such modern developments and need experience of their use during their engineering course.

A further school of thought (15,21-28) has grown up with the development of cheap and powerful micro and minicomputers and with the rapid progress of CAL. Some CAL supporters believe that laboratory experiments in total are now so inefficient in comparison with CAL as to be of little practical use. They emphasise how powerful, flexible and inexpensive CAL is in comparison to more traditional methods.

In the light of these conflicting views, a decision was made to undertake a study and review of the departmental laboratory teaching philosophy. As a result of the study, it was concluded that many of the conflicts arise because the earlier studies were subject specific and related to a period of one academic course year only. Only a few attempts (2,9, 29-31) have been made to structure a coordinated laboratory programme covering all subjects and lasting the full duration of the course. These attempts however lack the integration of computers into their programmes of work.

From our review it was also concluded that all the different schools of thought on laboratory teaching objectives and methods do in fact contain complementary rather than conflicting objectives. Moreover, it was felt that in order to produce engineers fully capable of meeting the demands of today's rapidly developing engineering world, experience must be provided in all the above areas.

The results of our study have suggested that it is possible to structure a coordinated laboratory programme, lasting the duration of an engineering course which will satisfactorily cover these areas. In this paper the common objectives of the various aspects of laboratory teaching are identified and the different approaches to laboratory exercises described. The basic structure of a coordinated programme to achieve most of these objectives is also outlined. An attempt is made to illustrate how to combine 'traditional' and 'modern' laboratory techniques to the students' best advantage.

AIMS AND OBJECTIVES OF LABORATORY WORK

Before any attempt could be made to plan a coordinated programme of laboratory teaching,the objectives of such a scheme had to be defined clearly. Most papers on laboratory teaching have made some attempt to define objectives (1-13) and several authors have pointed out the value of differentiating between aims and objectives (11). 'Aims' are defined as a desire to move in a particular direction e.g. to develop communication skills and 'objectives' are a target to be reached e.g. teaching the ability to use a certain instrument. However, for the sake of simplicity, the word objectives will be used collectively to describe both aims and objectives as defined above.

The number of objectives listed in any one paper varied from a listing of over 20 (11) to blanket statements such as 'teaching experimentation' (1,13). In general, the broader the objectives, the fewer they were in number. After careful study of the various lists in the literature and discussion with departmental staff, the following were agreed as objectives to be met by our overall laboratory teaching programme over the three years. This list represents a distillation of the majority of the possible objectives from the published literature.

i) To supplement and support lecture material.

ii) To familiarise students with engineering equipment, techniques and materials.

iii) To develop in students the skill of experimentation.

iv) To develop engineering 'nous' in students.

v) To develop oral and written communications skills.

vi) To develop in the student a 'feel' for good technical judgements so that a balance between theoretical and experimental techniques of analysis and design can be made.

vii) To inspire, excite, interest and motivate students towards engineering.

The above objective listing is not in order of priority, and the objectives listed have been selected as those most likely to develop in students the final goal of engineering education - the skill of problem solving.

TYPES OF LABORATORY EXERCISES

Having identified the teaching objectives for the laboratory programme it became necessary to examine the different types of laboratory exercise in order to identify how best to satisfy them. It is generally agreed (5, 6, 11, 29, 32) that although laboratory exercises can be considered as covering a continuous range of types, they can be conveniently categorised into three main groups.

Controlled Assignments

These are well defined experiments of short duration (about 2 hours) in which an 'instruction' sheet is provided. This contains a clear statement of the problem, its origin and significance; lists of apparatus and a brief description of their application, limitations and accuracy. It also describes the steps to be followed and the range in which measurements are to be taken with appropriate reasoning. A guide to the discussion and conclusions to be drawn is also given.

This type of experiment has been widely criticised (33, 34) on the grounds that the student will lack any sense of achievement and hence interest in such experiments, and so derive no benefit from them.

Against this, it is believed however, that a limited number of such experiments should be included in any programme of laboratory work at the initial stages - they serve to give the student confidence in using the laboratory. As this kind of experiment leaves little room for 'floundering' on the part of the student, disappointing and confidence eroding results are avoided. In fact the student gains positive instruction in good working habits giving confidence for later experiments.

Open-ended Assignments

These tend to be longer exercises or problems, normally set by the tutor, which can extend over more than one laboratory period. These experiments should encourage the student to accept responsibility for investigations and to work independently. The student can plan and execute his own experimental work supervised by the tutor.

These kind of assignments provide the student with a challenge and the opportunity to meet it with previously gained knowledge. The degree of 'open-endedness' can be varied greatly over a wide spectrum of applications. Often a certain degree of 'open-endedness' can be introduced into controlled experiments without the need to modify equipment. Simple modifications of instructions to the student achieve this open-ness e.g. "test engine at specific conditions" changes to "determine how the performance of the engine varies with mixture strength". These assignments also lend themselves to analytical or computational interpretations which require experimental data.

Project

These are assignments designed to give the student opportunities to practise problem solving of the type he will face during his engineering career. It follows that these problems should be as realistic as possible within a reasonable time limit (usually about 9 hours). The problem should contain an element of analytical or numerical analysis and be as original as possible.

Projects can be the most challenging and interesting type of laboratory exercise the student meets, providing him with the opportunity to develop his own originality and organisational abilities. The student is given complete charge of the planning and execution of the project while the supervisor keeps a minimal profile. The student is expected to give an oral presentation of the work and complete a formal report at its conclusion. To introduce an extra element of the real engineering world, the student can be asked to write the report as if it were for presentation to an identified recipient, e.g. an immediate manager, the Board of Directors etc.

THE PROPOSED CO-ORDINATED PROGRAMME

As this paper must be limited in length, only the underlying rationale and a brief outline of the programme structure will be described here. The following is a summary of the structure of the laboratory programme over the duration of our three year degree course in Mechanical and Production Engineering. The whole of this course is overseen by a single coordinator. Although the subject leaders maintain responsibility for their own subject laboratory classes, the coordinator is responsible for ensuring that the overall programme objectives are met. This can result in requests to subject lecturers to introduce extra elements into their laboratory classes to satisfy the overall course needs. To assist in this task, subject teachers are asked to identify the teaching objectives of each of the individual laboratory exercises for which they are responsible. (The students are also informed of these particular objectives before they start any exercise).

One of the stated objectives of this programme is to supplement and support lecture material. Experimental exercises are unsatisfactory in themselves in achieving this aim. CAL techniques, however, provide a more effective, flexible and efficient way of supporting lectures. Therefore, CAL exercises are used in place of the more traditional tailor-made experiments used to illustrate lecture material. Consider for example, the effects of exhaust heat exchanger characteristics on the overall system performance. This is best illustrated by the use of a simulation package for two reasons. Firstly, it allows a more complex and therefore, more realistic system to be studied. Secondly, the speed and flexibility of the simulation provides the opportunity to study the system more comprehensively.

Although the programme is coordinated for all subjects the examples given will be in the Thermodynamics and Fluids area as this is the area of this conference.

1st Year

In the first term of the first year the programme starts with a structured course of experimentation lasting 12 hours including lectures and demonstrations. The syllabus of this course is given in Appendix I.

The student is also given additional lecture notes on the subjects covered in this course and a list of reading which will help him throughout the laboratory programme. The course is intended to equip the student with the necessary tools to gain good results from the laboratory programme and also instil good experimentation skills. The student also follows a computing course in parallel with the experimentation course. It consists of familiarisation with the basis of two computer languages, one compiled (e.g. FORTRAN) and one interpreted (e.g. BASIC).

During the remainder of the first term and throughout the second term students continue their laboratory work with controlled experiments. These are subject specific but are coordinated within a general framework to provide applications for material studied in the experimentation course. By taking this radical approach it is felt that experiments become more realistic and interesting for students, as opposed to the more usual case when students perform specially designed exercises to illustrate aspects of the experimentation course. This approach also provides the opportunity for more comprehensive discussions of the topics covered in the course. An example is an engine test. This programme invites students to examine and discuss how the engine is instrumented, and to comment on the different measurement techniques used and to estimate errors in measurements, etc. The emphasis falls on how the test rig is designed and the experiment planned and executed to best achieve the stated objectives of the test. This is in strong contrast to the normal approach of asking students to determine the engine performance under different running conditions.

In parallel with these experiments students are taught how to use a multi-user mainframe computer and a microcomputer. They are also taught computer graphics and the use of graphics libraries.

By the end of the second term it is expected that the students will have gained confidence, developed good work habits, and have the necessary experimentation and computational skills to progress to the next 'stage'.

During the third term assignments progressively contain more and more elements of open-endedness. They encourage the student to accept more responsibility and provide opportunity for more independent experimentation. Students are also encouraged to experiment with CAL packages e.g. losses in pipes or one-dimensional heat conduction.

Throughout the first year students work in groups of 4 or 5 taking it in turn to lead and coordinate the group effort and write the final report. Each member of the team keeps a log book regularly examined by the laboratory supervisor. All members of the group share equal marks for reports which count as 60% towards the total laboratory mark. The remaining 40% is gained by individual oral assessments at the end of the year.

2nd Year

During the second year the number of laboratory exercises is reduced and their duration increased (typically 6-8 hours). Assignments become more open-ended and the student assumes full responsibility for the planning and execution of work. Experiments frequently include a limited amount of multi-disciplinary work but in general are subject specific.

In addition, some semi-controlled exercises are included in the programme, including automatic data acquisition and analysis. We feel that it is important that students are exposed to these new techniques during the course, to learn and appreciate their advantages and their limitations.

In order to avoid 'automating' laboratory work, automatic data acquisition is used only where it increases the potential benefits from an experiment. For example, the calculation of lift and drag characteristics from the pressure distribution around an aerofoil. The students are required to do one experiment 'manually' and to process the data themselves for one angle of incidence. This done, the computer is now used to repeat the experiment at different angles of incidence. This, of course, allows the full characteristics of the aerofoil to be established during the course of one laboratory session.

In parallel to the laboratory exercises, more sophisticated CAL packages are used, e.g. potential flow, 2-D steady and transient heat transfer, performance of Hydraulic machines etc.

Again group work is done, this time with smaller groups (2/3). Students again complete log books and formal reports are presented for each experiment. The recipients of these reports are identified to students in advance and via co-ordination a variety is provided. Students are also asked to give short oral presentations of their work at the end of each assignment.

3rd Year

By the third year assignments take the form of mini research projects whose objectives are the solving of realistic problems. These contain an element of original work as far as the student is concerned. The projects are subject specific but may contain significant amounts of multi-disciplinary areas of work. The problem is given to the student and no preference of approach, either analytical (computational) or experimental investigation is indicated to them. However, projects should contain both elements.

The student is allowed to develop and use his 'feel' for technical judgement. He decides which approach to use and when to use it in order to procure the most efficient solution. He has to learn to adopt an unbiased approach, using the one best suited to the occasion. He must also be able to produce a co-ordinated approach to achieve the best and most efficient solution.

An example of a realistic problem in which a computer is appropriately used to assist in the development of a solution is the determination of the optimum thickness of insulation for a steam pipe. In this problem, saving in fuel as a result of thicker insulation is offset by the cost of the insulation itself. The solution depends on fuel cost, insulating material costs, life of insulation, diameter of pipe, temperature of steam, conductivity of insulating material and many other variables.

The solution to such a multivariable problem can be achieved by repetitive calculation which could be performed by programming a micro-computer. The heat transfer data, e.g. Nu vs Re relationship has to be obtained via experiment, The student is also faced with a technical decision regarding approximations, e.g. whether the radiation part of the heat transfer from the outer surface of the insulation is important or whether convection is more dominant.

Each student is now working individually and is expected to complete 2 or 3 of these types of project. In each case they are required to present individual formal reports.

SUMMARY

It has been shown that a programme for laboratory teaching can be coordinated between different subjects over the duration of an engineering course. Such a programme contains all the elements required to develop in the student the necessary skills for solving Engineering problems.

In addition to the experimental exercises this programme also includes and develops the use of modern instrumentation and CAL techniques so as to cover most of the objectives of laboratory work.

An example of such a programme is currently undergoing development in our Department and our experiences will be published at a later date.

REFERENCES

1. Wood, W. G., "The Aims of Laboratory Teaching", Bull. Mech. Engg. Educ., 1969, Vol.8, pp.115-117.

2. Smith, A. G., "Laboratory Work - Aims and Achievements" Bull. Mech. Engg., 1969, Vol. 8, pp.105-114.

3. Williams, H. W., "Remarks on Laboratory Instruction", Proc. ASEE Annual Conference, 1981, pp.515-518.

4. Sturley, K. R., "Is Laboratory Work Really Necessary?", Int. J. Elect. Engng. Educ., 1967, Vol. 5, pp.63-66.

5. Berthoud, L. A., "A First Year Electrical Engineering Laboratory to Satisfy All Requirements?", Int. J. Elect. Engng. Educ., 1977, Vol.14, pp.301-306.

Carter, G., Lee, L. S., "A Study of Attitudes to First Year Undergraduate Electrical Engineering Laboratory Work at the University of Salford", Int. J. Elect. Engng. Educ., 1975, Vol.12, pp.278-289.

7. Longer, G., "The Role of Experimental Work in the Teaching of Fluid Mechanics", Proc. Conference on Teaching Thermo-Fluids, 4 December, 1979, I. Mech. E., London.

8. Kapplin, J. O. et al., "Use of Self-demonstrations in an Undergraduate Laboratory Program", Int. J. Elect. Engng. Educ., 1967, Vol. 5, pp.203-213.

9. Dubey, G. K., "Some Ideas on Laboratory Instruction in Engineering Education", Int. J. Elect. Engng. Educ., 1967, Vol.5, pp.215-219.

10. Boud, D. J., "The Laboratory Aims Questionnaire - A New Method for Course Improvement?", J. Higher Educ., 1973, Vol. 2, pp.81-94.

11. Carter, G., et al., "Assessment of Undergraduate Electrical Engineering Laboratory Studies", IEEE Proc., September, 1980, Vol. 127, Pt.A, No.7.

12. Fishenden, C., and Markland, E., "A System Approach to Elementary Laboratory Instruction", European J. of Engng. Educ., 1980, Vol. 4, pp.293-302.

13. Ernst, E. W., "A New Role for the Undergraduate Engineering Laboratory", IEEE Transactions on Education, May 1983, Vol.E-26, No.2.

14. Graham, A. R., "Instrumentation for the Undergraduate Laboratory", ASEE Annual Conference Proc., 1982, pp.204.

15. Nuttall, H. E., and Mead, R. W., "The Use of a Combination of Data Acquisition and Simulation as a Cost Effective Laboratory Teaching Aid", ASEE Annual Conference Proc., 1979, pp.375-379.

16. Wright, E. H., et al., "Automatic Data Acquisition for Undergraduate Fluid Flow Studies", ASEE Conference Proc., 1982, pp.603-609.

17. Pellei, D. J., "Computer Requirements for Electrical and Mechanical Engineering Curricula", IEEE Trans. on Educ., November, 1982, Vol.E-25, No. 4.

18. Crowe, D. A., and Durisin, M. J., "A Low Cost Microcomputer System for Process Dynamics and Control Simulation", ASEE Annual Conference Proc., 1982, pp.658-662.

19. Booker, S. E., et al., "Use of Micrcomputer Control to Enhance Learning in a Fluid Mechanics Laboratory", ASEE Annual Conference Proc., 1981, pp.794-799.

20. Douglas, G. W., "Basic Instrumentation for Mechanical Engineering Undergraduate Laboratory", ASEE Annual Conference Proc., 1982, pp.211-213.

21. Gosman, A. D., "The Impact of the Computer on Teaching of Fluid-Mechanics and Heat Transfer", Proc., Conference on Teaching Thermo-Fluids, 4 December, 1979, I.Mech.E., London.

22. Sommer, H. T., et al., "Computer-Aided Engineering Education", Mechanical Engineering, December, 1982, Vol.104, No.12, pp.38-40.

23. Mitin, A. I., and Pashkin, E. N., "Methods and Types of Computer-Aided Instruction", Moscow University Comput. Math. Cybern., 1982, Part 4, pp.44-50.

24. Jocquot, R. G., et al., "Interactive Microcomputer programs for Teaching Engineering Systems", ASEE Annual Conference Proc., 1982, pp.663-666.

25. Schuermann, A. C., and Hammertzheim, D., "Microcomputer Use in the Engineering Classroom", ASEE Annual Conference Proc., 1982, pp.667-671.

26. Carlson, D. A., and Work, C. E., "Application of Computer Graphics to Engineering Mechanics Courses", ASEE Annual Conference Proc., 1982, pp.763-769.

27. Swisher, G. M., "Simulation Laboratory for Mechanical Engineering Students", Model Simul. Proc. Ann Pittsburgh Corp., 1979, Vol. 10, Part 5, pp.2123-2127.

28. Moore, W. R., and Baxter, R. L., "Interactive Computer Graphics: A Powerful Tool for Technical Education", ASEE Annual Conference Proc., 1982, pp.34-37.

29. Breckell, T. H., and Manning, A. S., "Experimentation for the Mechancial Engineer in a Modular Degree Course", Int. J. Mech. E. Education, Vol.11, No.2, pp.71-82.

30. Aloroni, H., and Cohen, A., "Manpower for Engineering Laboratory Courses", Int. J. Elect. Engng. Educ., 1977, Vol. 14, pp.293-300.

31. Bedore, R. L., et al., "Development of an Engineering Systems Laboratory Course Sequence", ASEE Annual Conference Proc., 1982, pp.621-626.

32. Carter, G., and Lee, L. S., "A Sample Survey of Departments of Electrical Engineering to Ascertain the Aims, Objectives and Methods of Assessing First-Year Undergraduate Laboratory Work in Electronic and Electrical Engineering", Int. J. Elect. Engng. Educ., 1981, Vol.18, pp.113-120.

33. Lee, L. S., and Carter, G., "A Sample Survey of Departments of Electrical Engineering to Determine Recent Significant Changes in Laboratory Work Pattern at First Year Level", Int. J. Elect. Engng. Educ., 1973, Vol.10, No.2, pp.131-135.

34. Martin, D. G., and Lewis, J. C., "Effective Laboratory Teaching", Bull. Mech. Engng. Educ., 1968, Vol. 7, pp.51-57.

APPENDIX

Syllabus for Experimentation Course

Measurement:

A1.1 Measurement system (transducer, signal conditioning, recording or displaying).

A1.2 Characteristics of measurement systems:

<u>static</u>: Accuracy, sensitivity, linearity, resolution, threshold, repeatability, hysteresis, drift, zero stability, dead band, readability, range, ... etc.

<u>dynamic</u>: system response, frequency response, ... etc.

A1.3 Measurement of fundamental parameters:
each section should deal with: methods and techniques, units, standards, errors and suitability of techniques to applications. The sections are:

- Dimensional and angular measurements.
- Strain measurement.
- Measurement of time, speed, acceleration and frequency.
- Measurement of force, torque and power (to include dynamometers).
- Measurements of pressure.
- Measurements of flow.
- Measurements of temperature and heat flux.

A2. Instrumentation:
Brief introduction to electrical measurements of mechanical parameters, meters and oscilloscopes, recorders, introduction to automatic data acquisition.

A3. Errors:
Random and systematic errors (brief description of statistical techniques of data processing), propagation of errors, dealing with errors.

A4. Technical report writing:
The basic skill of written and graphic communication.

DEVELOPMENTS IN A FIRST COURSE IN THERMODYNAMICS USING TAPE/SLIDES

B.L. Button* and B.N. Dobbins*

Department of Mechanical Engineering
Coventry (Lanchester), Polytechnic
*Now Dept. of Mechanical Engineering
Trent Polytechnic

SYNOPSIS

A first course in thermodynamics, based on aims and objectives and using tape/slide programmes has been used for seven years. Every year a formative evaluation has been carried out using a student attitude questionnaire. The structure and development of the course are described. Developments include an advance organiser, a learning style inventory, a greater variety of learning tasks and some reading guides. Proposals for future developments and computer based learning within the course are discussed.

INTRODUCTION

A decade ago five members of staff collaborated in the preparation of a set of tape/slide programmes. The reasons for choosing tape/slides and details of how the collaboration was achieved can be found in Button et al. (1980). Over a three year period thirteen tape/slide programmes were prepared which replaced traditional lectures for teaching the fundamental knowledge of first year thermodynamics.

Each tape/slide programme consists of an audio tape with up to 20 minutes continuous playing time, a carousel of up to 40 slides and a handout containing some or all of the following: instructions, a pre-test, interactive text, post-test, summary sheet, answer sheet and tutorial sheet.

The course, using the tape/slide programmes has retained the conventional elements of laboratory demonstrations and end-of-year examinations. The staff time released has been used to provide more tutorials and more marked course-work assignments.

Thermodynamics like other core subjects in Mechanical Engineering is compulsory in the first two parts of the degree course and optional in the third. Part one is equivalent to half a subject in terms of time allocation and until this year was time-tabled for 45 hours in two-hour periods spread over 28 weeks. Within this time one hour was allocated to an introductory lecture, eight for tape/slides, 26 for tutorials, seven for laboratories and three split between the mid-sessional and end-of-year examinations. By comparison, in a conventional lecture based framework, 16 hours would have been allocated for lectures and 19 for tutorials. For a class size of about 60 students, lectures were given by one member of staff. For tutorials and laboratories the class was split into four groups, with a tutor for each group. As a result of using the tape/slide framework a 37 percent increase in student tutorial time was achieved for an eleven percent increase in staff time.

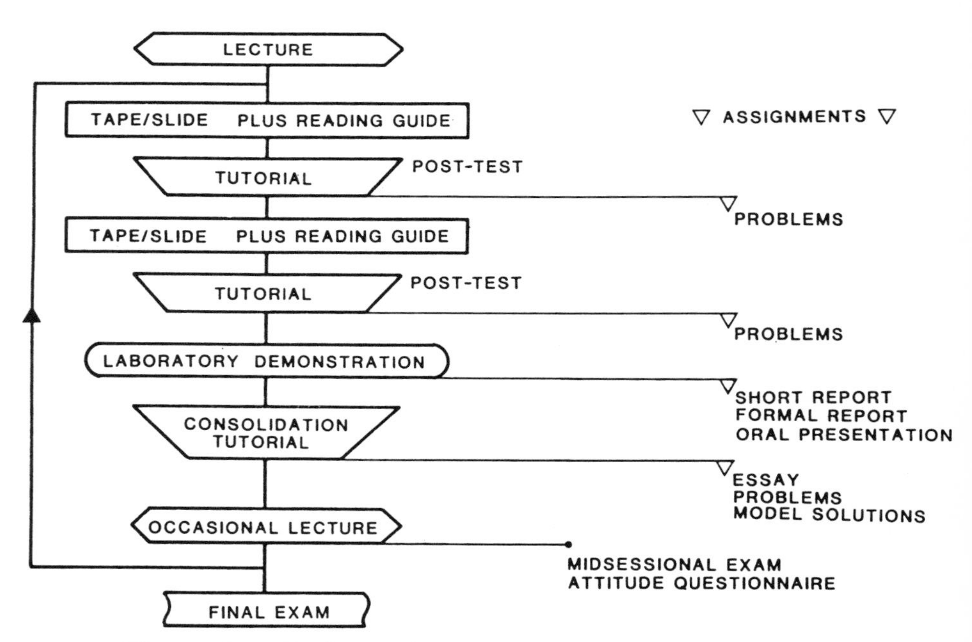

Figure 1 Typical element of course structure.

This year the organisation, content and hours allocated to each activity have been changed to take account of a number of developments detailed below. Fifty-three hours are now allocated to the course. The relationship between the different activities and the course-work assignments in the course is shown in Figure 1.

It should be noted that the typical cyclic element shown is repeated six times in the current course. The course aims and developments, which have led to this structure, will now be discussed.

AIMS FOR THE COURSE

Since the course was originated, significant changes in the perception of the role of engineers and engineering education have arisen from the Finniston report (HMSO 1980). In addition, for each year the tape/slide framework has been used, student feelings about, and their attitudes to the course have been elicited using an attitude questionnaire. A summary of the findings for the first five years is reported in Button (1983), results for this year are discussed below. The final element leading to change has been the work carried out in the last nine years in the field of educational psychology, as discussed in Button and Dobbins (1983).

These influences, with our original aims, have led us to formulate eight aims for the students on the course.

1. To learn the fundamentals of thermodynamics and to apply them to engineering equipment.

2. To gain first hand experience of thermodynamic equipment.

3. To learn how to solve problems.

4. To use different methods of communication.

5. To work as a group.

6. To use different methods of learning.

7. To train for the exmaination at the end of the course.

8. To learn how to find information.

These aims do not, nor do they attempt to, represent a synthesis of the recommendations of any of the above influences, but they may reasonably be dealt with by the thermodynamics course operating within a modular degree structure. We realise that all the aims cannot be fully satisfied by thermodynamics alone, but we believe that the more general aims should permeate all courses and that their introduction in individual courses will enhance student perceptions and expectations.

DEVELOPMENTS

Our philosophy has been to provide our students with the best course and learning medium possible, not to provide a vehicle for educational research. This has resulted in many changes to the tape/slide programmes and their administration being implemented as and when the need has been perceived. As the changes have been extensive, it is impracticable to catalogue them all here. The following are the major changes, most of which were implemented this year.

After the first two years students were provided with an advance organiser at the beginning of the course; this told them what they had to do, when it was time-tabled and the content and submission deadlines for course-work. Student attitudes towards learning from tape/slide programmes compared with lectures and tutorials, and about the organisation of the course, improved markedly after this, although changes in the administration of the questionnaire may account for some of the difference.

More, and more varied, laboratory equipment has been introduced. In four laboratory sessions, eight pieces of equipment are now demonstrated, one piece is used twice, and four are split between groups. For the laboratory involving split groups, the laboratory report has been replaced by an oral presentation.

To supplement the first hand experience gained in the laboratories and to convey the scale and costs of some industrial thermodynamic equipment, students visited a power station. The assignment for the visit was to take some readings make observations and report on the efficiency of the steam turbine plant.

In order to encourage different methods of communication, assignments have been based on a formal report, an essay question and a problem solution layout. These are in addition to the oral presentation, short laboratory reports and normal assignment questions. The problem solution layout was supplemented by a handout which described an approach to solving problems in thermodynamics. Also discussed was the relationship between tutorial and examination problems and 'real' engineering problems.

Our interpretation of the students' attitudes, supported by work in educational psychology (see Button and Dobbins 1983), indicates that students prefer to learn in different ways related to preferred methods of acquiring and processing information. Predispositions towards learning in one particular way have been termed 'learning styles'. Numerous tests and techniques have been used to assess an individual's learning style, but because they either take too long to administer or demand considerable experience to interpret, we felt unable to allocate the necessary time from the thermodynamics course. In consequence during their introductory lecture, students were asked to complete a learning style questionnaire derived from Kolb's learning style inventory (1976); this inventory is not so well related to psychological theory as others we had considered. However the reason for administering it was to introduce the students to the concepts, variability and consequences of learning styles; we also hoped to derive information about learner types from it.

Two further changes this year were designed to support a wider range of learning styles and also to draw further on the presumed advantage of the advance organiser; these changes were the introduction of five lectures and the provision of reading guides.

The five lectures were used to provide an overview of the course material both in its entirety and in parts. Additionally the lectures were intended to consolidate previous learning and to motivate the students. Lectures were given before and after groups of tape/slide programmes as illustrated by Figure 1.

Reading guides were introduced halfway through the course. These were a supplement to the tape/slide handout and covered the same content as the tape/slide by reference to four text-books, by chapter, and by section and page number.

The introduction date was chosen since we wanted to ensure that students experienced using the tape/slide programmes. At the same time as the reading guides were started, a multi-choice question post-test replaced the previously self-administered and marked post-test. This multi-choice test was given in the tutorial following the relevant self-study period and included two questions on the study method used. The multi-choice format allowed in-class marking, with immediate feedback of results and correct answers for discussion in the tutorials.

The attitude questionnaire had remained unchanged since the introduction of the tape/slide programmes. We were aware that some of the questions were ambiguous while others had become redundant. So to remove the ambiguities and to extend its scope, the questionnaire has been revised this year. To do this, six students were asked open-ended questions about various aspects of the course and based on their replies 24 new questions were written. These covered new aspects of the course and novel perspectives on some established aspects. In addition, eight questions were added by the staff, either to broaden the scope of the questionnaire (four questions) or to remove perceived ambiguities. Seventeen redundant questions were deleted. The revised questionnaire contained 59 questions compared with the original 45. As in the original questionnaire, the final section of the revised questionnaire gave the students the opportunity to comment on those factors of the course which impressed them most and least favourably. The questionnaire was given out during the last consolidation lecture period of the course which was three weeks before the examination. Thirty-one out of 49 students voluntarily completed the questionnaire this year.

EFFECTS OF DEVELOPMENTS

In establishing the effects of our developments, the results of the revised questionnaire will be used to provide information about overall student attitudes. The results from the questionnaire have not yet been analysed statistically in any way, but the percentage of students either agreeing or disagreeing with each of the items in the questionnaire have been calculated, and where appropriate are compared with responses obtained in previous years. Where a result is quoted this will be the percentage of students who agreed with the item. The numbers in brackets refer to the item number in the questionnaire, all items referred to in the text are listed in Appendix 1.

It has not been possible, at this time, to correlate individual student response profiles with either the results of the learning style questionnaire or any objective measure of performance. We are, however, still hoping to obtain these correlations.

Based on the responses to the attitude questionnaire, this year stands out compared with all previous years, in that very strong positive attitudes have been shown towards the thermodynamics course and the tape/slide method (e.g. Q18, Q41, Q43 and Q49). These and other related aspects will now be discussed.

Organisation of the Course and Comparison of the Learning Methods

Over 90 percent of students said that they enjoyed thermodynamics (Q28), did not find it the most difficult subject (Q38) and thought that the course was well organised (Q18). This latter point was reinforced by a large number of students in the free response section of the questionnaire; the course timetable, revision questions and solutions and summary sheets were particular items provided by the course which students associated with these comments. Negative feelings were less in evidence than in previous years (e.g. only 48 percent were tempted to 'bash on through the tape/slide' (Q2) compared with 68 percent averaged over the last six years). However, a majority (74 percent) still feel that being made to 'think more and work problems out' (Q40) would help understanding, and in free response students thought that more emphasis on applications and problem solving was desirable. Eighty-seven percent thought that the self paced aspect of the tape/slide allowed them to get ahead (Q24), while the availability of the tape/slide for repetition received favourable comment.

Also for the first time since the introduction of the tape/slide programmes the majority of our students now feel they learn more from the tape/slide than from tutorials (Q33), and still more than from lectures (Q23, Q34 and Q55).

Consolidation Lectures and the Reading Guides

Only 50 percent of the students thought lectures were very efficient as a method of conveying information (Q30) compared with the previous six-year average of 74 percent. This may have been affected by poor attitudes towards the consolidation lectures; only 26 percent thought they were useful (Q25), while 63 percent felt they didn't teach anything (Q22). We may infer that the difference may be due to a perception that the overviews, consolidation, and guidance on learning to learn were of no value. We would, however, contend that these aspects are likely to have been instrumental in promoting the marked improvement in attitudes described above.

The reading guides have received a cool reception from students, based on a 70 percent return of post-test questionnaires the reading guides were used on less than one in ten possible occasions, although 15 percent thought the reading guides were about right given the alternatives; they were too complicated (15 percent); too long (19 percent); trivial (13 percent) or that there was insufficient time to read the books (30 percent). Students were apparently evenly split as to whether a full week should be allowed between study-periods and tutorials to provide some extra time (Q48). Sixty-four percent of students thought that the reading guides were well thought out (Q59), similar numbers thought that they involved too much work (Q50) and should have been started in the first term (Q1 and Q36). Free responses indicate that a lack of emphasis on the reading guides in tutorials may also have contributed to their lack of use.

Laboratories and Course-Work

Ninety percent of students thought that laboratory work conveyed some idea of relevance (Q37); all students responding found the visit to the power station very interesting (Q53) although 40 percent did not feel that it helped their understanding of thermodynamics (Q44).

Very clear views were shown towards course-work assessment with about 70 percent of students preferring more emphasis on course-work (Q4 and Q12) this is probably because these same students felt they had more control over their course-work mark than final examination mark (Q21). Eighty-one percent of students felt that course-work was a fairer method of assessment than examination (Q9).

In seeking help with their course-work 93 percent would ask a friend (Q17), 87 percent their tutor (Q19) and 77 percent would refer to a book (Q6).

Learning Styles

The intention in administering the learning style inventory was to interest the students in their own learning and to encourage them to think about alternative methods of learning. It had also been hoped that it would be possible to establish relationships between learner types as categorised by the learning style inventory and their relative performance in differing tasks. Relationships between learning style and laboratory, oral presentation and essay course-work marks, final examination mark and use of the reading guides have not yet been found. This may be due to the informal administration of the learning style questionnaire during the first week of term and inadequate checking of the validity of the data supplied by the students. We have however gained some experience of the difficulties involved in the use and interpretation of a learning style inventory. This experience will enable us to have more confidence in the administration and interpretation of next year's inventory.

Benefits for the Staff

During the years in which the tape/slide programmes were being prepared, discussions between the collaborative authors resulted in a common view and approach to thermodynamics teaching. In addition, the skills and experience of those members of staff, two of whom have since left the department, have been retained and staff time is spent in tutorials rather than in preparing lectures.

THE FUTURE

We should identify and be flexible in the methods we use for assessment. These methods should be linked to what we want the students to learn, both for the thermodynamics course but also in the more general context of the degree course as a whole.

From our experience with the learning style inventory, we realise more care should go into the choice and validation of the instrument we use to 'measure' learning styles, and we should spend more time in planning and designing our evaluations.

We judge that the use of learning style inventory has the potential to provide meaningful information to both students and staff. We anticipate using more extensive and reliable learning style inventories, which may allow students to better understand and manage their own learning needs, and enable staff to better understand and plan the resources needed to provide the most appropriate learning material to identifiable groups of students.

More and different types of learning material, matched to subject content and assessment, should be developed. In particular computer controlled tape/slides or video systems will provide a more individualized and demanding resource for some students. The use of computer based self learning assessment modules will permit students to independently assess their level of competence.

Our reading guides have been little used by students this year, perhaps for the reasons discussed above. The analysis carried out for them is however a necessary part of the preparation needed if we hope to develop materials for a computer based thermodynamics tutor. To implement this requires us to develop a model of the structure and inter-relationships between facts, concepts and principles of thermodynamics; such a structure can be called a topic map, entailment mesh or petri net. If this structure could be linked to a model of the learner, perhaps based on a learning style inventory, and to a resource for study (tape/slide programme, section of a text-book, video, etc.), then a computer system which has been called an intelligent tutoring system would be with us. Such a system would be capable of interacting independently with a student, evaluating their present knowledge and assisting the student to choose the sequence and content of the subject matter to suit his or her individual needs.

CONCLUSIONS

We have presented the developments in our thermodynamics course using tape/slide programmes. We believe it is the underlying philosophy of approach that is important, other learning materials could, and should be included in the course.

Our developments are taking us into, and giving us experience in, the fast developing fields of educational technology and educational psychology. Over the years the course has been run, and this year in particular, we have sought to emphasize learning rather than teaching. This is because we believe that if we are to maintain standards of student learning we must provide students with the responsibility, skills and resources to learn independently. Staff time will be freed from providing information and can be used to support more tutorial time and more marked assignments than would otherwise be possible.

We cannot, at present, detect any improvement in the course using objective measures of performance, despite our efforts so to do, but we believe our developments are changing student perceptions of engineering and of their own learning.

Thermodynamics has provided a test-bed for the development of a course design, which we believe could be used for any course dealing predominantly with the knowledge and comprehension levels of learning. Unlike many others who try alternative methods of teaching, we have reported the developments of our course. We hope that by doing we others may benefit from our experience, and we may benefit from their criticism and comment.

This year students have revealed the most positive attitudes towards the thermodynamics course since its introduction. Such attitudes can only be beneficial, both for the students' study of thermodynamics and for their learning in general. The positive feedback from students encourages us to continue to use, develop and promote our course.

ACKNOWLEDGEMENTS

We acknowledge the support given by Mr. J. D. Wallis, Head of the Department of Mechanical Engineering. Also to the authors and co-authors of the tape/slide programmes for their comments on the reading guides (Dr. D. Wilcox and Mr. D. A. Yates) and permission to modify the tape/slide handouts (Mr. P. R. Koch and Dr. P. R. S. White). Our thanks also to Mr. S. M. Cox for his support and assistance in preparing the revised attitude questionnaire, and in other educational aspects of the work, and finally to our students for their cooperation.

REFERENCES

1. Button, B. L., "A Five-year Summary of Student Attitudes to and Performance in a Thermodynamics Course Taught Using a Tape/slide Framework", In: Ninth International Conference on Improving University Teaching. Dublin, 1983, III, 658-666.

2. Button, B. L. and Dobbins, B. N., "Computer Assisted Instruction and Learning Strategies - Which Way to Go?" In: Ninth International Conference on Improving University Teaching. Dublin, 1983, III, 648-657.

3. Button, B. L., White, P. R. S. and Wilcox, D., "The Use of Tape/slides in the Teaching of Thermodynamics to Engineering Students". European Journal of Engineering Education, 1980, 4, 303-317.

4. H.M.S.O., "Engineering our Future" : Report of the Committee of Inquiry into the Engineering Profession. H.M.S.O., London, 1980.

5. Kolb, D. A., "Learning Style Inventory: Self-scoring Test and Interpretation Booklet", McBer and Company, Boston, Massachusetts, 1976.

APPENDIX 1

SUMMARY OF SELECTED QUESTIONNAIRE RESPONSES 1977 TO 1984

		Year of entry						
		83	82	81	80	79	78	77
LECTURES		TRUE FOR ME / percent						
30*	Lectures are very effective as a method of presenting information.	50	69	83	70	67	78	83
34	I believe we learn more in tutorials than in lectures.	81	86	88	89	88	67	87
55	I learn better from lectures than from a tape/slide programme.	17	50	38	46	32	54	65
23	I would prefer a straight lecture to the tape/slide programme.	17						
25	Consolidation lectures were useful.	26						
22	Consolidation lectures did not teach me anything.	63						
TUTORIALS								
33	Most of what I learned about thermodynamics came from tutorials rather than the tape/slide.	41	59	67	61	61	85	96
TAPE/SLIDES								
2	You are tempted to bash on through the tape/slide rather than stopping or going back when necessary.	48	67	72	68	61	72	64
49	I felt out on a limb watching the tape/slide sequence.	3	33	58	46	25	57	43
43	Tape/slide information does not sink in very well.	37	63	64	64	60	61	68
40	If we were asked to think more, and work problems out whilst watching the tape/slide it would help our understanding.	74	83	84	89	91	77	90
24	With the tape/slide you can get ahead by going at your own pace.	87						
READING GUIDES								
59	The reading guide is well thought out.	64						
36	The reading guides should be started in the first term.	58						
50	Using the reading guides involves too much work.	68						
LABORATORY DEMONSTRATIONS								
37	The lab. work helps to convey some idea of relevance (e.g. turbines, thermometers).	90						

44	The visit to Rugeley helped me understand thermodynamics better.	59						
53	The Rugeley visit was very interesting.	100						
COURSE-WORK								
12	I would prefer a heavier emphasis on course-work assessment rather than final exams.	68	97	83	81	74	70	74
4	There could be more emphasis on course-work assessment as a proportion of our final marks.	73	97	88	90	75	79	86
9	Course-work is a fairer method of assessment than final exams.	81						
21	I have more control over a final exam mark than of my course-work mark.	26						
6	When I find the course-work difficult, I look in a book for help.	77						
17	When I find the course-work difficult, I ask a friend for help.	93						
19	When I find the course-work difficult, I ask my tutor for help.	87						
COURSE ORGANISATION								
41	We could have done with the same method of organisation in a couple of other subjects beside thermodynamics.	93	53	46	53	60	50	26
38	Thermodynamics is the most difficult subject.	3						
28	I enjoy thermodynamics.	97						
18	The thermodynamics course is very well organised.	97	77	85	87	80	37	41
48	There should be at least one week between the tape/slide and the tutorial.	48						
1	There should be more emphasis on using the books in the first term, otherwise it is too easy to rely totally on the tape/slide programmes.	61						
45	The study guides contain all the notes that you need for revision.	55						
	<u>Attitude questionnaires</u>							
	Number distributed	47	61	49	61	58	44	62
	Number returned	30	30	36	48	36	38	23
	Percent returned	64	49	73	79	62	86	37

* The numbers refer to the sequence of items as they appear in the revised questionnaire. Only items addressed in the text have been included in this appendix.

VIDEODISC: A NEW DEVELOPMENT IN VISUAL AIDS

T. R. Haynes

University of Cambridge

INTRODUCTION

This paper describes the present and future application of the interactive videodisc for teaching and learning. At the workshop the author demonstrated the interactiveness of the videodisc and discussed some problems with the videodisc production techniques, the authoring (courseware) procedures and the costs involved.

There are many educators nowadays producing their first videodisc and we at the Audio Visual Aids Unit (the University of Cambridge) and Mr. John Sayer, the Managing Director of Interactive Pictures, Ltd., are presently producing a videodisc to illustrate these newer technologies for this conference. A pilot videodisc containing material from the National Physical Laboratories Research, the Rivers Video Project (RVP) collection at Cambridge University and the University of London Audio Visual Aids Centre was made early in 1984. The equipment needed is a videodisc player controlled by a microcomputer with a suitable interface. The output from the disc player is displayed on a domestic teletext television screen (Figure 1).

Programs running within the microcomputer act as a software interface to both Computer Assisted Instruction (CAI) author and the student. The videodisc authoring involves a number of complex functions, the planning and preparation of the visual information, and production of the computer software for programme control. There are often special effects for superimposing animation and text information and the planning and preparation of sound tracks.

We hope from this project to encourage some further activity in the field, demonstrating the potential of the interactive videodisc for teaching and learning in higher education.

The author was first introduced to the videodisc player at the "National Audio Visual Conference" in Sydney, Australia in 1974 and there was great interest and discussion by the educators that were there. There were claims that the videodisc player would be one of the greatest innovations of educational technology made so far this century. It was not until 1982 in Australia, because of technical problems in producing the master videodisc, that educators had a chance to evaluate its usefulness in education. The disc was still in the American television standard (NTSC) and this created a problem of colour compatibility to resolve good colour from the disc player onto a normal colour television monitor. At this time the author was briefly involved in the evaluation of the disc and used it on three occasions to give lectures at the University of Adelaide. On these occasions the videodisc player was connected to a large video projector screen for large audiences. It produced favourable comments on quality and its ability to execute swift functions. These include, for example, its ability to scan from the first to the last picture on the disc within four seconds.

At about the same time General Motors-Holden was using the video disc (NTSC system) successfully throughout all its branches in Australia to teach their staff salesmanship, mechanical maintenance and other inhouse training programmes. In the last few months I have been carrying out research on the effectiveness of the computer controlled videodisc in education.

All the above is really to show that the videodisc player has been around for a long time in the American NTSC system and now there are a number of different makes of videodisc players in the PAL video system available to evaluate.

How far other audio visual and media departments have advanced with the evaluation of its usefulness in education is difficult to measure. In a recent survey of media and learning centres in New York State Colleges and universities, it was found that 84 percent of the respondents reported a positive attitude towards interactive video. Only 25 percent have already incorporated it into their Media Centres although 58 percent plan to do so in the future (D. M. Gayeski, D. V. Williams 1984). This shows some awareness and the use of interactive video in America.

THE OPTICAL DISC

Philips have two videodiscs available for use with optical disc players, the constant linear velocity and the constant angular velocity disc (CAV). The latter only is discussed here. The CAV is the most suitable videodisc format for interactive video because it is capable of random access

Figure 1 The Videodisc equipment with Computer Controller and Monitor

of many modes and special effects which will be described later on. The CAV laser head scans one revolution of the videodisc. Each revolution is called a 'track' and is equal to a frame. Each side of the disc contains 54,000 of these tracks and the playing time is 36 minutes if run at normal speed (25 frames/second).

First of all let us look at the videodisc itself (Figure 2). This is a 365 mm diameter glass disc optically ground, polished and spotlessly clean. It is coated with a

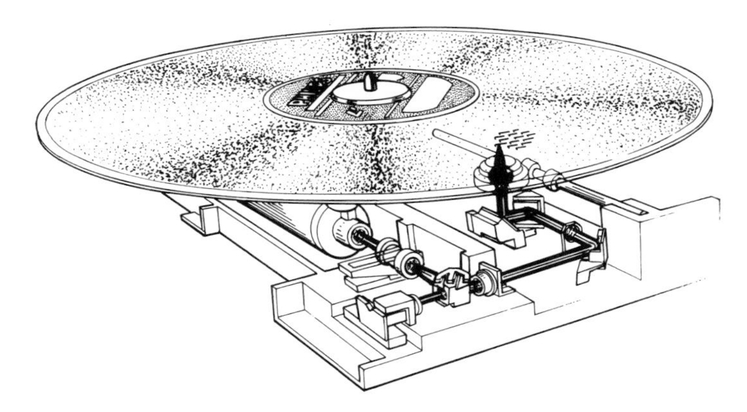

Figure 2 The light path of the Optical Laser Disc

1 mm layer of photoresist material evenly distributed by a centrifuge. This material forms a substrate for the recording process. The recording laser is modulated by the signal from the master video tape and writes a pattern on the photoresist material. This pattern is photographically developed to generate the final pit structure. Thus the "master disc" is created giving a positive plate carrying the actual pit structures required for the finished discs. The master disc is silvered and electroplated with nickel. When plating is separated from the glass master it forms a metal "negative", known as the "metal father", from which the copy discs may be pressed. In the player the laser scans the pits on this disc.

THE VIDEODISC PLAYER

The advantage of the laser beam is its ability to be focussed accurately onto the reflecting rotating disc. This laser beam pathway is directed by a few lenses and a prism. The laser beam is projected onto the reflected surface of the disc (Figure 3) where there are billions of little pits. They are reflected by this surface and then transferred into an electrical digital signal by the potential layer of the plastic. Scratches and dust will be kept out of the light beam, which tracks the pits precisely. The information that

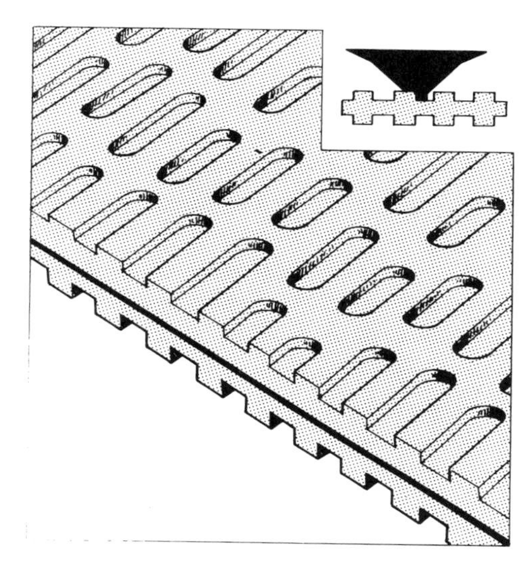

Figure 3 Placement of the Pits on the Disc. The insert shows the laser beam scanning the disc.

one disc contains is in the row of pits which is very compact and tens of miles in length.

The laser beam scans this spiral of pits from the middle of the disc. A single picture frame is kept still by the light source scanning over that one spiral track for an indefinite time. If you want the picture to go backwards push a reverse still frame button and the laser beam goes backwards onto the next frame and a still picture is achieved. When the disc is going either forwards or backwards there is no interruption to the picture i.e. there is no objectionable "noise" at all on the screen, in either fast or slow motion.

Each picture is designated with a picture or frame number and selection can be extremely fast. For example the new Philips videodisc player can jump 13,500 frames in 1 second. Also this does not have to be just still pictures; it can be a high speed or slow speed film, video tape sequences of high quality or a page of text from a book catalogue or card system. The player has two audio tracks which may be selected either one at a time, giving single mono channel (e.g. for bilingual capability) or both for stereo. A computer program can be dumped onto the usused audio track of the videodisc.

This program controls the disc for teaching purposes. The player can be controlled in one of three ways, manually, remotely or by a computer.

The manual mode of operation of the videodisc player uses the buttons located on the front of the panel; the latest professional machines only have a panel of indicator lights and must, therefore, be controlled remotely or by computer. The remote control uses an infra red hand-held key pad. The third method of controlling the videodisc player is by the microcomputer. This is the most versatile of all methods as it brings in the mechanism for human-videodisc-computer dialogue which is an interactive video system. There are many commands that may be used to effect the following operation.

RANDOM ACCESS

Precise and accurate access is either manual, keyed in on the remote control panel or by the computer.

At any time the picture can be stopped at a pre-selected frame and the read-head scans that frame continuously. This function has a distinct advantage over the conventional video cassette players because when they are in the pause mode or freeze (still) frame the video scanning heads are continuously moving and the video tape is being minutely worn away; also there is mechanical wearing of the video head assembly. This could damage the video tape for further use. The cassette still picture is unstable i.e. it jitters. Although some video cassette players have a reasonable still mode it is not as good as the videodisc. For example on the University of London pilot videodisc there is a film of a nitrogen bubble bursting its way through a tank of water and the selection of freeze frames at any interval of the bubble pathway can be studied in detail. The displacement of the bubble on explosion as it reaches the surface of the water has a great impact visually and what you are seeing is many effects that can be analysed in far greater detail. In comparison to this when the author was producing a video tape on a similar engineering project i.e. watching flames go down glass tubes, measuring time and colour, size, shapes etc. the freeze or stills were rather crude. With this kind of material on videodisc it would have a more visual impact with the freeze frame, giving adequate time to make observations, with a slow motion effect in either forward or reverse pictures from the 25 frames per second down to a single picture every four seconds. If single stepping after the freeze frame is needed, then it can be switched either backward or forward one frame at a time. The fast motion of the disc can be viewed in either reverse or forward up to three times its normal speed again without any interference to the picture. The indexing operation has auto stopping and repetitive frame skipping and all these functions used effectively can help the student using the disc to learn at his or her own pace.

THE RIVERS VIDEO PROJECT (RVP)

When the University of London were making an experimental disc they offered Mr. Martin Gienke from the University of Cambridge AVA Unit space to fill one fifth of one side of the disc, room for 10,000 slides. Under the auspices of the RVP within the Department of Social Anthropology the slides were sorted and put onto 16 mm film before transferring onto master video tape. The disc was produced with text and an authoring program to make an interactive video system. The test videodisc containing material from the National Physical Laboratory, Shell Research, the Rivers Video collection at Cambridge University and the University of London Audio Visual Centre was shown at a conference at the Gatwick Hilton Hotel 3rd-4th November, 1983. Perhaps delegrates from this Thermodynamics Workshop, 1984, could also pool resources and share in the production of teaching material for interactive video systems.

OVERLAYING TEXT ONTO THE PICTURE AND MODIFYING THE PROGRAMME AS REQUIRED

Some videodisc players have a built in teletext encoder which enables text from the computer to be placed onto the television monitor screen. There must be a decoder in the television set. Teletext can be stored on the disc and then displayed over the videodisc pictures and teletext can be generated by a computer and then displayed over the picture onto the television screen. If teletext is not available or if you want good graphics it is then necessary to have a genlock circuit board fitted into the microcomputer. This will enable overlaying of disc picture by text and graphics which are stable and clear.

THE OPERATION OF THE BBC MICROCOMPUTER

The BBC microcomputer is connected to the videodisc player's RS232 connection via its RS423 socket. Commands from the computer are sent out to the videodisc player, and operate in the same way as the functions on the front panel; with certain commands the player will inform the computer when it has carried them out. The BBC microcomputer is used because it is very suitable and readily available.

THE STUDENT'S ACTIVE INVOLVEMENT

There are a number of communication channels available through the videodisc computer system, for example, video (image production), high resolution animation and graphics, sound (on either one of the two sound tracks or both) and the keyboard.

Students interact with the videodisc system answering questions, following different lines of inquiry, making comments on text, branching off at various points in the program, changing sequences, constructing arguments and presenting answers in text and, on arriving at the end of the program, filling in the questioning sequences provided. There could then be an assessment made on the student's comprehension.

AUTHORING

There are many kinds of authoring (software) languages and it is difficult to make a choice of the most appropriate one. Each has its own characteristic and advantage. Philips have their own software called PHILVAS which is suitable for the professional laser disc. Many others are also suitable and there is a good discussion on this subject set out in a series of articles in the British Journal of Educational Technology by Barker et al 1982. They have produced a book to be published in 1985 called "Introducing CAL" which examines, reviews and goes into more detail on this subject. There is also a consultancy service provided by Interactive Pictures Ltd. of Cambridge.

"Microtext" authoring software from the National Physical Laboratory (NPL) has two versions. Level two version is a basic, cheaper version and there is a level three version. This is an updated version from level two and it has more facilities and is available in ROM to allow more room for memory.

The Microtext language, level 3.1 was used for the RVP project test disc and a few problems were encountered. The latest version (3.3) has solved the most important of these and also provides commands for writing to and reading from the RS232 port.

PROBLEMS OF THE FUTURE

The videodisc production involves three basic steps: the assembling of suitable material, pre-mastering and the reproduction from the master of copies for replay.

The requirements for signals on the master video tapes used for cutting the videodisc are that they are of broadcast quality PAL specification.

Philips require two crucial standards to be met and they are that the colour sub-carrier phase be continuous throughout the whole tape and field dominance must be preserved.

The pre-mastering of the videodisc material required for pressing has to be on a master video tape such as one inch "C" Helical video tape or any broadcast quality two inch quad video tape recorder. Alternatively a 16 mm cine film can be made which is then transferred onto either of the broadcast formats. The videodisc is not cheap to produce and at Philips 1984 prices there is a charge of approximately £1900 per side to produce a videodisc master and then £24 each for the first 10 copies and £6 each for every 1000 wanted.

INTERACTIVE PICTURES LTD. (IPL)

IPL offers a full consultancy service to Education Departments, Schools, Universities, businesses and other organisations, from the first planning discussions including demonstrations to show what can be done with interactive video, right through advising on preparation of film, slides sequences, scripting, to production of the final one inch master tape and the pressing of the discs themselves. They will also advise on marketing, market area potential, produce computer software, and so on.

Everybody always wants to have some idea of what it is going to cost them and this is of course the most difficult thing because the cost depends entirely on what sort of production you want, the desired picture quality, what sort of material you have and duration etc. Maximum duration for an interactive videodisc is approximately 108,000 stills, roughly 54,000 each side, or a total playing time of 36 minutes per side and from this you can deduce the number of film sequences or slide sequences you can fit onto the disc, the calculation being based on 25 frames per second to achieve a motion sequence.

A videodisc containing say 50 minutes of motion picture sequence and 15 to 20,000 still pictures will have an interactive experience, that is to say a real teaching time, running into hundreds and hundreds of hours if indeed it is ever to be fully exploited. So we can afford to think in terms of using parts of a disc, or only processing one side of a disc and so on. Cost per unit, that is per film section or per slide put onto disc, are always formed as a proportion of the total pressing costs and to these must be added the production costs of the particular sequence you have done. They can be as low as 50p, or as high as £2 or £3 per picture. It depends how much grading is necessary - the state of your slides in the first place. It is possible to envisage sharing production costs on a disc or slotting in. For example, a disc in current production with IPL contains a University Department's set of 48 slides and the cost of putting these on would be £50. Educators may think this very expensive but as in the case of the RVP disc there were other universities involved in sharing the costs. The more involved in sharing, the more realistic the budget.

How will we find ourselves in the future, say in the next decade, when the smaller electronic news gathering (ENG) videodisc recording system is predicted to be the accepted format? At the 1984 NAB Conferences in America, Matsushita released the first compact videodisc recorders which will have a terrific impact on video recording and studio production. The new videodiscs will reduce editing time by a considerable amount and will be much cheaper. If this is so then educators will have an excellent opportunity to make their own videodisc productions with more emphasis on educational values and less on the technical judgements. (Haynes 1982).

In summary the advantages in using videodisc in the future are that it can be used to link audio visual and computer databases, with facilities for overlaying text onto the picture and modifying as required, and providing a stable high quality picture with precise and rapid random access. The active involvement with students allows them to be self-paced throughout the learning experience with various routings through a programme to suit different needs and allows feed back of the users' comprehension.

ACKNOWLEDGEMENTS

The author is indebted to Mr. David Hurworth, Technical Manager of the University of Cambridge AVA Unit, for facilitating the equipment for research; to Jenny O'Dell for help with the preparation of this paper; to Philips Laservision for copies of their graphics used for illustrations and to John Sayer and Val Loose of Interactive Pictures Ltd., for their technical support.

The author is from the Advisory Centre for University Education (A.C.U.E.) at the University of Adelaide, which exists primarily to assist in the development of effective educational and teaching practices within the University including the provision of audiovisual production. He is currently on a six month job exchange with Mr. M. A. Gienke, the Head of the Audio Visual Aids Unit at the University of Cambridge.

REFERENCES

1. Barker, P. G. and Singh, R. for their series of papers in the British Journal of Educational Technology, 1982.

2. Gayeski, Diane M. and Williams, David V. "Interactive Video in Higher Education", Chapter 3. Video in Higher Education, 1984.

3. Haynes, T. R. "Video Production in Australian Higher Education", Journal of Educational Television 8, 2 119-26 (1982).

THERMODYNAMICS FOR THOSE WHO THINK WITH THEIR FINGERS

C. S. Sharma

Department of Mathematics
Birkbeck College
University of London

INTRODUCTION

The author's ideas on teaching thermodynamics arise from the following unusual experience: he first learnt thermodynamics as a chemist and was so baffled by δQ, dS, d^2f and so on, that he learnt mathematics properly and later had to teach thermodynamics to mathematicians, and mathematics (particularly that of Pfaffian forms) to Ehrenberg's physics students whom Ehrenberg claimed to be teaching to think with their fingers. As a result of the way thermodynamics is taught to them, the overwhelming majority of experimental scientists get a very distorted view of both thermodynamics and mathematics. Consequently they begin to think that the language of mathematics is similar to that of politicians, where words are interpreted and reinterpreted till they lose all real meaning. On the other hand, even if one were to succeed in teaching an experimental scientist to understand dx, in one dimension, as that linear operator which maps $\frac{d}{dx}$ to 1, there is no chance that one would be able to reconcile $d^2f = 0$ of mathematics with the rather formidable expression for d^2f given by Margenau and Murphy. The author proposes that experimental scientists should be taught non-mathematical descriptions of both the phenomenological and probabilistic ideas behind thermodynamics and should be given tables of useful thermodynamic formulae. They should also be trained to search the literature for other formulae they might need. A mathematical account of the subject should be an option for only those exceptional students who have a distinct aptitude for theory.

As a chemist the author learnt thermodynamics from his teachers and two books (2, 5), the first one by Glasstone and the other one by Saha and Srivastava. The subject was full of objects like dS, dQ, dV, dP and so on; all of these without exception were 'infinitesimals' and most were 'perfect differentials', although dQ was definitely not. The author was, to say the least, much puzzled by both these concepts: infinitesimals and perfect differentials. He asked his teachers many relevant questions and was always given the same reply: the teachers expected the author to find out the answers for himself and then explain them to his teachers in the fullness of time!

In due course the author did a Master's degree in mathematics. He did not meet any infinitesimals in the process though he had a brief encounter with perfect and imperfect differentials in the course on differential equations. Apparently perfect differentials were things which could be integrated and imperfect ones could not be. This was precisely what his chemistry teachers had told the author: $\frac{dV}{V}$ was a perfect differential and $T\frac{dV}{V}$ was not - T is a function of both P and V and the integral is with respect to V alone, so there is bound to be some difficulty. Nevertheless, that dQ could not be integrated to yield Q seemed very surprising, even though one could see from the equation for the isothermal expansion of a perfect gas

$$\frac{dQ}{T} = R\frac{dV}{V} \tag{1}$$

that if $\frac{dV}{V}$ was a perfect differential $\frac{dQ}{T}$ was also one and as $\frac{TdV}{V}$ was not a perfect differential dQ could not be. Unfortunately this argument does not really work: for isothermal processes, T is surely a constant and a perfect differential multiplied or divided by a non-zero constant is also supposed to be a perfect differential. The integral of dV was V, of dx was x and of df was f, but integral of dQ was not Q. The author was uneasy, unhappy and unsure. The author did notice that some books did not use dQ but δQ or q instead. The trouble with δQ is that in usual old fashioned notation $dx = \lim_{\delta x \to 0} \delta x$ and in general δx is much larger than dx, but this surely did not apply to δQ. Writing q instead of dQ confused the issue even further because $\int \frac{dQ}{T}$ then became $\int \frac{q}{T}$ and at that stage the author was not familiar with an integral without a d of something or other after it.

Since then, the author has learnt much, become wiser and accepted the fact that there are many - far too many - unanswered questions, that there are many different levels of understanding and that no one, absolutely no one, knows enough about anything to be able to answer every legitimate question that can be asked about that particular thing. On the other hand, scientists take pride in describing themselves as 'logical' and 'rational' and the situation about dQ seemed to be so irrational, to the author at any rate, that it needed an answer. This gave him motivation to learn about both infinitesimals and differentials. He has pursued these matters to the point where he has found answers which, though not absolute, satisfy him. With hindsight he can say that none of his teachers or the books mentioned above could have given a meaningful definition of the concept 'infinitesimal' with the prerequisities they could take for granted. The concept 'perfect differential' on the other hand could have been explained in a way which could have satisfied this author at least.

In summer 1963 after the author had agreed to teach mathematical methods to physicists, he was instructed to go and discuss the course with the late Professor W. Ehrenberg who was then the Head of the physics department at Birkbeck. The instructions which the author received from Ehrenberg were curt and precise. Ehrenberg was teaching his students to think with their fingers and he did not mind in the least what mathematics the author tried to teach them, as long as anyone, who could pass his physics examination, did not fail the degree because of a failure in the mathematics examination. At the placid response of the author, Ehrenberg mellowed a little and added as an afterthought that he would like his students to know enough about Pfaffians to be able to have some appreciation of thermodynamics. The author agrees with both the edicts in the instructions he received: firstly that though it is important for physicists to know some mathematics, it is not so important that they should fail their examination for a degree in physics if they failed their examination in mathematics, and secondly that thermodynamics is really important. (This was many years before 'energy' became such a dominant issue.) Whether it is right and proper to require a junior lecturer under probation (which is what the author was then) to cook the examinations so that everyone passed is better not discussed in this work. However, the author would like to point out that in a good educational system, educators should have clear ideas as to what precisely they are trying to teach, and why, and the examination is merely a means of finding out how well this objective has been achieved in teaching each individual student. It is sadly the case that a majority of teachers have no clear idea of why their students should learn a particular thing in a particular way and examination systems have become so rigid that no one has any idea whether examinations do actually examine anything worth examining.

INFINITESIMALS

The author bought two books, (1, 3) both called 'Infinitesimal Calculus' in the hope that they would enlighten him about the concept of the infinitesimal. The first one was by Dieudonne, a member of the famous Bourbaki school. The term 'infinitesimal', as far as the author can make out, does not occur in the text. In the preface Dieudonne says something whose essence is that infinitesimal calculus is the name of traditional calculus and is the art of distinguishing between what is 'large', what is 'small', what is 'dominant', and what is 'negligible'; the philosophy being summarized by the three words: majorize, minorize and approximate. He then goes on to say that all this does not imply that he would sacrifice rigour to convenience and reduce Infinitesimal Calculus to a series of recipes.

In all applied sciences the art of distinguishing between 'small' and 'large', 'dominant' and 'negligible', 'relevant' and 'irrelevant' is precisely what makes these subjects alive and in touch with reality. It is for mathematicians to find out the precise reasons for treating some things as small and negligible and other things as large and dominant. Heuristic and plausibility arguments are enough for most physicists and engineers, but they ought to know that the lack of rigour renders their arguments fallible. In any case, the author did not learn a thing about infinitesimals from either of the two books he bought: the second one by Horace Lamb did not seem to contain anything of interest to the author.

An infinitesimal is often defined as a number which, in absolute value, is smaller than every real number and yet is non-zero. This is in direct contradiction with one's intuition of the real line as a continuum with no gaps between one real number and another, which is not occupied by other real numbers. In this picture there is simply no room for infinitesimals. One often hears that infinitesimals exist and their existence can be demonstrated rigorously by mathematical logic. Most people who make this assertion have no idea what this demonstration involves. This author did actually supervise a Ph.D. thesis on non-standard analysis and after learning a great deal about semantics, formal languages, ultra filters, superstructures and such like, he does finally accept and understand infinitesimals: they do exist in a highly technical formalism of logic, though the author does not believe, contrary to what many authors say, that most people have a good intuition about infinitesimals. They, in the opinion of the author, are best avoided. However, defining differentials without infinitesimals makes it necessary to bring in the fancy formalism described in the introduction.

PERFECT DIFFERENTIALS

Though it is possible to do without infinitesimals, thermodynamics cannot be done without some understanding of perfect differentials. After reading the theory of differential forms the haze around perfect and imperfect differentials lifted from the author's mind. In this theory d is an operator, which among other things, turns functions into differential 1-forms, though not all differential 1-forms are the results of d acting on functions. A differential 1-form which arises as a result of d acting on a function is a perfect differential and one which does not arise in this way is imperfect. It is clear from this that what is written as dQ is not what results from an action of d on a function Q. Thus the notation dQ (or δQ) is meaningless and should be avoided. The crux of the matter here is that Q is not a thermodynamic function, that is, not a function of just two variables (any two of P, V, T and S), it is a function of more variables in addition to the thermodynamic ones such as its value at some initial state and the path from the initial state to the final state. The thermal energy of a system is measured by T and not by Q. The situation here is much like the balance sheet of a multi-national company: the U.K. profits do not account for all of its sterling balances. Heat like sterling is just one of many convertible forms of the same commodity. If it had been emphasized that Q is not a thermodynamic function, that is, its value is not determined merely by a specification of the thermodynamic state of the system and if d had been explained even vaguely as inverse of integration, the author might have shown greater inclination to accept what is meant by imperfect differential. At this point it becomes necessary to discard both dQ and δQ in favour of q of Glasstone (2) and for once we find how necessary the modern notation for integration is where $\int f$ is used for $\int f dx$. In the new notation $\int \frac{q}{T}$ makes sense even though there does not happen to be a d after the integral sign.

THE CHAIN RULE

The author came across the following in one of the books:

$$dP = \left(\frac{\partial P}{\partial V}\right)_T dV + \left(\frac{\partial P}{\partial T}\right)_V dT \tag{2}$$

Since dP = 0 for isobaric processes

$$\left(\frac{\partial P}{\partial V}\right)_T \left(\frac{\partial V}{\partial T}\right)_P + \left(\frac{\partial P}{\partial T}\right) = 0 \tag{3}$$

or

$$\left(\frac{\partial P}{\partial V}\right)_T \left(\frac{\partial V}{\partial T}\right)_P \left(\frac{\partial T}{\partial P}\right)_V = -1 \tag{4}$$

For isobaric processes P is constant and if P is constant, the author would argue that $\frac{\partial P}{\partial V} = 0 = \frac{\partial P}{\partial T}$ and getting (4) involves dividing by 0 which is meaningless. It can be argued that if dV and dT in (2) are in such a ratio that dP = 0, then the process would be isobaric and the ratio would be the value of $\left(\frac{\partial V}{\partial T}\right)_P$. The trouble with this argument is that it uses a novel definition of partial derivatives. All such confusion can be avoided by using the chain rule with which most students are familiar. If a coordinate transformation from variables (x_1, x_2) to (y_1, y_2) is made then according to the chain rule

$$\frac{\partial f}{\partial x_1} = \frac{\partial f}{\partial y_1} \frac{\partial y_1}{\partial x_1} + \frac{\partial f}{\partial y_2} \frac{\partial y_2}{\partial x_1} \tag{5}$$

Taking $x_1 = T$, $x_2 = P$ and $y_1 = T$, $y_2 = V$ gives

$$\left(\frac{\partial f}{\partial T}\right)_P = \left(\frac{\partial f}{\partial T}\right)_V + \left(\frac{\partial f}{\partial V}\right)_T \left(\frac{\partial V}{\partial T}\right)_P$$

If we take f = P, $\left(\frac{\partial f}{\partial T}\right)_P = 0$ because P is constant and f is P and we get (3). Thus a correct procedure for getting the correct result also uses fewer new concepts.

THE SECOND DIFFERENTIAL

The author found the following formula in Margenau and Murphy (Ref. 4).

$$d^2f = \frac{\partial^2 f}{\partial x^2}(dx)^2 + 2\frac{\partial^2 f}{\partial x \partial y}dxdy + \frac{\partial^2 f}{\partial y^2}(dy)^2 + \frac{\partial f}{\partial x}d^2x + \frac{\partial f}{\partial y}d^2y \qquad (6)$$

The theory of differential forms asserts that one of the definitive properties of the operator d is that $d^2 = 0$ which implies $d^2f = 0$. The author has not been able to find a modern theory of Pfaffians which will reconcile this contradiction.

CONCLUDING REMARKS

One of the author's classmates during his undergraduate days had a slightly warped mind as is shown by the following remark seen jotted in pencil on the margin adjacent to a problem in one of his mathematics text books: "One gets the right answer in this problem by adding 2 to the answer one gets by one's calculations". No one can dispute that by using this arbitrary totally meaningless procedure the author's friend was able to get the correct answer. Had it been a joke, it was definitely not a bad one and one could merely laugh at it. However, the author and others decided that there was something wrong with the mind of the discoverer of the artifice just mentioned because the discoverer solemnly, sincerely and persistently proclaimed his absolute faith in the validity of the artifice. When one considers some of the mathematical arguments fed to students in the experimental sciences - chemistry, physics, engineering and biology, even though the arguments produce the right answers, it is not difficult to demonstrate that the arguments are based on strategems which are hardly better than the one used by the author's friend. Some students are fooled, but the author gets the feeling that contrary to what their teachers believe, the majority of students become sceptical and begin to think that either advanced mathematics is not what it pretends to be (logical and precise) or the teachers do not really understand their own arguments. Such training cannot possibly be of much use to anyone. However, one must concede that it is not infrequent that great discoveries are inspired by fallacious theories and one of the best examples of this comes from thermodynamics: Joseph Black (1759-63) Professor of Medicine and Lecturer in Chemistry at Glasgow University had based his theory of heat on the assumption that dQ is indeed a perfect differential: the conclusions of the theory were obviously and apparently wrong, but the theory is reputed to have inspired the discovery of the steam engine by Watt. It was at least 35 years after Black had gone to his grave that anyone seriously disputed his theory.

Thermodynamics is an important subject, not just because it is the dynamics of the therm in an era when energy has become an issue of public even political concern, but much more so because it enhances the intelligence of man. In this context intelligence is the capacity of seeing interrelations between things and these interrelations are what give the world around us a structure and make it organised. Some of the interrelations one discovers through thermodynamics, such as between the bulk modulus of elasticity in an isothermal process and the coefficient of volume expansion in an isobaric process, and between latent heat of isothermal expansion and the rate of increase of pressure with temperature at constant volume are some of the most surprising and unexpected interrelations of which the author is consciously aware, and he would have been deprived of the joy of seeing these exciting and unexpected interrelations without thermodynamics. These delights have more than compensated the author for the agony he has suffered because of his difficulties with infinitesimals, differentials and partial derivatives. However, quite apart from satisfying intellectual curiosity and demonstrating the beauty of the structure of physical reality, these interrelations are of considerable practical use to engineers and physicists. Thermodynamics should be taught to all scientists - but, how should one do it? Thermodynamics is essentially a mathematical subject and the mathematics which is a minimum prerequisite even for a rudimentary appreciation of the subject is far from trivial. In most courses for experimental scientists, thermodynamics is taught using arguments which yield the correct results and yet these arguments are very tenuous and in many cases are demonstrably dubious or even absurd.

The task of writing this essay has forced the author to concentrate his mind on some of the issues on which he still owes answers to his teachers. A surprising discovery was that writing differentials and integrals without a, d preceding any other symbol in the expression was not just a fad propagated by mad mathematicians in the name of modern mathematics, but the requirement is forced on one once the symbol d is given a precise meaning. The author believes that modern mathematics is simpler than traditional mathematics and much more precise and powerful. However, it is so very difficult to teach it because in their formative years most of the students are given a very distorted account of modern mathematics by teachers who pretend to know what is clearly beyond their grasp. Once the damage has been done, it is almost impossible to undo it. Perhaps one day when teaching mathematics becomes a profession which is duly rewarded and attracts men and women of the right calibre, it will be possible to train young men and women to think clearly and to realize that the essence of the elegance of modern mathematics is its extreme simplicity and then it will be possible to teach all experimentalists thermodynamics with a moderate amount of relevant mathematics.

Even mathematicians find the mathematics of thermodynamics hard and confusing. It is unrealistic to expect that engineers can be taught thermodynamics with rigorous mathematics. Since thermodynamics is important, perhaps a middle way can be found, where as far as possible thermodynamics is taught without using mathematical arguments which are quite obviously absurd, but unfortunately this can be done only if the teachers have an understanding of the subject at a level much higher than the one at which they are teaching. The author believes that the prescription proposed in the introduction of this article is more realistic. Engineering students in general should be taught non-mathematical descriptions of both the phenomenological and probabilistic ideas behind thermodynamics and taught recipes for their sums. A mathematical account of the subject should be reserved for only mathematically inclined engineers.

REFERENCES

1. Dieudonne, J., "Infinitesimal Calculus", Kershaw, 1973.

2. Glasstone, S., "Textbook of Physical Chemistry", Macmillan, 1956.

3. Lamb, H., "An Elementary Course of Infinitesimal Calculus", Cambridge, 1956.

4. Margenau, H. and Murphy, G. M., "The Mathematics of Physics and Chemistry", Van Nostrand, 1956.

5. Saha, M. N. and Srivastava, B. N., "A Treatise on Heat", Indian Press, 1958.

THE USE OF SMALL COMPUTERS FOR TEACHING

D. H. Bacon

Department of Mechanical Engineering
Plymouth Polytechnic

SYNOPSIS

Engineering thermodynamics, work transfer and heat transfer are closely allied subjects and the paper contains examples in all fields. As in other subjects, undergraduate applications of the principles presented in lectures have often in the past been limited by the time taken to solve equations or optimise solutions. The advent of inexpensive computing facilities (e.g. 48 K Spectrum £130) means that the limitations are being changed. New consideration must be given to the presentation of material in lectures and to the applications in laboratories, tutorials, seminars and projects. Consideration of examination techniques is also required.

A new style of text book[1] has been produced in which both the worked and unworked examples are aimed at integrating the use of computers into simple problems. The results of pilot trials with small student samples are discussed. The theme is further developed to include the next stages of the work involving extended programs in which each member of a group of students contributes to the solution of a problem which embraces wider principles than the short examples presented in the book.

In this way the student should be helped to an awareness that an engineer solves real problems rather than short examination examples and that he must be able to work as part of a team which may use computers for modelling, design or development. The student must also be prepared to combine various 'subjects', e.g. heat transfer and costs, engine performance and vehicle dynamics.

One of the great advantages of this method is that to write a successful program the student must fully understand the problem and the method of solution. The computer must be programmed with the correct analysis.

Three examples follow, both in simple form and extended form for groups of students.

ENGINEERING THERMODYNAMICS

The processes in the working substance of a four stroke compression ignition engine can be modelled by incrementing the crank angle and calculating the pressure and temperature after each increment from a knowledge of the pressure and temperature before the increment and the energy transfer during the increment. The processes can be considered in two forms; valve open and valve closed. The simplest solution for the single student to consider is an adiabatic process with air as the working substance for the valve closed period. The results of the program can be compared with those given by the reversible adiabatic relationship. For a group, the program can be extended to allow for heat transfer, delay and energy release during combustion and flow of working substance during the open valve period. The results may be compared to an indicator diagram obtained on a real engine.

WORK TRANSFER

The single student can conduct a laboratory trial on a spark ignition engine and process the results by computer program including, if possible, a graphics routine to plot a power-speed characteristic with specific fuel consumption as contour. The group approach is to develop a program to fit the engine to a vehicle which has to meet a given design specification for acceleration, economy and 'driveability'. This will involve choice of gears, estimating or measuring inertia values, etc. to deal with the acceleration dynamics and then optimising the variables to meet the specification.

HEAT TRANSFER

The single student considers a simple heat transfer process in which water flows through tubes held at constant temperature. The length of tube required for various combinations of diameter and numbers of tubes used can be determined for a given heat flux. The pressure drop in each combination can also be calculated. The group of students can then advance to the design of a heat exchanger to perform a given duty and to optimise the design on a cost basis (capital, running and maintenance).

REFERENCE

1. Bacon, D. H., "BASIC Thermodynamics and Heat Transfer", Butterworth, 1983.

RESOURCE-BASED LEARNING IN THERMODYNAMICS

D. J. Buckingham

Department of Engineering Science
University of Exeter

INTRODUCTION

The second-year course in Thermodynamics and Fluid Mechanics in the Department of Engineering Science at Exeter is presented largely through self-instructional programs based upon guided reading and tape-slide. The S.I. programs are directed through "keystone lectures" and by feedback through small-group tutorials.

OBJECTIVES

The objective of the scheme is to shift staff-student contact time from lecture-style exposition toward an interactive and informal tutorial contact. No claims are made that a resource-based presentation through tapeslide, or other means, is "better" or "worse" than any other style of teaching. The merit of the scheme is that, by easing the term-time burden of formal teaching, more time is available for small-group contact, and it is claimed that this permits a better environment for learning.

THE RESOURCE PACKAGES

The purpose of the resource material is to present information in a way that permits the student to generate reference notes, and leaves him in no doubt as to the engineering purpose and examinable features of that material. The scheme began in 1969 with tape-slide packages to replace blocks of lectures, and has developed since then into a mixture of tapeslide and guided-reading formats. In the laboratory exercises tapeslide, guided-reading, video, computer graphics and computer-controlled selected-answer routines are used in a variety of mixes.

KEYSTONE LECTURES

While the resource packages are well suited to delivering hard information they do not necessarily provide motivation. Therefore a number of "keystone" lectures are retained to introduce and conclude blocks of self-instructional material. They can be used to emphasise the relevance of syllabus material to engineering applications, and short video sequences can be employed to add impact. Because the purpose is motivational there is little formal note-taking, and they can afford to have the style of a "performance" rather than a "lecture".

SMALL-GROUP TUTORIALS

The primary aim of the self-instructional programs is to improve the time and enthusiasm given to interactive staff-student contact. At each session there will normally be discussion of assigned work and examples set up by the preceding resource material. The tutorial becomes the main mechanism for discussing engineering applications.

SCOPE FOR EXPANSION

While the scheme permits a reduction of formal teaching commitments during term it does require a considerable increase in staff time devoted to course preparation and organisation during vacations when new programs are being developed. It is unlikely that many academic staff would consider such effort worthwhile if at the expense of research and other professional -development activities. The possibility of resource exchange between Departments is attractive in that it should enable an easing of the burden of program construction, but attempts to do this seem to result in a destructive de-personalisation of the programs. The best hope for program interchange is to keep the resource units small so that another user could select and mix existing software.

CONCLUSION

A resource-based Thermofluids course has been operating at Exeter in various forms for the past 14 years. The objective of this style of course is to shift the staff rôle in staff-student contact from declamation toward discussion. The system may, or may not, be a better style of teaching but it can sponsor a better environment for learning.

REFERENCES

1. Matthews, J. J., and Buckingham, D. J., "Resource based learning: a pragmatical approach", Studies in Higher Education, 1976, Vol. 1.

2. Buckingham, D. J., "Engineering Education at Exeter: Three years on", Brit. J. of Educational Technology, 1974, Vol. 5.

3. Austwick, K., and Harris, N. D. C. (Ed.), "Aspects of Educational Technology VI", Pitman, 1972, p. 181.

A LABORATORY APPROACH TO TEACHING THERMODYNAMICS

D. W. Pilkington

Department of Mechanical Engineering
Manchester Polytechnic

INTRODUCTION

Faced with the problem of introducing Thermodynamics to a group of technicians, it seemed very likely that the usual emphasis on the more academic aspects would now have to be slanted in a more practical direction. Many students and not only technicians, can easily lose sight of the subject aims in a mass of formulae and definitions, thus causing considerable loss of motivation. It was considered that a 'hardware' or laboratory approach would overcome some of the problem of relevance; it would also help to maintain interest when the topics with less obvious practical application were introduced.

GENERAL OUTLINE

The initial plan envisaged an introduction to the subject by coordinating simple theoretical aspects with a detailed description of an experiment designed to illustrate the topic. The two areas selected were 'Optimisation of Energy Usage' and 'Properties.' After the first group of experiments had been completed it was planned to have a consolidation period in the classroom followed by the introduction of a second laboratory stage.

Overall, the aim was to provide each part of the theory with an industrial application and a laboratory experiment. The emphasis being laid on the need for all three parts of the process to occur at the same time.

STAGE ONE

The laboratory equipments chosen for this stage were, in order: Diesel and petrol engines to cover the 'energy' aspects; the steam calorimeter and refrigeration plant to cover 'properties'.

In more detail, the Diesel engine was used to demonstrate a specific fuel consumption - load characteristic with the engine governed to a constant speed. The graph obtained clearly shows a rapid improvement of specific fuel consumption at the initial stage of increasing the external load, followed by a range of economical operation and finally deterioration on overload conditions. This characteristic leads to a discussion on the effects of internal friction and combustion efficiency; it also allows for comments on its operational range and comparative efficiency. The use of the specific fuel consumption term provides some practice in the manipulation of units in the conversion of kJ into kWh. The test on the petrol engine performs a similar function and shows that variable efficiency is not solely a characteristic of the Diesel.

For the 'property' experiments students were strongly encouraged to sketch the p-h and block diagrams for the plant to be tested. In addition to gaining familiarity with the properties of vapours, the refrigeration plant helped to clarify the idea of a thermodynamic cycle. In addition, by plotting the appropriate state points on a large scale property diagram, it became possible to comment on the actual and theoretical coefficients of performance.

STAGE TWO

This was based on the first law with particular reference to the steady-flow energy equation. This is usually considered to be a classroom exercise, but it was now possible to refer back to the laboratory experience and clarify ideas which might otherwise appear to be purely academic. The laboratory equipment used in this stage was steam turbine, reciprocating air-compressor and a gas calorimeter.

Consolidation of the experimental results reinforced the first law concepts, and provided a lead in to the second law. Finally, the association of reversibility, entropy and the Clausius statement were considered for their practical implications.

CONCLUSIONS

In practice the scheme worked even better than initially hoped for, the students proved themselves to be interested and much more competent than expected. To some extent the scheme evolved as the session progressed and a revised version is planned in the light of experience gained. This is intended to expand the section on 'Optimisation of Energy Usage' to four experiments by the inclusion of two based on Fluids topics. Although the scheme was planned with technicians in mind, the intention now is to extend its range of application.

DISCUSSION IN SESSION 2 "INNOVATIVE METHODS OF TEACHING"

Edited by D. Walton and J. D. Lewins

On the paper by D. R. H. Jones, "Materials Thermodynamics for Engineers"

Dr. Jones agreed that the availability concept was both more fundamental and an easier concept to grasp than the Gibbs function, so that materials scientist and chemists might perhaps use availability in preference. To get his own back, as it were, on engineers, the author expressed some concern that concepts of order and disorder were not more widely referred to in engineering thermodynamics.

On the paper by J. D. Buckingham, "Resource Based Learning in Thermodynamics"

In the discussion, Dr. Buckingham acknowledged that the preparation time for a tape-slide teaching model was very substantial. The students coped well with the mechanics of using the machines. Assessment of student progress was made indirectly in overall assessment rather than in respect of specific programs. The authors general view was that good students got to the same standard by such techniques as they would from conventional lectures but that the weaker students who persevered raised themselves to a higher standard than they reached conventionally. The tail is pulled up because students at the bottom can go back for program instruction without the program ever getting annoyed and putting the student off.

On the paper by C. R. Stone, "Thermodynamics in a Broad Based Course"

Professor Turner enquired of the difficulties in maintaining collaboration with 40 companies. The author explained that a division of labour amongst his colleagues meant that he dealt directly with a few companied that he would have a natural interest in anyway and he found this desirable. The collaborating companies changed from time to time, some ceasing to sponsor students.

On the paper by P. Foss and D. R. Croft, "Maximising the Teaching of Experimental Thermodynamics"

The authors emphasised the importance of maintaining the presence of the engineering lecturer in the student demonstration laboratory and not just before a blackboard. Dr. Croft spoke of the difficulties of achieving this in a Polytechnic where all too often the laboratory work was placed with unqualified lab. assistants.

A comment from the audience drew attention to a slide in which the microprocessor employed to run the engine and provide the data was positioned so that the student could see the microprocessor but not the engine!

On the paper by M. R. Heikal and T. A. Cowell, "Engineering Laboratory Teaching: a Case for Coordination"

Dr. Lewins had pointed out that the drift in previous discussion had been unsympathetic to microprocessors in the engineering laboratory. There seemed two reasons in favour however. The first was that if microprocessors are used in professional laboratories then we have an obligation to expose our students to professional practice. The second is that thermodynamics or at any rate the thermal part of mechanical engineering has to compete for students with other engineering topics; if students are attracted to modern developments from microprocessors up to information theory, then we must offer some comparable opportunities in our own topic areas.

Dr. Heikal emphasised the question of judgement in using microprocessors appropriately. He had also found that it was desirable to require a manual experiment and hand calculaton before, for example, allowing the remaining data to be obtained and processed automatically, a view supported by Mr. D. G. Walton who emphasised the importance of planning and coordinating the whole laboratory course to see computers took their right place.

On the paper by B. L. Button and B. N. Dobbins, "Developments in a First Course in Thermodynamics Using Tape/Slides"

In reply to a point from Professor Woods, Professor Button explained that the questionnaires were constructed by a technique in which one tenth of the students were questioned in depth about the course and their comments formed the questionnaire completed anonymously by the whole class in the form of a range of 'support-don't support' values from 1 to 6. The answers have been given anonymously largely at the request of most students.

The authors speculated about the advantages of a teaching team as available in the Open University where educationalists and psychologists for example might be available to prepare and process such questionnaires more professionally. Most of the audience did not seek such team teaching however.

On the paper by D. H. Bacon, "The Use of Small Computers for Teaching"

Dr. Burnside had raised the desirability of making thermodynamics into more of a design oriented study, not accepting the common equating of computer aided design courses with mechanical design. The author felt that this raised problems. Of course the major purpose of engineering was design in order to create and he had some sympathy with the desire for a second year engineering course that might be entitled 'thermal design and thermal systems'. But to be realistic, there were already tremendous pressures on the syllabus. Perhaps some time could be saved by not lecturing on some of the steady cycle theory and leaving this to be absorbed in a design project.

On the paper by R. Haynes, "New Developments in Visual Aids"

The lecturer presented a demonstration video-disc under microcomputer control. In comparison with video tape, the disc offers quick access, long life and lack of wear. It provides perfect stills if the CAV format is used although admittedly not the long play that at present is only available with CDV format.

On the paper by Professor C. S. Sharma, "Thermodynamics for Those Who Think With Their Fingers"

There was a variety of opinion expressed in the discussion of this paper as to what extent thermodynamics was necessarily a mathematical subject and to what extent it could be taught without the use of mathematics.

It was generally agreed that the engineering students would not find highly advanced mathematical methods suitable and the teacher had an obligation to teach within the capabilities of his students. The Chairman would support a view that favoured the physical concepts above the mathematical techniques. Much could be done in pictorial or topological terms, moving about a three-dimensional surface, without forcing Pfaffians upon experimentalists.

Professor Gurney then pointed out that some of the difficulties the author drew attention to would be resolved if the partial differentials giving the slopes of elementary areas in the topological space were correctly interpreted as functions and not as something that happened necessarily during the process itself. Other speakers commented upon the graphical interpretation of the thermodynamic equations. Sir Brian gave us an impeccable defence in quoting from the first edition of "Statistical Physics" by Landau and Lifshitz: "No attempt is made at mathematical rigour, since in physics this is always illusory".

On the paper by D. W. Pilkington, "A Laboratory Approach to Teaching Thermodynamics"

The question was raised in discussion of the logistics of phasing laboratory sessions with lecture/tutorial periods where large classes exist. Mr. Pilkington replied that it was very difficult, if not impossible, with a large number of students to get the right experiments done at the right time, since there would be insufficient equipment for all to participate at the same time. On the type of course with which he is dealing they tend to have smaller classes and so this method of teaching is possible. He admitted that it would be difficult to achieve the required level of integration with a large class".

On the book review by B. L. Button and R. J. Munton

In response to a variety of enquiries, the authors explained that they limited themselves to thermodynamics for mechanical engineers, excluding heat transfer as such. In their view, the question of whether a text was intended for students to read or for their teachers to select from was highly germane and quoted discussion with an author at the meeting who agreed entirely that his recent text book was meant to inspire teachers and was not directed to the student directly.

Views on the use of classics, available perhaps as reprints, differed. A personal favourite of Professor Gurney was the Bridgman text: "The nature of thermodynamics". Some speakers recommended keeping away from modern text books altogether! The Chairman however was on the side of the modern author and quoted Willard Gibb's own text book as entirely inappropriate for today's engineering student.

Editor's note: in view of the interest in the survey of books, it is published as an Appendix to these proceedings with a supplement for chemical engineers and a short list of classical reprints which at least may be useful to the teacher.

Written contribution to the paper by Dr. D. R. H. Jones, from J. D. Lewins

The exponential weighting $\exp(-\delta\mu^*/kT)$ can be made plausible without recourse to statistical thermodynamics by use of the Clausius-Clapeyron equation. Making usual assumptions of constant molar enthalpy of evaporation ($\Delta\tilde{h}$), $\tilde{v}_g >> \tilde{v}_\ell$ and an ideal gas, we have $p(T)/p_o = \exp(-\Delta\tilde{h}/\tilde{R}T)/\exp(-\Delta\tilde{h}/\tilde{R}T_o)$ along the saturation curve. Writing $\tilde{R} = N_o k$ then $\Delta\tilde{h}/N_o$ is the evaporation energy per molecule and the activation energy, $\delta\mu^*$, for this process. Hence $p(T) \sim \exp(-\delta\mu^*/kT)$.

In turn we can consider the pressure at different temperatures to be proportional to the molar density in the gas at varying pressure, $p(T)/p(T_o) = n(T)/n(T_o)$ and hence to the probability of finding that proportion of molecules having energy above the activation energy. It is certainly easy to

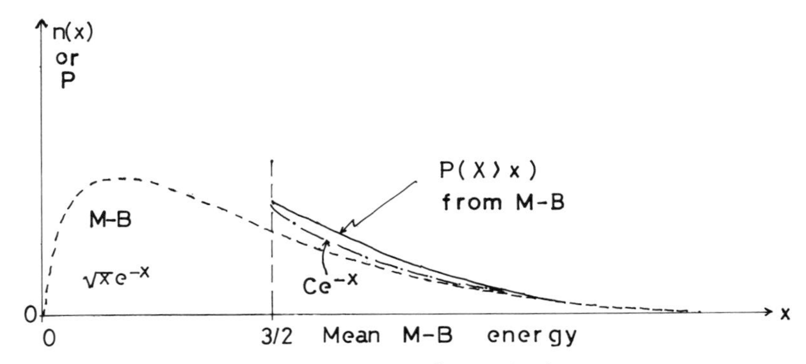

Fig. 1. Validation of $\exp(-\delta\mu^*/kT)$ approximation

compare the Maxwell-Boltzmann distribution, $n(E) \sim \sqrt{E/kT}\exp(-E/kT)$ above the mean energy $3kT/2$ to see that the right-hand tail falls off in approximately exponential form, although the normalization of the function is not particularly accurate.

SESSION 3

COMPUTER ORIENTED AND LABORATORY ORIENTED DEMONSTRATION

Joint Secretaries: T. Hinton and B. R. Wakeford

REAL GAS EFFECT AND GAS LIQUEFACTION

T. H. Frost

Department of Mechanical Engineering
University of Newcastle upon Tyne

A laboratory scale Linde plant has been used routinely to demonstrate to engineering students the essential features of the process for the liquefaction of a gas whose inversion temperature is much higher than ambient temperature i.e. nitrogen. In particular the effect of supply pressure and the effectiveness of the heat exchanger on the yield can be observed and can be explained on the basis of the thermodynamic properties of nitrogen and of real gas effects.

INTRODUCTION

It is important that one section of a practical course in Thermodynamics should be devoted to exposing the student to real gas effects. Obvious traditional examples are the phase changes which occur in boilers, steam turbines and refrigerators; a more unusual and unexpected example is the Linde plant for the liquefaction of the 'permanent' gases. For many years the ability of a proprietary mini-cooler* (Fig. 1) to liquefy nitrogen has been demonstrated to 2nd and 3rd year undergraduate students in Mechanical Engineering.

PRESENTATION TO STUDENTS (Part II Level)

In our particular undergraduate course lectures and laboratories are not integrated. Thus groups arrive in sequence in the laboratory over the whole period of our course.

---

* Hymatic Model No. MAC108. Hymatic Eng. Co., Redditch, Worcs., England.

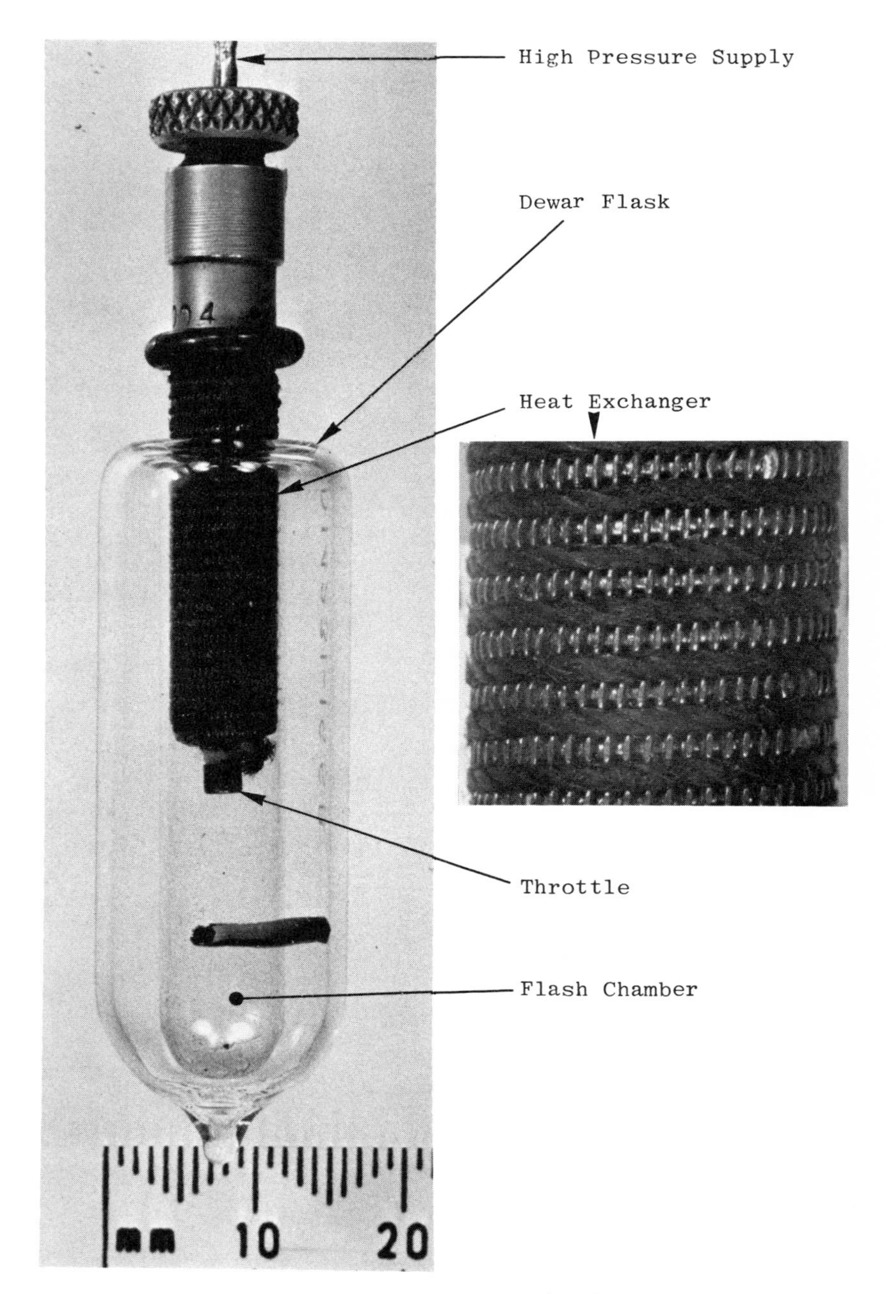

Fig. 1 HYMATIC MINI COOLER (70% Immersion)

At second year level some will have met Maxwell's equation and the Joule Thomson coefficient but none will have met as yet heat transfer, a topic which at the moment is only introduced in the final year. Thus it is necessary to tailor the presentation of an experiment on this mini-cooler according to the detailed background and ability of each group as it arrives. Therefore the students are introduced to the experiment in easy steps. (a) The operation of the equipment is demonstrated. (b) The variables which are under their control, in particular the supply pressure to the cooler and the immersion in the vacuum flask, are discussed. (c) They are asked to quantify the effects of these two variables on the time for liquefaction to actually commence and the time to collect a fixed volume of liquid. (d) They are asked to examine the construction of the mini cooler and to deduce how, in fact, it works.

TYPICAL RESULTS AND INITIAL INTERPRETATION

Effect of Supply Pressure. The students are left to determine the pressure at which they will carry out the experiment subject to the maximum pressure which the system can supply which is to the order of 120 bar. Under this condition it takes about two minutes from placing the mini cooler in the vacuum flask before liquid starts to form and a further minute to collect a fixed volume of liquid which is of the order of 0.2 cc. As the supply pressure is progressively reduced students find that both these times increase quite sharply and eventually at a supply pressure of 80 bar or lower, the cooler will not in fact produce any liquid whatsoever. (Fig. 2).

Effect of Immersion. The above figures are for full immersion of the mini-cooler in the vacuum flask. Similar trends occur as the immersion is reduced at constant supply pressure. The plant fails to produce any liquid once the immersion is less than 50%. The good student, by this time, will have appreciated that the two important features in the plant are (a) reduction in temperature due to throttling from the supply pressure to atmospheric pressure. (b) The construction of the of the mini cooler which gives intimate contact between the cooled exhaust gas (at atmospheric pressure) and the high pressure incoming gas (at ambient temperature) leading to a progressive reduction as time proceeds in the temperature of the high pressure gas before

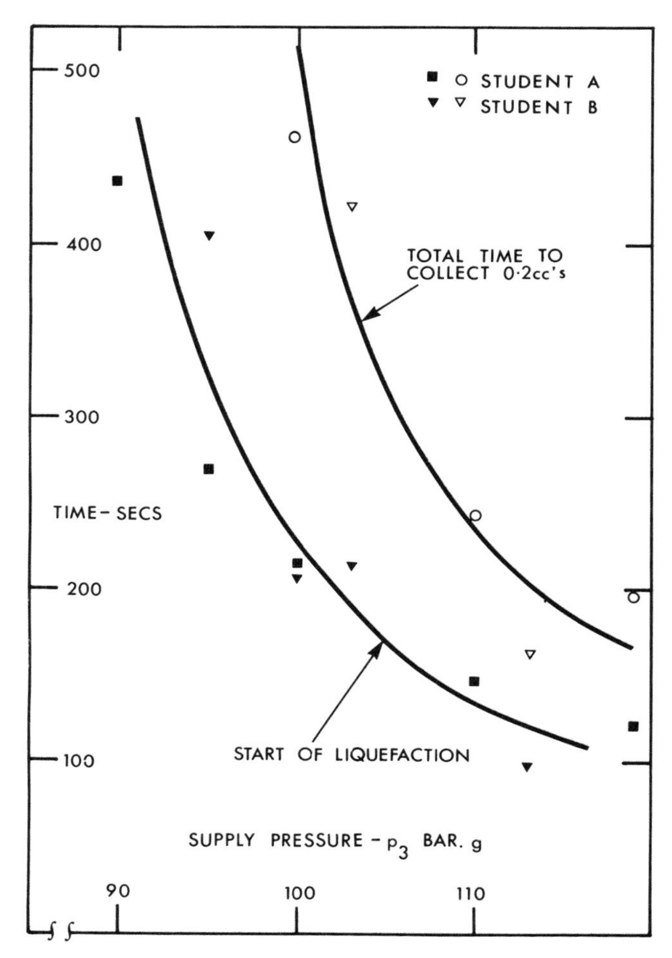

Fig. 2. Effect of Supply Pressure

it is throttled (Fig. 3). Eventually the throttled temperature is low enough for some liquid to form and separate from the remainder of the gas.

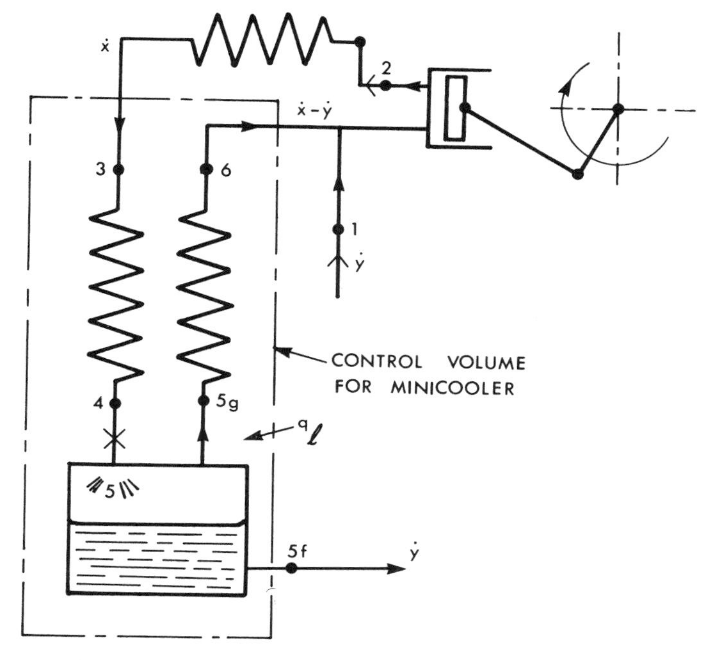

Fig. 3. Linde Plant for Gas Liquefaction

THEORETICAL DEVELOPMENT

Control Volume Analysis. At this stage I ask the students to write down an energy balance for the control volume surrounding the complete mini-cooler. By Part II level all the students will have met the steady flow energy equation, but only the good students manage to cope with this control volume where in fact there is one fluid entering and two fluids leaving. The correct expression is shown in equation AI.2, Appendix I and this forms a basis for discussing the conditions under which we might expect to get the maximum yield from the plant.

T-S Diagram. All students are provided with a copy of the T-S diagram shown in Fig. 4. The author explains the

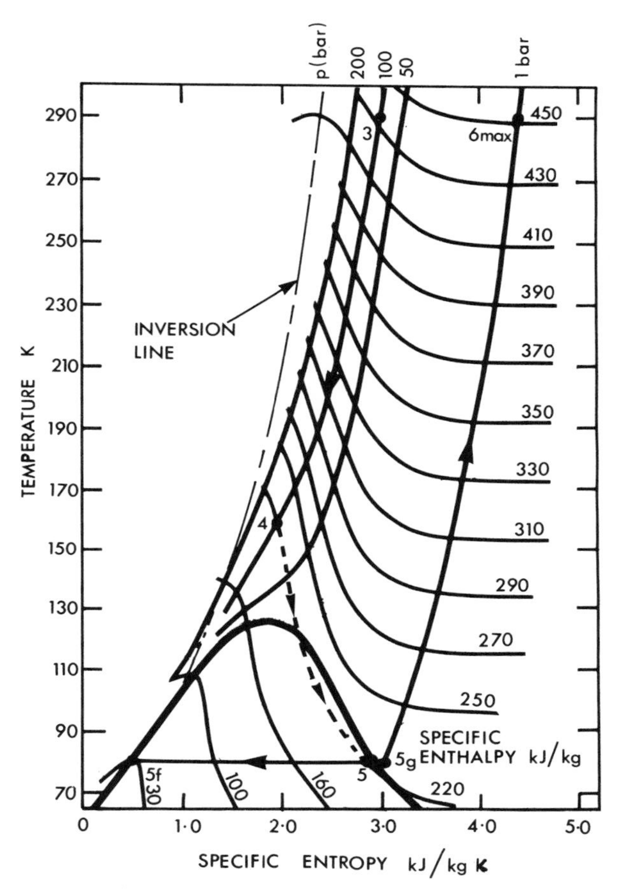

Fig. 4. Temperature Entropy Diagram for Liquefaction of Nitrogen

self-cooling ability of the gas on throttling to be due to the effect of the intermolecular force attraction (Appendix II) but this only occurs when the supply temperature is well below the inversion temperature which for nitrogen is round about 600 kelvin. Thus the shape of the constant enthalpy lines on the T-S diagram can be easily explained and appreciated by the students. It is then a simple matter for them to calculate the maximum yield for a fixed supply pressure under ideal conditions; typically for a supply pressure of

100 bar this comes out to about 5%. The observed effect of both supply pressure and immersion on the performance of the plant can now be explained to the students. In particular, the inability of the plant to work at all below a certain pressure and immersion can be attributed to (a) the thermal inertia of the pipework of the mini-cooler and (b) the inevitably small but finite heat leakage into the system through the vacuum flask from the atmosphere $(\dot{q})_\ell$. Two further topics may then be introduced;(a) the concept of an exergetic efficiency, Equation AI.3;(b) the problems of liquefying a permanent gas whose inversion temperature is much lower than ambient temperature, i.e. hydrogen and helium.

PRESENTATION TO STUDENTS (Part III Level)

The equipment is used purely for demonstration purposes in the final year course on Thermodynamics. Firstly, however, the full theory for inversion temperature and inversion line is developed as summarized in Appendix II and for the predicted behaviour of the plant as outlined in Appendix I. In fact the derivation of the exergetic efficiency for this plant is a very good example of the use of availability analysis.

CONCLUSIONS

This experiment has been used successfully along the lines indicated for about 12 years at Part II level and recently for one year at Part III level. In addition it is quite often used as a demonstration during schoolboy visits. It has many attractive features in practice. Firstly, it never fails to work provided that the supply pressure and immersion are at the required levels. In fact, it is the most reliable piece of equipment in the whole of our Thermodynamics Laboratory. Secondly, it can be demonstrated within five minutes and specific tasks can be given to the students in a very short time. Thirdly, it is completely silent, which again is unique in our laboratory. Fourthly, it provides a very stimulating experience for students, with whom in fact it is very popular, and supervisors alike as there are so many practical and theoretical facets which can be explored.

NOTATION

$\dot{x}$ Mass flow rate of high pressure gas.
$\dot{y}$ Mass flow rate of liquefied gas.
$\dot{q}_\ell$ Rate of heat transfer due to conduction.
β Coefficient of volume expansion (isothermal).

All other symbols (i.e. p, v, T, R, u, h, s, c_p) as S.I. Standards.

Subscripts

in Inversion
in_0 Inversion (temperature) at low pressures
a.b. Constants in equation of state of a Van der Waal's gas.

ACKNOWLEDGEMENTS

The Author would like to thank the Technical Staff in the Department for maintaining the equipment and the Clerical Staff for the preparation of the manuscript.

REFERENCES

1. Frost, T. H., "Temperature Changes due to Throttling of a Van der Waal's Gas", Bull. of Mech. Eng. Education, Vol. 11, No. 1, March 1972.

2. Haywood, R. M., "Analysis of Engineering Cycles", Pergamon (3rd Ed.) 1980.

APPENDIX I

YIELD AND EXERGETIC EFFICIENCY OF THE LINDE PLANT

The energy balance for a control volume containing the heat exchanger, throttle and flash chamber is given by

$$\dot{x}h_3 + \dot{q}_\ell = \dot{y}\, h_{5f} + (\dot{x} - \dot{y})h_6 \qquad (AI.1)$$

The yield $\dot{y}/\dot{x}$ is thus given by

$$\frac{\dot{y}}{\dot{x}} = \frac{h_6 - h_3 - \dot{q}_\ell/\dot{x}}{h_6 - h_{5f}} \qquad (AI.2)$$

The conditions for maximum yield are

a. $q_\ell/\dot{x}$ as small as possible.

b. $h_6 - h_3$ to be a maximum

(i.e. when $T_6 = T_{amb} - T_3$)

Using the Availability Function the Exergetic Efficiency η_{ex} is given by

$$\eta_{ex} = \frac{\dot{y}}{\dot{x}} \; \frac{h_1 - T_a s_1 - h_{5f} + T_a s_{5f}}{h_1 - T_a s_1 - h_3 + T_a s_3} \qquad \text{(AI.3)}$$

APPENDIX II

TEMPERATURE CHANGES DURING ADIABATIC THROTTLING OF A REAL GAS

The following presentation is heavily condensed as it can be found in Reference (1) and in standard texts. Ignoring velocity terms:

$$h_4 = h_5 \qquad \text{(AII.1)}$$

Using Maxwell's 4th relation $\left(\frac{\partial s}{\partial p}\right)_T = -\left(\frac{\partial v}{\partial T}\right)_p$ (AII.2)

$$\mu_h = \left(\frac{\partial T}{\partial p}\right) = -\frac{v}{c_p}(1 - \beta T) \qquad \text{(AII.3)}$$

where $\beta = \frac{1}{v}\left(\frac{\partial v}{\partial T}\right)_p$ (AII.4)

Since δp is negative in a throttling process
then: δT is negative if $(1 - \beta T)$ is negative
δT is positive if $(1 - \beta T)$ is positive
δT is zero if $\beta T = 1$

The last case is satisfied:-

a. for a gas obeying the ideal equation of state
b. for a real gas at the inversion temperature.

THROTTLING OF A VAN DER WAAL'S GAS

$$\left(p + \frac{a}{v^2}\right)(v - b) = RT \qquad \text{(AII.5)}$$

Starting with AII.4 and AII.5:-

$$\beta T = \frac{1 - b/v}{1 - \frac{2a}{RTv}\left(1 - \frac{b}{v}\right)^2} \qquad \text{(AII.6)}$$

Putting $\beta T_{in} = 1$ into AII.6 gives

$$T_{in} = \frac{2a}{Rb}\left(1 - \frac{b}{v}\right)^2 \qquad \text{(AII.7)}$$

The inversion temp T_{in_0} at low pressures is given simply by

$$T_{in_0} = \frac{2a}{Rb} \text{ since } \frac{b}{v} \text{ is small at low pressure.} \qquad \text{(AII.8)}$$

The inversion line is given by substitution of AII.7 into AII.5.

$$p_{in} = \frac{2a}{bv}\left(1 - \frac{3}{2}\frac{b}{v}\right) \qquad \text{(AII.9)}$$

The maximum inversion pressure $(p_{in})_{max}$ is then given by

$\frac{a}{3b^2}$ at which condition $T = \frac{4}{9} T_{in_0}$

COMMENTS

The first correction term $\frac{a}{v^2}$ in AII.5 represents a correction for the intermolecular attractive force. Ignoring the term (b), the molecular volume, it can be argued that since v increases during throttling the temperature must fall. Correspondingly if the a/v^2 term is ignored equation AII.1 becomes:-

$$RT_2 + u_2 - (RT_1 + u_1) = (p_1 - p_2)b \qquad \text{(AII.10)}$$

hence, since $p_2 < p_1$, $T_2 > T_1$

At the inversion temperature as calculated above these two effects cancel out.

THE USE OF TV AND AUDIO-CASSETTES IN TEACHING THERMOFLUID MECHANICS

W. K. Kennedy

Faculty of Technology
The Open University
Milton Keynes

The paper outlines the structure of the Open University 2nd Level (half-credit course):T233 'Thermofluid Mechanics and Energy' (see Session 6), first presented in 1982. The course comprises an integrated multi-media teaching package consisting of 16 'unit' texts, a home experimental kit (HEK), 8 T.V. programmes and 4 audio-cassettes.

INTRODUCTION

The various course components of the Open University course T233 'Thermofluid Mechanics and Energy' are fairly closely integrated with the students being required to cover a specified amount of text and audio-cassette material before the T.V. broadcast linked with each Block. In this way the T.V. component acts to some extent as a 'pacer' to the student. In general the main teaching objective of the T.V. programme is to reinforce and supplement the teaching material presented by other means. Naturally there are fluid-flow phenomena that can only be shown and taught adequately by means of laboratory and/or visual representation. Some of the T.V. programmes involve experimental work with students required to analyse and comment on the results and submit their work as part of their assessed assignments.

STRUCTURE OF THE INTEGRATED TEACHING PACKAGE

Block 1. Unit texts (1 and 2): Introduction to energy and thermodynamics; the first law. Audio-cassette: 'Perpetual Motion' machines (linked to an in-text work sheet); T.V. programme: The Hydraulic Ram - a 'T.V. experiment' with energy analysis procedure.

Block 2. Unit texts (3 and 4): The second law, available energy, entropy and experiments (use of HEK). Audio-cassette: Property diagrams - introduction to P - V diagrams; the laws of thermodynamics (revision). T.V.: Carnot and Stirling Cycles.

Block 3. Unit texts (5 and 6): Modelling fluids; similarity analysis and dimensionless groups. Audio-cassette: Home experimental kit (introduction to the HEK flow table experiment); similarity analysis. T.V.: Modelling fluids: canals and other examples.

Block 4. Unit texts (7 and 8): The first and second laws for flow processes. Audio-cassette: Properties of real fluids (use of steam tables, etc.). T.V.: 'Inputs and Outputs' - the control volume procedure (mass, energy and entropy balances).

Block 5. Unit texts (9 and 10): Fluid Mechanics: energy and momentum. (Bernoulli's equation as special case of the previous energy balance analysis). Audio-cassette: None. T.V.: Bernoulli's equation (with experimental component).

Block 6. Unit texts (11 and 12): Water wheels and turbines. Audio-cassette: experiments with water turbines, water turbine selection (with in-text work sheets). T.V.: Water turbines: experimental work.

Block 7. Unit texts (13 and 14): Heat transfer analysis - introduction to conduction, convection and radiation closely linked to HEK work. Audio-cassette: back-up to HEK work, with in-text worksheets. T.V.: 'Looking at Heat'.

Block 8. Unit texts (15 and 16): Power station cycles; revision of course. Audio-cassette: gas turbine cycle; Revision: preparing for the examination. T.V.: Gas and Steam Turbines.

CONCLUSIONS

The above teaching package which has been developed at the Open University has been shown to be a viable and successful means of teaching thermofluid mechanics to distance learning students. The course is now in its third year of presentation (some 700+ students in total) and the standard attained by the majority of students is commendable.

COMPUTER ORIENTED AND LABORATORY ORIENTED DEMONSTRATIONS

T. Hinton

University of Surrey
Guildford, England

B. R. Wakeford

Heriot-Watt University
Edinburgh, England

INTRODUCTION

A valuable part of the meeting consisted of demonstrations of various laboratory and computer oriented techniques briefly reported in what follows together with some of the discussion engendered

The demonstrations provided by participants showed a wide range of uses to which the computer had been put, to aid the teaching of important concepts in thermodynamics. They illustrated well that low cost microcomputers can be a valuable aid to teaching provided that a clear objective is established as to the role the computer is to play and the teaching/learning activity is integrated into a course. The materials demonstrated were essentially simulations of complex situations for which CAL is well suited and for topics which had been found difficult to teach or understand. In all cases the materials were supplementary to well established courses or an enhancement to laboratory work.

Brief outlines of the CAL materials demonstrated are listed below to provide a useful resource to the reader who may wish to discuss in more detail with the particular author how the materials were used and the kinds of learning gains that have been found. A list of CAL materials relevant to thermodynamics written by various authors not involved in this workshop is included at the end of this section, for information.

COMPUTER ORIENTED DEMONSTRATIONS

Author: J. D. Lewins
Program: Thermoquiz
Microcomputer: BBC with colour monitor

Thermoquiz has been written for a BBC micrcomputer with a colour monitor to support the teaching of thermodynamic diagrams for pure substances. It is used after the relevant lectures and is intended to allow a student to confirm his knowledge.

Author: J. J. Quirk
Program: Van der Waals Condensing Fluid
Micrcomputer: BBC with colour monitor

The van der Walls gas model with constant specific heat at constant volume can be combined with Maxwell's lever rule to predict the condensation 'catastrophe' and thus model a two-phase as well as one-phase fluid.

This program performs the calculations of the saturation curves and then allows plots of the characteristic energy surfaces, such as g(P,T), to be shown.

Author: S. H. Darwoodi
Program: Stefan's Problem
Microcomputer: BBC with colour monitor

This is a solution to a heat transfer problem involving condensation of a liquid to solid and a moving interface. It therefore is relevant to casting problems.

Analytic solutions to Stefan's Problem are known but tedious. This finite element program in one-dimension speedily visualises the behaviour of the casting.

Author: A. J. Organ and G. Weatherall
Program: Stirling Simulator
Microcomputer: BBC with colour monitor

The Stirling Simulator shows the pressure volume behaviour in the two reservoirs of a Stirling engine.

Author: B. Davies
Program: Computer Aided Gas Cycles
Microcomputer: BBC with colour monitor

A simple CAD program that accepts a block design for a gas turbine system permitting interchanges and multiple compressors.

The calculations are based on the perfect gas model and allow isentropic efficiencies or heat exchanger effectiveness to be specified. Simple block diagrams are drawn and thermodynamic diagrams crudely plotted.

The user can quickly modify his design and see the effect on efficiency.

Author: J. P. Packer
Program: Composition and Thermodynamic Properties of the Combustion Products of a Hydrocarbon-Air Reaction
Microcomputer: 48K Spectrum with colour monitor

The program is of particular use when considering the topics of chemical equilibrium and single-zone and multi-zone models of combustion in reciprocating I.C. engines.

There are two options:

1. Temperature sequence; in which the user specifies initial pressure, equvalence ratio and fuel composition. The program then calculates the equilibrium composition of the mixture and plots this against temperature from 750K to 3000K.

2. Conservation of energy; is fundamentally a single-zone of a combustion model. Here the user has a choice of two fuels, either methane (CH_4) or $C_{16}H_{30}$ which could be considered to be an "ideal" diesel fuel. Both are considered to be gaseous and to take part in a single step combustion process. The user specifies the initial zone mass, equivalence ratio, pressure and temperature. The program uses the chemical equilibrium section together with a First Law of Thermodynamics approach to determine the final composition of the zone. This information is output in tabular form.

The program which considers 11 chemical species of combustion products is based on work by Way[1] and Benson[2]. In addition a video showing fuel injection and combustion in a d.i. diesel engine was demonstrated.

<u>Author</u>: R. I. Crane
<u>Program</u>: Simulation of Phenomena in Convergent-Divergent Nozzles
<u>Microcomputer</u>: BBC

A one-dimensional treatment of compressible flow in varying-area ducts is commonly given in second-year under-graduate thermodynamics courses, as a pre-requisite for studies of turbomachinery. In the limited time usually available for teaching convergent-divergent nozzle flows, there is a tendency to concentrate on pressure variations and to curtail the description of off-design phenomena in choked nozzles, shock waves often being treated only qualitatively. The average student may acquire only a hazy notion of such features as the dependence of shock wave location and strength on back pressure, the magnitude of the associated irreversi-bility, and variations in temperature etc.

To supplement such a course, a simulation of inviscid, adiabatic convergent-divergent nozzle flow of a perfect gas has been written in which nozzle shape is fixed, but throat area, inlet state and back pressure are specified by the user. Graphs of pressure, temperature, density, velocity, Mach number and stagnation pressure may be requested; mass flow rate and entropy change are also displayed, together with an indication of the existence of oblique shocks or expansion waves beyond the exit plane. Stored tables of design choked flow parameters and of shock location as a function of exit plane pressure allow plots for off-design conditions to be computed in acceptably short times. Given a guide to the pressure ratios required to choke this nozzle and to achieve the design isentropic expansion, students may quickly explore all possible flow regimes.

---

1. Way, R. J. B., *Methods for Determination of Composition and Thermodynamic Properties of Combustion Products for Internal Combustion Engine Calculations*, Proc. I. Mech. E., Vol. 190, 1976.

2. Benson, R. S., *Advanced Engineering Thermodynamics*, 2nd Edition published by Pergamon Press, 1977.

Author: J. M. Jervis
Topic: The Use of Microcomputers in Thermodynamic Laboratories
Microcomputer: Apple II and Sage IV

(a) The microcomputer is used to replace the conventional engine indicating equipment, and can sample an in-cylinder pressure transducer when triggered by an optical shaft-encoder. The computer carries out real-time data processing and creates pressure-crank angle and pressure-volume diagrams, calculating the indicated power etc. The equipment operates from a spark-plug mounted pressure transducer, which gives a damped response of sufficient accuracy for undergraduate laboratory work.

(b) The computer requests data for a particular test during an investigation (e.g. engine fuel loop) and then processes the data. The processed data is then presented to the student in graphical or tabular form, thus freeing the student from tedious calculations. This allows him to concentrate on the required investigation and the equipment on which it is to be carried out. The student engineer is thus placed in a decision-making role whereby he decides what tests to carry out, considers (with the help of the computer) the validity of the data after each test, and then, using the graphically presented results, he decides upon the next step.

Author: N. Collings, H. Daneshyar and A. J. Organ
Topic: Automated Instrumentation of Internal Combustion Engines
Microcomputer: An EECIV Commercial

A micro-processor is used to control fuel injection and the timing of a spark ignition engine using in-cylinder flame front sensors and ionisation pressure detectors. To obtain the necessary feedback control high-speed data acquisition is used which is easy to set up with low cost microcomputers and analogue to digital convertors.

Author: D. H. Bacon
Topic: Use of Computers in Problem Solving
Microcomputer: 48K Spectrum

The objective is to introduce students to more realistic problems than provided by the traditional 'short examination problem" and how to use a computer in their solution. Students work in a team (4) on a project to solve a problem set to them in the form of a detailed problem specification. Solution of the problem requires a detailed analysis and the use of a microcomputer to solve the problem numerically. To aid this task a number of utility programs are available upon which the student can build a complete solution.

A source of these programs is a recent book by the author Bacon[1]. The underlying theme of this approach is that to write a successful program the student must fully understand the problem and the method of its solution before the computer can be programmed to carry out the correct analysis.

Three examples are available:

1. Engineering Thermodynamics in which a four stroke compression ignition engine is modelled by incrementing the crank angle and calculating the pressure and temperature after each increment from a knowledge of the pressure and temperature before the increment and the energy transfers during the increment. The processes can be considered in two forms; valve open and valve closed. The simplest solution to consider is an adiabatic process with air as the working substance for the valve closed period. The results can be compared with those given by the reversible adiabatic relationship, and extended to allow for heat transfer, delay and energy release during combustion and flow of working substance during the open valve period. The results may be compared to an indicator diagram obtained on a real engine.

2. Work Transfer in which a student can conduct a laboratory trial on a spark ignition engine and process the results by computer program including a graphics routine to plot a power-speed characteristic with specific fuel consumption as contour. A more extensive problem is to develop a program to fit the engine to a vehicle which has to meet a given design specification for acceleration, economy and 'driveability'. This will involve choice of gears, estimating or measuring inertia values, etc. to deal with the acceleration dynamics and then optimizing the variables to meet the specification.

3. Heat Transfer in which a student considers a simple heat transfer process in which water flows through tubes held at constant temperature. The length of tube required for various

---

1. Bacon, D. H., *BASIC Thermodynamics and Heat Transfer*, Butterworth, 1983.

combinations of diameter and numbers of tubes used can be determined for a given heat flux. The pressure drop in each combination can also be calculated. An extension is to design a heat exchanger to perform a given duty and to optimise the design on a cost basis (capital, running and maintenance).

Author: H. K. Zienkiewicz
Program: Statistical View of Entropy
Microcomputer: BBC Model B, 48K Spectrum with colour monitor

Statistical/information-theory view of entropy is illustrated by a microcomputer simulation of the approach to equilibrium of a hypothetical isolated system. The simulation is similar to that described by Reynolds and Perkins[1].

The system consists of a number of distinguishable particles each of which may be in any one of five energy levels. The particles are allowed to interact at random, exchanging energy according to the appropriate transition probabilities. At the start of each 'run', or series of interactions, the system is arranged to be in a specified (most usefully the least probable) state. Population of the energy levels and the corresponding entropy are monitored and can be plotted after every interaction. To simulate the presence of many particles a run can be repeated any number of times. As the number of interactions and runs increases, the population distribution and the entropy are seen to tend towards their equilibrium values corresponding to the maximum of entropy at the given total energy. By a suitable choice of the total energy, an inverted population distribution can be established, illustrating the concept of negative absolute temperature.

Author: M. Reed and D. Robson (Video-Slide Project, Southampton).
Topic: Computer Based Video-Slides
Microcomputer: BBC model G with colour monitor

A valuable video-slide system has been produced at the University of Southampton which can replace or compliment overhead project transparencies or conventional slides.

---

1. Renolds, W. C. and Perkins, H. C., *Engineering Thermodynamics*, McGraw-Hill, Kogakusha, 1977, 177-182.

With this package users can create 'slides' of coloured information and store them on disc for later display. Each screen is referred to as a Video-Slide and conforms to the teletext standard. This provides 24 lines of 40 characters with 8 colours and simple graphics. Each slide may also have associated overlays in a manner akin to overhead projector transparencies. Slides are stored on disc in 'slide show' files. Parts of Video-Slide manage the ordering of slides for display and allows libraries of slides to be created.

Having prepared the slide-show files in advance, the presenter simply executes the displayer program. From then on only single key-depressions are needed to move on to the next slide, move backwards and highlight information within a slide.

In addition there is an auto-pilot display mode so that, once initiated, Video-Slide can cycle through slides at preset intervals in a continuous unattended presentation. This gives the effect of an electronic noticeboard.

Video-Slide is composed of a master program and three main utility programs. All are menu driven with associated help information. All programs are written in BBC Basic.

This facility was brought to the attention of the workshop through a written contribution by Dr. J. D. Lewins and is clearly a valuable facility if BBC microcomputers are readily available.

Author: J. C. Sayer (Interactive Pictures Ltd.)
Topic: Computer Controlled Videodisc
Microcomputer: BBC with Teletext TV and Philips Videodisc Player

A demonstration of an interactive computer controlled videodisc system was shown which represents one of a range of hardware available from different manufacturers varying in price and specification. Basically, a system consists of a videodisc player, interface, computer (with a monitor), and a Teletext colour television. This sytem will present real pictures on screen, either individually, or in motion picture sequences, intermixed with computer generated graphics and/or overlaid text or Teletext. Overlaid text, Teletext or graphics may be changed or updated locally by the tutor whilst still retaining the real photographic images from the disc.

The student may interact with the system answering questions which then route him to particular sequences of static or moving images. The intervention of the microcomputer provides for text (teletext mode) to be written to the TV screen, and for the videodisc mode to be switched off and normal microcomputer mode to be activated.

With a teletext printer television in use relevant teletext information may be printed out.

The major use of the videodisc facility to date has been for Computer Based Training in which visual images of real industrial situation are invaluable as part of the training material.

Interactive videodisc systems are claimed to be one of the most user friendly information/learning systems available, and where no previous computer experience at all is required.

A useful facility for any potential user wishing to explore this technology has been provided by the University of London Audio-Visual Centre who have prepared a Laser Vision videodisc for use as a developmental tool for interactive video programming. It is made up of several independent segments, some of which are extracts from previously-recorded programmes. The others consist of original material specially devised to exploit the properties of the videodisc system: in their diversity they illustrate the range of new options provided by interactive videodisc technology.

Topics covered are:

Micrometer Mantissa
(how to use a micrometer, from the National Physical Laboratory)

Call Yourself a Manager
('Trigger tapes' for management training)

The Rivers Video Project
(10,000 still photographs)

The Nitrogen Bubble
(a piece of research film for image analysis, from Shell UK)

Amphioxus
(an excerpt from a teaching film showing swimming and burrowing of Amphioxus).

Moving the Acute Spinal Injured Patient.

Swift vs. Greenshire County Council
(excerpt from a programme for law students).

Jupiter from Voyager 2
(a selection of NASA images).

Earth from Meteosat 1
(A selection of ESA images).

Earth from Meteosat 2
(one month's weather over the whole globe and other sequences from IPIPS).

Elementary Tabla Technique

Histologic Slides

CAL CATALOGUE

A list of CAL programs in the area of thermodynamics that have been produced by various authors, but who were not present at this workshop, is included for completeness. These materials have not been reviewed and therefore no comment is made as to their availability or suitability. They have not been written specifically for use on microcomputers but for use on a terminal linked to a multiuser computer.

Author: G. Reece, University of Bristol.
Name: CHEAT
Computer: PRIME (FORTRAN)
Publisher: Engineering Science Program Exchange (ESPE), Queen Mary College, University of London.

This is a program for heat conduction problems. It is interactive and can be used for solving a wide range of problems in heat conduction and for illustrating the technique of finite-difference solution of analytically intractable differential solutions.

Author: P. Ayscough, University of Leeds
Name: A710
Language: STAF Author Language

The first Law of Thermodynamics: An application of Hess's Law is the topic of a program used by first year students in the experimental laboratory.

Author: P. Ayscough, University of Leeds
Name: T711
Language: STAF Author Language

Basic Thermodynamics in a self-contained tutorial package for students of Chemistry.

Author: J. Hunter, Glasgow University
Name: HEAT
Language: FORTRAN IV

A program on heat and thermodynamics for students of Physics.

Author: B. Jones, Wolverhampton Polytechnic
Name: ENGIN
Language: Hewlett-Packard BASIC

Investigates the efficiency of an ideal Stirling engine.

Author: B. Jones, Wolverhampton Polytechnic
Name: PATHS
Language: Hewlett-Packard BASIC

Integration of temperature and work on a PV diagram.

Author: L. F. Walker, Wolverhampton Polytechnic
Name: SSPR
Language: FORTRAN IV

Enables the user to obtain property values of superheated steam from a terminal rather than a Mollier chart or tables. SI and British units available.

Author: Dalhousie University, Canada
Name: RRHO
Language: FORTRAN

Calculates the thermodynamic properties of a perfect gas in the rigid rotor, harmonic oscillator approximation, given the vibrational frequencies and cartesian coordinates.

Author: Dalhousie University, Canada
Name: BINVAP
Language: FORTRAN

Plots a boiling point diagram for an ideal miscible binary liquid mixture with positive deviations from Raoult's law.

CONCLUSIONS

The CAL materials demonstrated by the workshop participants had in general not been expensive of resources to produce for a number of reasons.

For most writers these materials were the results of their first attempts at microcomputer programming and at writing CAL materials. Because the materials were written in this way they had been tailored to particular needs which is less time consuming in design and development than writing generic materials that can be adapted for different teaching situations. Although the materials would clearly transfer technically from one institution to another, it is not certain how readily they would be acceptable for integration into existing courses at other institutions. This raises the issue of the true economic cost of such materials if other than marginal costing is taken into account. It is instructive to consider the true cost of such CAL materials if a professional programmer has to be employed to create the courseware to a detailed brief given by the teacher. In a number of projects in which the cost of transferable CAL material has been estimated, a development ratio of 200-500 hours per hour of courseware has been established from which a real cost can readily be calculated.

It is to be hoped that such cost considerations do not inhibit the exploration of new ways of using microcomputers to aid the teaching of complex subjects such as thermodynamics nor impose a restrictive approach to the style of CAL materials which are so often limited more by the imagination of us the teachers, than by the computer itself.

LABORATORY ORIENTATED DEMONSTRATIONS

Demonstration by D. J. Buckingham;
"Resource Based Learning in Thermodynamics"

The paper, which explained the use of tape-slide presentation of subject matter to be followed by small group discussion tutorials, was presented in session 2 and the discussion on the paper is covered there. The demonstration during session 3 consisted of a number of displayed photographs to illustrate the equipment and methods used. Individual discussions with Dr. Buckingham took place during session 3.

Demonstrations by B. L. Button and B. N. Dobbins;
"A Self Learning Assessment Module for the Laws and Definitions of Thermodynamics" and "Developments in a First Course in Thermodynamics using Tape/Slides".

The first of these was a particularly good demonstration of the use, in interactive mode, of computer based teaching. The Law or definition to be learned was clearly stated. The next stage in the program was to ask questions relating to the statement with multi-choice answers. A correct answer was received by a courteous "Yes - well done" appearing on the screen followed by another question or step to the next definition. An incorrect answer was countered with "No" followed, most impressively, by an indication of why the answer was wrong and further explanation of the statement of the Law or definition. This seemed to be coming as near to personal tuition as is possible with a machine.

The second demonstration was complementary to the first. A full paper covering experience in using the method was presented in session 2 and the discussion on the paper is covered there. The demonstration in session 3 illustrated the type of tape/slide material used which appeared well thought-out. The method is used to teach the more extensive applications of the Laws and definitions. The method appears to have some advantage over the formal lecture in that the student can obtain an immediate or later replay of any material which he has not understood. Consequently he has a better chance to note points which he feels the need to discuss during the more extensive time available for tutorials.

During the demonstrations session Professor Button and Mr. Dobbins discussed their material with individual members of the Workshop.

Demonstration by T. H. Frost
"Real Gas Effects and Gas Liquifaction".

This was a most interesting demonstration, the theory and procedure for which is adequately covered in the accompanying paper. The apparatus involved is very small in size, yet worked reliably through the session during which individual demonstrations and discussions were held. This is probably a tribute to its simplicity.

Liquifaction of gases is a topic which is given little attention in many undergraduate Mechanical Engineering courses. This simple experiment can much enhance the teaching of properties of real gases, particularly as they approach the two-phase region. The rapidity with which the liquid Nitrogen evaporated when stirred with a small screw driver blade was rather amusing and provided yet another demonstration of energy transfer.

Demonstration by T. Bright and M. Wright;
"Use of TV and Audio Cassettes in Teaching Thermodynamics".

This demonstration was of material used by the Open University in teaching material related to Thermodynamics. In view of the large proportion of "teaching at a distance" which must be carried out by the Open University, such material has had to be developed by them to an advanced stage. Mr. Bright was unfortunately not able to attend the Workshop and in his absence the demonstration was run by Mr. Wright and Miss R. Armson. Individual discussions with other members of the Workshop took place during the demonstration session.

Display by Tec Quipment International:
"Computer Controlled Engine Test System".

Messrs. Tec Quipment displayed their TD 2000 system which received considerable interest from members of the Workshop. Unfortunately it was not possible to provide facilities to run the engine and so demonstrate the system fully, but it appeared to be a well thought out system which could be used to introduce students, in the later stages of their course, to modern methods of testing engines.

WRITTEN CONTRIBUTIONS TO THE DISCUSSION

J. D. Lewins raised the following in written contribution to Session 3.

Statistical Thermodynamics and Entropy

In seeking to 'understand' entropy, it is natural to embark upon statistical mechanics and be led into discussions of order. McGlashen (1979) gives a salutory warning, pages 111-115, parts of which I reproduce here, against associating entropy as defined in (classical) thermodynamics with 'order'. His first example involves the acceptance of the metastable state as a proper study of thermodynamics having, in particular, a defined entropy, Operationally, metastable states can last a human lifetime; they are operationally an equilibrium state and entropy changes, defined by reversible heat transferring processes, are operationally measurable.

We might agree with McGlashen then that they are a proper subject of study. Now, he says, consider a supersaturated liquor in a metastable state that can be induced to crystalize to a well-ordered crystal and a depleted liquor. Entropy increases as this happens to an isolated system but visible order increases not decreases and is not accompanied by disorder therefore.

And if metastable states do not attract you, consider an argon-hydrogen mixture and the use of a palladium membrane. The hydrogen will escape outwards into an evacuated space. The ordering of the system would appear to increase by the separation of pure hydrogen from the mixture in an isolated system for which the entropy increases. It is true of course that the pressure decreases but what is that to do with apparent 'order'.

The warning then is that a naive recourse to statistical mechanics applied to the special cases of a perfect gas or an ideal liquid may leave the tyro with totally misconceived interpretations of entropy.

H. K. Zienkiewicz replied in writing:

Statistical Thermodynamics and Entropy

Few people at Exeter, with any interest in thermodynamics, can be unaware of McGlashan's salutary warning aginst the mistaken interpretation of entropy as a quantitative measure of disorder (McGlashan 1965); other writers (Wright 1970, Brostow 1972) have uttered similar strictures. Neither in my teaching nor in the naive computer simulation under discussion do I identify entropy with disorder; indeed one of the main points of the simulation is to illustrate how entropy is related to appropriate probabilities rather than to any vague notion of disorder.

Landsberg (1978, 1984) shows how one <u>can</u> define a disorder function

$$D(n) = - \sum_{i=1}^{n} p_i \log_n p_i ,$$

where n is the number of accessible microstates. D(n) is simply proportional to the statistical entropy

$$S(n) = - k \sum_{i=1}^{n} p_i \ln p_i$$

<u>only when n is constant</u>. Otherwise, when S increases less rapidly with n than $k \ln n$, one can have decreasing disorder even though entropy increases.

Incidentally, McGlashan would regard much of what has been written for and said at the workshop as nonsense, for he asserts that the whole of thermodynamics is contained in

$$\Delta U = W + Q, \quad dU^{\alpha} = -p^{\alpha}dV^{\alpha} + T^{\alpha}dS^{\alpha} + \sum_i \mu_i^{\alpha}\, dn_i^{\alpha}, \quad dS(=\sum_{\alpha} dS^{\alpha}) \geq 0 \quad (U,V\text{const}),$$

the rest is easy algebra and there is no such thing as a thermodynamic theory.

McGlashan, M. L. 1965. 'The use and misuse of the laws of thermodynamics'. An inaugural lecture, Exeter University Press.

Wright, P. G. 1970. Contemp. Phys., <u>11</u>, 581.

Brostow, W., 1972. 'Between laws of thermodynamics and coding of information'. Science, 178, 123-126.

Landsberg, P. T. 1978. 'Thermodynamics and statistical mechanics',pp.366-7. Oxford.

Landsberg, P. T. 1984. 'Can entropy and "order" increase together?" Physics Letters, 102A, 171-173.

M. J. French wrote the following comment on the above contribution by J. D. Lewins:

It is no doubt true that a naive recourse to statistical mechanics may leave the tyro with a slightly misconceived view of entropy, rather than the totally confused one which is the common result of the classical approach. However, we do not have to be so naive as to say 'disordered' when we mean 'probable'.

In the second example of the gas mixture, when hydrogen escapes not all the molecules are to be found inside, i.e., we have a more probable state - and a more disordered one too. Incidentally, this is a metastable case, since no pressure difference can exist in a classical 'stable equilibrium', and the first steps in the 'single axiom' derivation fall if this point is not adhered to. In the truly classical case the argon escapes, too, so there is no paradox anyway.

The first example involves an increase in order associated with a more probable state, which can be illustrated by a solitaire board with loose marbles on it being shaken gently.

SESSION 4

PRINCIPLES OF THERMODYNAMICS

Chairman - N. Hay

Secretary - N. Zienkiewicz

TEACHING THERMODYNAMICS TO FIRST-YEAR STUDENTS BY THE SINGLE-AXIOM APPROACH

R. W. Haywood

University of Cambridge

SYNOPSIS

The paper shows, with short, simple proofs, how non-cyclic and cyclic statements of both the First and Second 'Laws' may be deduced as corollaries of the Law of Stable Equilibrium. It also points out that an equally simple proof establishes the State 'Principle' as a further corollary of that Law. Simple 'family tree' diagrams enable a direct visual comparison to be made between the development of ideas in the Single-Axiom Approach and the more conventional presentation put forward by Keenan in 1941. Further advantages of this new approach are also briefly discussed.

INTRODUCTION

Hatsopoulos and Keenan[1] presented the Single-Axiom Approach to equilibrium thermodynamics in a book which was not suited to an undergraduate course. My own book[2], published in 1980, was designed to meet that requirement. It goes well beyond the needs of a first-year student and, indeed, should serve him well throughout a three-year course. It was essentially written for teachers, in the belief that from it they could construct their own shorter, first-year presentation. This short paper indicates how this might be done. The single axiom on which it is based is termed the Law of Stable Equilibrium (LSE for short).

THE LAW OF STABLE EQUILIBRIUM (LSE)

The LSE relates to a Constrained System (Fig. 1) confined within a fixed bounding surface and subject to fixed physical constraints, both internal and external (e.g. subject to a specified gravitational field). It says that, from an initial specified state (described as an Allowed State), such a system can (and will) settle to one, and only one, Stable State while subject to no net interaction with its environment. This implies, of course, that if the initial state is itself stable, the state of the system will not change in these circumstances. This Law does not run counter to our everyday experience but remains a Law in the absence of proof.

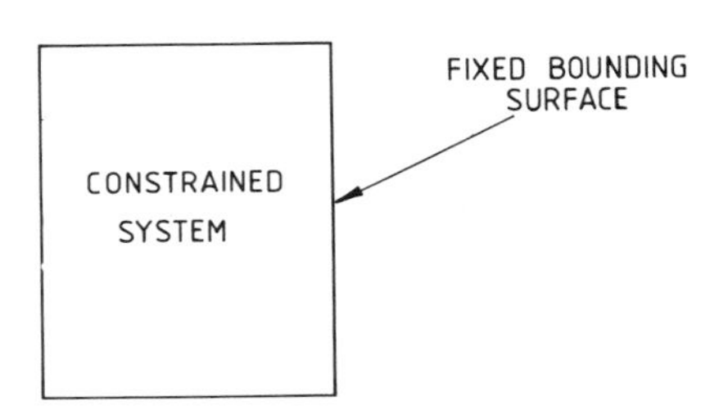

Figure 1 Constrained System

A Constrained System is one which is subject to fixed physical constraints, including confinement within a fixed bounding surface, and for which none of the constraints to which it is subject are altered during the process under consideration.

WORK INTERACTION

We start with the concept of a pure work interaction, in which work is the only form of interaction between a system and its environment, and we define such an interaction in the conventional way in terms of the change in level of a weight in a gravitational field.

COROLLARY 1 - FIRST 'LAW' (NON-CYCLIC STATEMENT)

As Corollary 1 of the LSE we can deduce the Non-cyclic Statement of the (so-called) First 'Law', namely:

<u>The work is the same for all adiabatic processes between two given Stable States of a system</u>

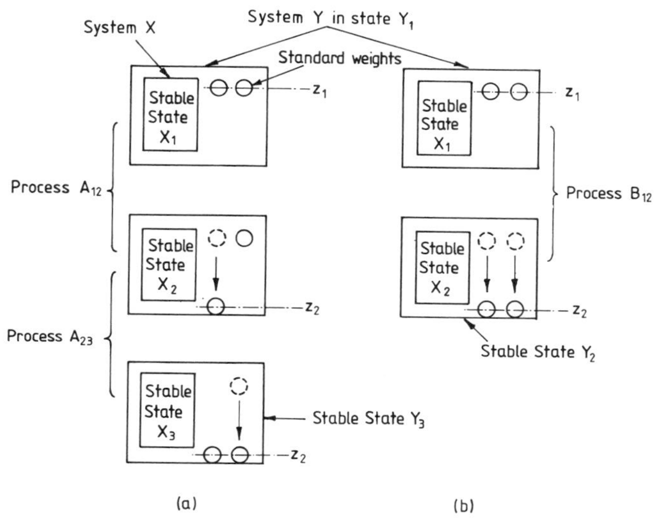

Figure 2. Proof of First 'Law' (Non-cyclic Statement)

For simplicity, we prove this first for the case of work <u>input</u>, leaving the case of work <u>output</u> as an exercise for the student. Thus, in Fig. 2 we consider the two alternative processes A_{12} and B_{12}. In (a), process A_{12} takes System X from Stable State X_1 to Stable State X_2 by the work input to the system provided by the work interaction arising from the descent of <u>one</u> standard weight. In (b), and contrary to Corollary 1, the work input resulting from the descent of two standard weights takes System X between the same two Stable States in process B_{12}.

To process A_{12} in (a), we now add process A_{23}, in which the interaction arising from the descent of the second standard weight will carry System X to some new Stable State X_3.

We look now at System Y. This suffers no interaction with its environment and yet, from the same initial state Y_1 in both (a) and (b), it settles to Stable State Y_2 in (b) but to a different Stable State Y_3 in (a). This contravenes the LSE, so Corollary 1 is proved for work input. The student can be left to prove it for work output.

The Non-cyclic Statement of the First 'Law' is incomplete because it relates to a work-only (adiabatic) type of interaction. Before we turn to the Cyclic Statement we need to define a heat interaction, but we first define the property energy.

ENERGY

Having proved the Non-cyclic Statement of the First 'Law', energy may now be defined explicitly by the following expression for energy change:

$$E_2 - E_1 \equiv -W^a_{12} \quad , \tag{1}$$

$$\text{or} \quad dE \equiv -dW^a \quad , \tag{1a}$$

where W^a_{12} is the work output (or the negative of the work input) that would result from taking the system by any available adiabatic process from Stable State 1 to Stable State 2. The unit of work being the newton metre, namely the joule, so therefore is also the unit of energy.

HEAT INTERACTION AND TEMPERATURE DIFFERENCE

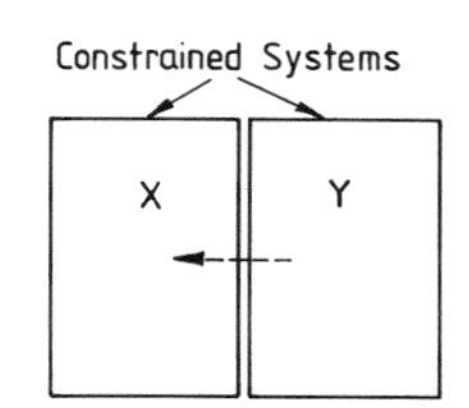

Figure 3. Heat Interaction

We define a heat interaction by considering the two Constrained Systems X and Y of Fig. 3, each initially isolated and each in a Stable State. If, when brought into communication with each other while remaining otherwise isolated, they suffer a mutual interaction which changes their states, that interaction is termed a pure heat interaction.

As we shall next see, the energy transfer between them during the interaction is described as heat. (Incidentally, if pressed by a student of suspicious mind, we can point out that it could not have been work, for one of the systems would then be behaving as a Non-cyclic PMM 2, whose impossibility of existence we shall shortly establish as a corollary of the LSE.)

From experience, we know that if the energy transfer is from System Y to System X of Fig. 3, then the (empirical) temperature of Y is greater than that of X, thus giving us the concept of temperature difference.

QUANTITY OF HEAT

We next need a definition for the quantity of heat. We obtain this by expressing it in terms of the energy change of either system in the heat-only interaction, namely

$$Q_{12} \equiv (E_2 - E_1)_{\text{heat only}} \quad , \tag{2}$$

$$\text{or} \quad dQ \equiv dE_{\text{heat only}} \quad , \tag{2a}$$

where Q refers to heat input to the system (or the negative of heat output). The unit of heat is then clearly the same as that of energy, namely the joule (≡ 1 N m).

For the discerning student who is a little suspicious of this step because energy difference was defined in terms of work, and here we are using energy difference to define heat quantity, one may refer him or her to Ref. 2, Chapter 5 for the simple proof of the (so-called) State 'Principle', as Corollary 2 of the LSE. This 'Principle' states that the Stable State of a Constrained System (of volume V, say) is fully identified when its energy E alone is specified, thus justifying the above step.

FIRST 'LAW' (CYCLIC STATEMENT)

Having quantified both work and heat, we can now turn to the more complete statement of the First 'Law', namely the Cyclic Statement. This is expressed in mathematical terms as

$$\oint (dQ - dW) = 0.$$

We prove this so-called 'Law' as a corollary of the LSE by considering the respective heat and work interactions depicted in Fig. 4.

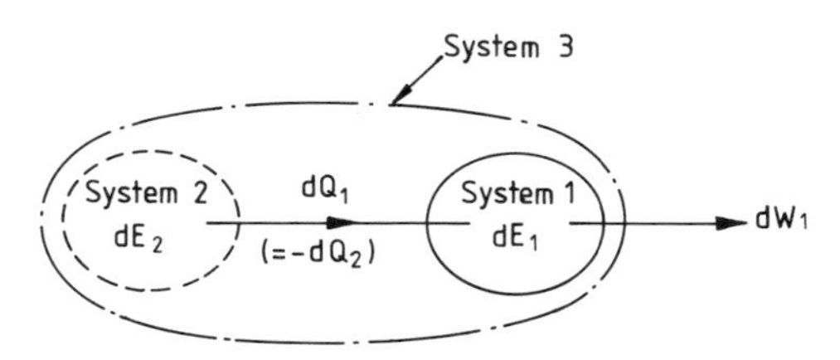

Fig. 4 Proof of First 'Law' (Cyclic Statement)

Then we have:

For System 2: From equation (2a) $dE_2 = - dQ_1$ (3)

For System 3: From equation (1a) $dE_3 = dE_1 + dE_2 = -dW_1$, (4)

Whence, for System 1: $dE_1 = dQ_1 - dW_1$,
or, in general, $dE = dQ - dW$ (5)

For a cyclic process, $\oint DE = 0$

$$\therefore \oint (dQ - dW) = 0 \qquad (6)$$

The Cyclic Statement of the First 'Law' is thus proved.

Our next concern is to prove the (so-called) Second 'Law' as a corollary of the LSE, for which we first need to define what we call a Non-cyclic Perpetual Motion Machine of the Second Kind (Non-cyclic PMM 2, for short). This hypothetical device was first given that name in Ref. 2.

SECOND 'LAW' (NON-CYCLIC STATEMENT)

A non-cyclic PMM 2 is depicted in Fig. 5. It represents a Constrained System going from some initial Stable State to a different Allowed State while the sole effect external to the system is the raising of a weight. The inverse of the paddle-wheel process would thus constitute a Non-cyclic PMM 2. The impossibility of such a process conforms to our every-day experience.

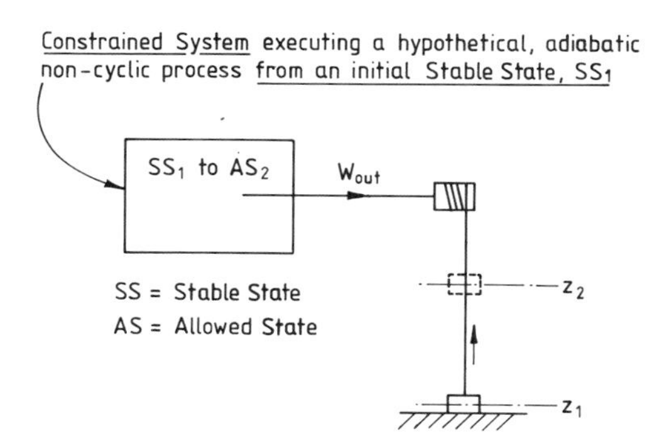

Figure 5. Non-cyclic PMM 2

We shall prove, as a corollary of the LSE, that it is impossible to construct a Non-cyclic PMM 2. The statement that this is impossible is what has been called in Ref. 2 the Non-cyclic Statement of the Second 'Law'.

To prove this, we simply note that the weight of Fig. 5 having been raised to level z_2 by the Non-cyclic PMM 2, could then be lowered to the initial level z_1 while feeding work locally into part of the Constrained System (e.g. by a paddle-wheel process), thereby bringing the latter into some non-equilibrium state. The overall process (from z_1 to z_2 and back to z_1) would be one in which the Constrained System went from an initial Stable State to some other Allowed State, yet in which there had been no net effect on the environment. This would contravene the definition of a Stable State, and therefore the LSE, since it is the LSE which says that Stable States exist.

In passing, we may note that, if it were possible to create a Non-cyclic PMM 2, it could be used to obtain directly from the oceans, for example, an almost unlimited amount of work.

Like the non-cyclic statement of the First 'Law', this non-cyclic statement of the Second 'Law' is incomplete because it involves only a work interaction.

SECOND 'LAW' (CYCLIC STATEMENT)

We now use the foregoing Non-cyclic Statement to prove the well-known Cyclic Statement of the Second 'Law', which states simply that it is impossible to construct a Cyclic PMM 2.

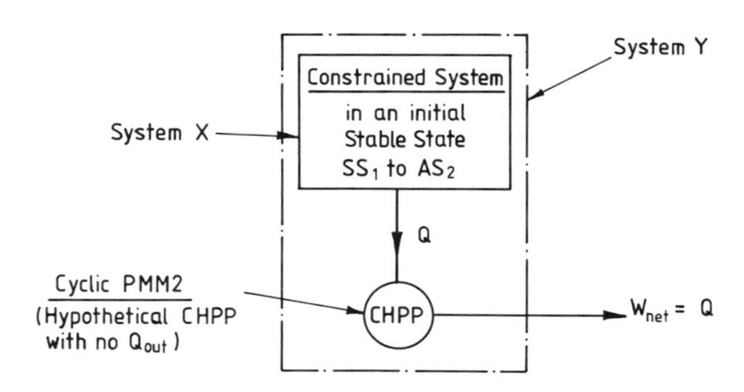

Figure 6. Proof of Second 'Law' (Cyclic Statement)

In Fig. 6, the hypothetical Cyclic Heat Power Plant (CHPP for short) is a Cyclic PMM 2, receiving heat Q from Constrained System X but rejecting no heat.

If we now look at System Y, we see that it constitutes a Non-cyclic PMM 2 (albeit containing a cyclic device within itself), which we have just shown to be impossible of realisation. The Cyclic PMM 2 represented by the CHPP of Fig. 6 must therefore itself be impossible of realisation. The proposition is thus proved as a further corollary of the LSE.

We may note in passing that, if the Cyclic PMM 2 were to exist, we could evidently obtain indirectly an almost unlimited amount of work by utilizing the oceans, for example, as a source of heat supply to a Cyclic PMM 2.

SUMMARY OF RESULTS AND COMPARISON
WITH THE CONVENTIONAL APPROACH

Fig. 7 summarises, in 'family-tree' form, the logical sequence of derivation of all these important corollaries of the LSE.

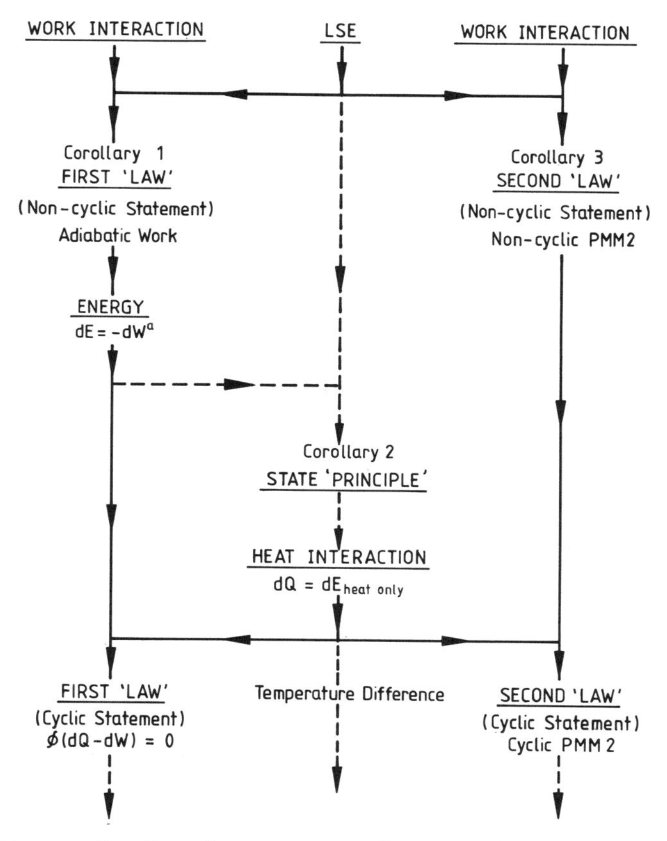

Figure 7. Development of Single-Axiom Approach

Particularly noteworthy is the evident symmetry as between the First and Second 'Laws', each with its own non-cyclic and cyclic statements. By contrast, Fig. 8 illustrates the conventional presentation following Keenan[3].

(Keenan, 1941)

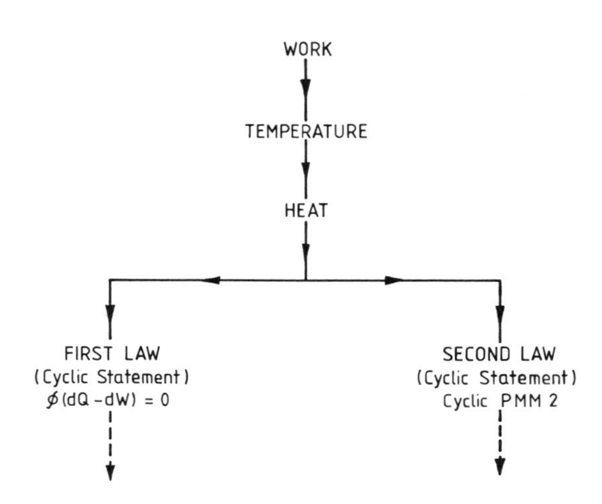

Figure 8. Sequence of Presentation in the Conventional Approach

In this conventional presentation, the student is asked to accept, as separate acts of faith, the cyclic statements of the First and Second Laws (at that time, truly Laws, because they had not been proved). Thermodynamics should be a science, not a religion. The Single-Axiom Approach proves these 'Laws', as well as the State 'Principle', as corollaries of the single basic physical law, the Law of Stable Equilibrium. This approach is surely much more satisfactory and, as here presented, well within the level of understanding of first-year students.

One of the greatest advantages of the Single-Axiom Approach arises from the fact that we start from consideration of non-cyclic processes, instead of cyclic processes (which, in any case, are not natural physical processes, but are constructs conceived by man). This enables us to lead naturally into the extremely important theorems of thermodynamic availability, which relate essentially to non-cyclic processes. Thus, after next proving the First Reversible-Work Theorem (relating to reversible gross work), we are able to use that to develop expressions for thermodynamic temperature and entropy without calling upon the Clausius Inequality. The latter, so ill-favoured by students, is in any case essentially itself a theorem in thermodynamic availability. That the latter topic has been so slow in gaining its due attention, both in the academic world and in industry, almost certainly results from the past fashion of starting the presentation of thermodynamics in terms of cyclic statements of the First and Second 'Laws'. By contrast, the concepts and theorems of thermodynamic availability (and therefore exergy) feature prominently at an early stage in Ref. 2.

The treatment of Simple Systems (usually described with unnecessary restriction as 'Pure Substances') also benefits from the Single-Axiom Approach through the early treatment of the State 'Principle', as also does the treatment of reactive Simple Systems and Open Phases. This, in turn, leads naturally into chemical thermodynamics, which receives thorough treatment in Part II of Ref. 2.

CONCLUSION

The paper has demonstrated the way in which the Single-Axiom Approach may be presented in a form readily assimilable by First-Year students. Its advantages over the more conventional type of presentation have also been discussed briefly.

This wide appeal has led to a Russian edition[4] of Ref. 2 which, inter alia, incorporates a revision to the argument presented in Ref. 2 in respect of a Cyclic PMM 2. A note on this revision for the English edition is included in the papers for the Proceedings.

REFERENCES

1. Hatsopoulos, G. N. and Keenan, J. H., "Principles of General Thermodynamics", John Wiley and Sons, Inc., New York, 1965.

2. Haywood, R. W., "Equilibrium Thermodynamics for Engineers and Scientists", John Wiley and Sons, Ltd., Chichester, England, 1980.

3. Keenan, J. H., "Thermodynamics", John Wiley and Sons, Inc., New York, 1941.

4. Хэйвуд, Р.У., Термодинамика равновесных процессов. Руководство для инженеров и научных работников, Москва, (Мир), 1983.

THE THERMODYNAMICS LAWS FROM THE LAW OF STABLE EQUILIBRIUM

P. H. Brazier

Mechanical Engineering Department
Imperial College

SYNOPSIS

The First and Second Laws of Thermodynamics and the Two Property Rule are derived from the Law of Stable Equilibrium. The treatment is simpler but not so rigorous as that of Hatsopoulos and Keenan (1) or Haywood (2). It is aimed at thermodynamics courses for engineering undergraduates.

INTRODUCTION

In a typical undergraduate thermodynamics course the First and Second Laws are usually presented as separate and unrelated laws. These laws are far from obvious and not related to common experiences or observations. They just have to be accepted as laws. However the Law of Stable Equilibrium, that all systems when isolated from their surroundings reach an equilibrium state, is intuitively correct and is frequently observed in everyday situations.

The derivation of the First and Second Laws from the Law of Stable Equilibrium would seem to offer therefore, a considerable advance in the teaching and understanding of Classical Thermodynamics. This derivation was first presented by Hatsopoulos and Keenan (1). Recently Haywood (2) presented the work in a far more understandable manner.

This work is concerned with the derivation of the First and Second Laws and the Two Property Rule from the Law of Stable Equilibrium. The objective was to achieve as simple and as direct a route as possible and as a result some loss of generality has been incurred. It is hoped that these ideas could form the basis of the first part of a thermodynamics course to engineering undergraduates. The remainder of such a course could follow traditional lines based on cyclic processes or the non-cyclic approach of Haywood (2).

DEFINITIONS

A SYSTEM is any collection of matter contained within a system boundary which may move but through which matter is not able to pass.

A system is ISOLATED when it is not able to have work or heat interactions with its surroundings. This restricts the system to a fixed volume.

The STABLE STATE of a system is the state to which an isolated system settles and from which it cannot deviate while the system remains isolated.

An ADIABATIC process is one in which there is only work interaction between the system and its surroundings.

THE LAW OF STABLE EQUILIBRIUM - LSE

The Law of Stable Equilibrium states that an isolated system will reach one and only one stable state. It cannot then deviate from this state while the system remains isolated. This law was first stated by Hatsopoulos and Keenan (1).

Some examples are shown in Figure 1. However many times each situation shown is repeated, starting from the same initial conditions the system will always settle to the same stable state. It will not deviate from this stable state once reached while the system remains isolated.

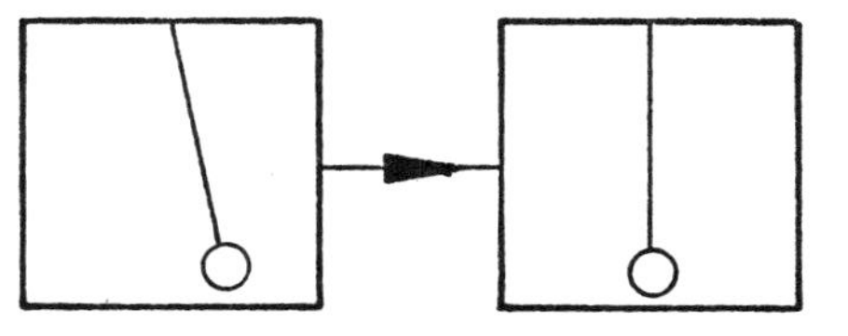

A rigid insulated vessel containing a fluid and a pendulum.

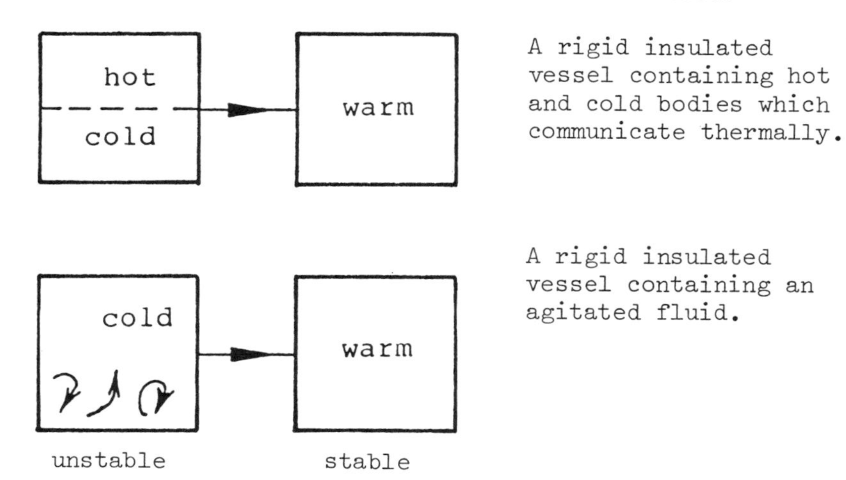

Figure 1. Examples of stable states

DEDUCTIONS FROM THE LSE REGARDING WORK TRANSFER PROCESSES

Work Input Processes. Figure 2a shows a vessel filled with a fluid. A paddle wheel in the vessel is driven by the descending weight so doing work on the fluid. We know from experience that if thermally insulated, the temperature of the fluid will increase.

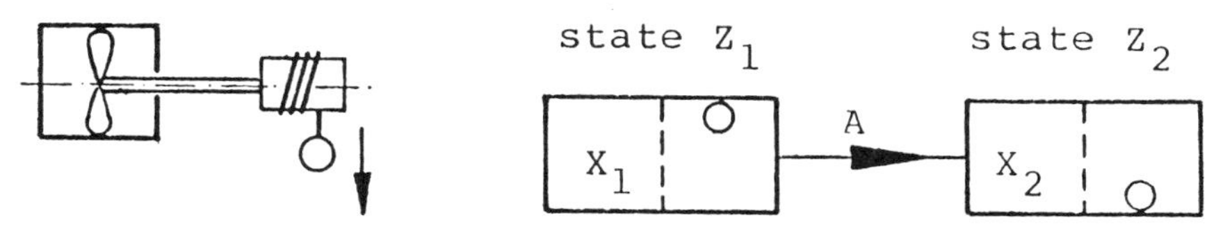

(a) A practical process (b) Symbolic representation

Figure 2. A system undergoing a work input process

The above paddle wheel arrangement can be shown symbolically as in Figure 2b. The isolated system Z comprises system X and the elevated weight. System Z undergoes process A in which the weight descends so doing work on system X which passes from state X_1 to X_2. The LSE states that the isolated system Z will pass from the initial state Z_1 to one and only one stable state Z_2. The components of system Z must also be in particular stable states, i.e. X in state X_2 and the weight on the floor of system Z. It follows therefore that system X at an initial specified state X_1 which experiences a specified work input will eventually end up in one and only one stable state X_2 regardless of the nature of the work input process.

Work Output processes. Consider first system X in the stable state X_1 as shown i Figure 3. System Z comprises system X and the adjacent weight which rests on the floor of system Z. System Z, which is isolated from its surroundings, is in a stable state since both X and the weight are in their stable states. From the LSE system Z cannot deviate from this stable state. Hence system X which is also in a stable state cannot raise the weight.

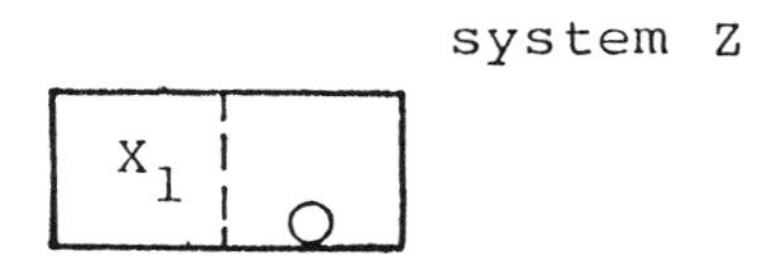

Figure 3. System Z in a stable state

Now consider the scheme shown in Figure 4a. Two fluids X and Y are separated by a piston. If the pressure of X is greater than that of Y the piston will move and can be made to raise a weight. This is shown symbolically in Figure 4b where the unstable system (X+Y) which comprises the mutually unstable systems X and Y, can raise a weight.

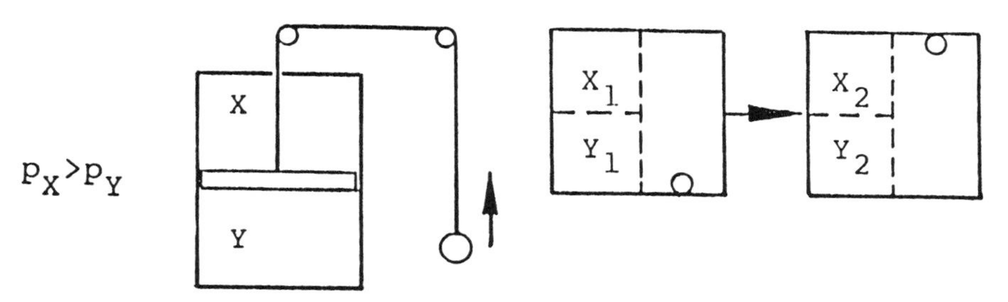

(a) Piston and cylinder (b) Symbolic arrangement

Fig. 4. System (X+Y) experiences a work output process

We therefore conclude that in accordance with the LSE a system in an unstable state can produce a positive work output as its sole effect on the surroundings while a stable system cannot. This is shown in Figure 5.

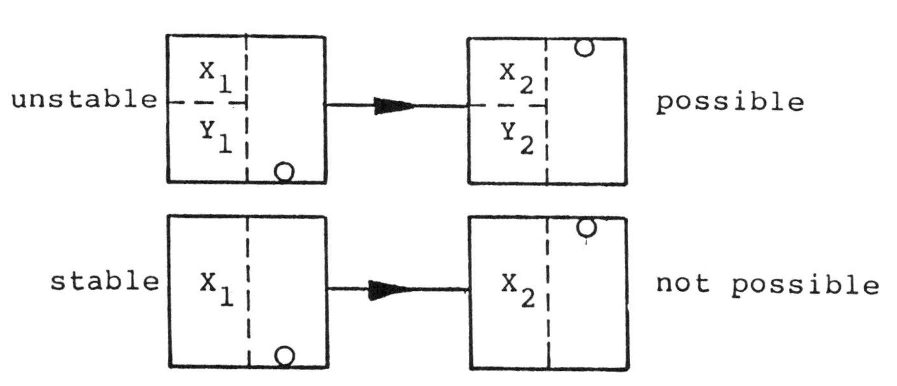

Fig. 5. The possibility of a system raising a weight

To find the magnitude of the work output of a system passing from an unstable state to another state, consider Figure 6. The isolated system Z consists of the unstable system (X+Y) and the weight which is on the floor. System Z can pass directly by process A to its stable state. Alternatively it could proceed to this stable state via processes B and C. In the previous section it was shown that the work input in process C depends only on the initial and final state of system (X+Y). Hence the work output for system (X+Y) in process B also depends only on its initial and final state.

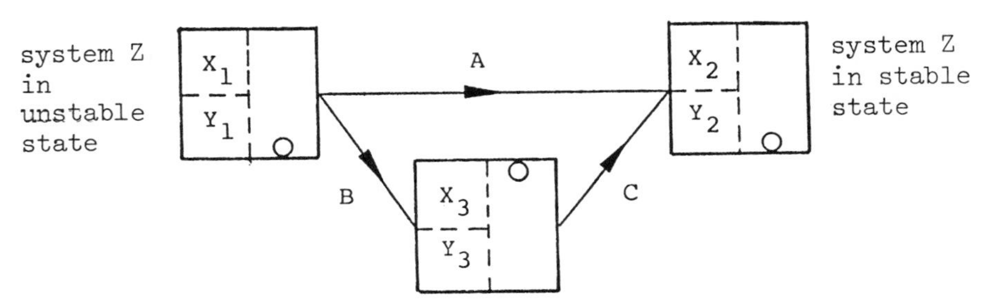

Figure 6. Unstable system undergoes a work output process

Work Corollaries of the LSE. In the previous two sections it has been shown that the LSE imposes certain restrictions when a system undergoes a work transfer process. These can be restated as Corollaries of the LSE as follows.

Corollary 1. The state of a system must change if it executes an adiabatic work process.

Corollary 2. It is impossible for a system in a stable state to execute an adiabatic process in which there is a positive work output.

Corollary 3. The work transfer to or from a system passing from an initial specified state to a final specified state is the same for all adiabatic processes.

ENERGY

From Corollary 3, the work transfer in an adiabatic process depends only on the initial and final states of the system. Hence this adiabatic work must be a property or the change in a property. Define the property E by

$$-dW^a = dE \qquad (1)$$

or

$$-W^a_{12} = E_2 - E_1 \qquad (2)$$

where W^a is the work transfer to the system. The superscript a indicates an adiabatic work only process. It is easily shown that E is the sum of the kinetic, potential and internal energies.

For an isolated system $W^a=0$. Hence when an isolated system passes from an unstable state to its stable state, its energy remains constant.

THE STATE PRINCIPLE

The State Principle can be stated as follows. The stable state of a system is fully identified by its volume and energy. This was first derived from the LSE by Hatsopoulos and Keenan (1).

To prove this statement consider system X and assume that, contrary to the State Principle, it can take up two different stable states X_1 and X_2 but which have the same volume and energy. Now states X_1 and X_2 could have originated from the same state X_0 via two different adiabatic work input processes as shown in Figure 7. From equation (2) the work input must have been the same in both processes and the initial elevation of the weight must also have been the same. The original state of the system comprising system X and the elevated weight is therefore identical for both processes. Now this system comprising X and the weight is isolated and from the LSE can reach one and only one stable state. Hence states X_1 and X_2 must be identical and the State Principle is proved.

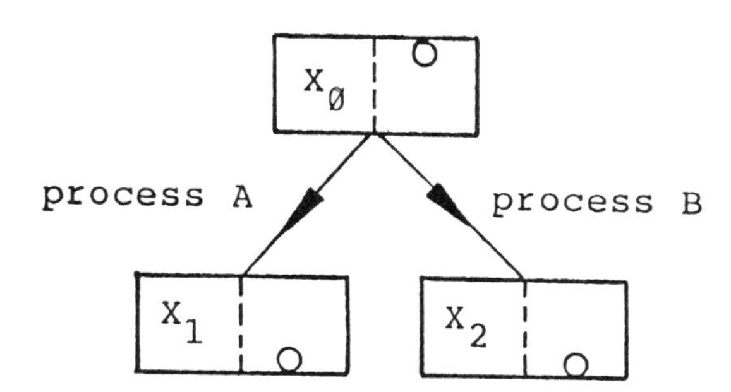

Figure 7. Equi-energy states from a common state

THE TWO PROPERTY RULE

From the State Principle, if the volume and energy of a system are known, the state is identified and all other properties are also identified. Hence for example, the pressure and temperature of a gas are given by

$$p=p(V,E) \quad \text{and} \quad T=T(V,E) \tag{3}$$

These may be solved simultaneously to give

$$V=V(p,T) \quad \text{and} \quad E=E(p,T) \tag{4}$$

Therefore the properties p and T identify the state. In general any two independent properties identify the state of a given system. This is known as the Two Property Rule.

HEAT INTERACTIONS

Definition of a heat interaction. When two initially isolated systems are brought into communication with each other, any interaction between the two that is not a work interaction is a heat interaction.

As an example consider two systems X and Y as in Figure 8. If X is initially warmer than Y an interaction will take place between the two. Eventually a stable state for the combined system will be reached. Is the interaction one of heat or work?

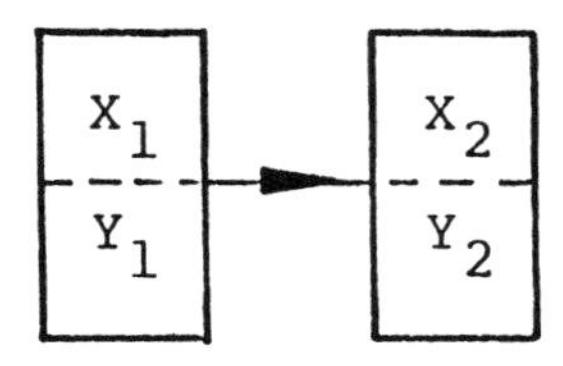

Figure 8. Heat or work interaction between X and Y?

System Y could have been taken from state Y_1 to the warmer state Y_2 by a work input process. However, in accordance with Corollary 2, system X could not have been taken from state X_1 to the colder state X_2 by a work output process and have no other effect. Hence the end states could not have been achieved by a combined process where a work interaction was transmitted from X to Y by say a rotating shaft as shown in Figure 9. The interaction is therefore not a work interaction and must be a heat interaction.

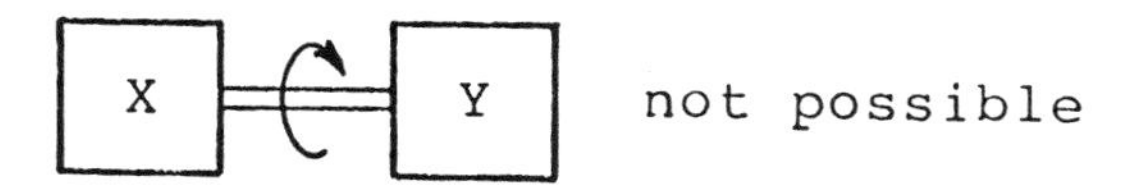

Figure 9. Combined work interaction process

Consider a system in a stable state which receives a heat only interaction. The state of the system will change and according to the State Principle the energy will change. Define the magnitude and sign of the heat interaction by

$$dQ = dE \qquad (5)$$

or

$$Q_{12} = E_2 - E_1 \qquad (6)$$

where Q is the heat transfer to the system.

THE FIRST LAW OF THERMODYNAMICS

It has been shown that the energy of a sytem is changed by work and heat interactions with its surroundings. For a small adiabatic work output process followed by a small heat input process the change in energy for the combined process is

$$dE = dE_{\text{work only}} + dE_{\text{heat only}} = -dW + dQ. \qquad (7)$$

hence for a cyclic process

$$\oint dQ - \oint dW = 0 \qquad (8)$$

since energy is a property. This is the usual statement of the First Law of Thermodynamics. Since it has now been derived from the LSE it is not really still a law.

THE SECOND LAW OF THERMODYNAMICS

The Second Law of Thermodynamics can be expressed in many different and apparently unrelated ways. They all however express certain one way restrictions on heat and work interactions. Corollary 2 is an example of this. A stable system can experience an adiabatic work input process but not an adiabatic work output process. The two most common statements of the Second Law and their derivation from the LSE follow.

1. The Clausius Statement: Heat cannot of itself flow from one body to a hotter body.

Proof. Consider the isolated system comprising of systems X and Y. System X is hotter than Y. There will be a heat interaction in which energy is transferred from X to Y until they are at the same temperature when they are in a stable state. The LSE states that while isolated the systems cannot deviate from this stable state. Hence the reverse process, that of an energy transfer from Y to X, cannot occur in an isolated system and the Clausius statement is proved.

2. The Kelvin-Plank Statement: It is impossible for an engine operating in a cycle to produce a positive work output if in communication with only one thermal reservoir.

Proof. Suppose that contrary to the above statement, the engine shown in Figure 10 can produce a positive work output while receiving heat from only one thermal reservoir. Let the engine complete one cycle in which it receives heat Q from system X and delivers work output W. Now consider system Z comprising of system X plus the engine. The sole effect of system Z on its surroundings is to raise a weight. This is contrary to Corollary 2 and therefore violates the LSE. Hence the Kelvin-Plank Statement is proved.

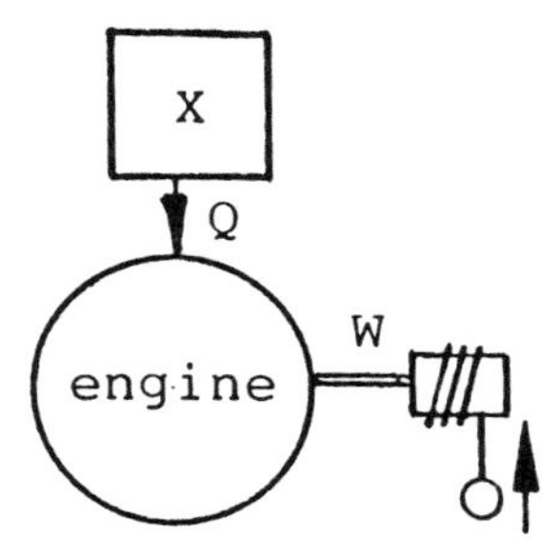

Figure 10 An engine violating the Second Law

CONCLUSIONS

The First and Second Laws of Thermodynamics have been derived from the fundamental and easily conceived Law of Stable Equilibrium. On the way the Two Property Rule has been proved and heat and energy defined.

For an engineering undergraduate course in Classical Thermodynamics this approach would appear to be a far sounder and more satisfying one than the traditional one which is based on three apparently unrelated laws. It could be followed by the traditional sequence based on cyclic processes or the non-cyclic approach of Haywood (2).

REFERENCES

1. Hatsopoulos, G. N. and Keenan, J. M., "Principles of General Thermodynamics", Wiley, 1972.

2. Haywood, R. W., "Equilibrium Thermodynamics", Wiley, 1980.

ENTROPY AND TEMPERATURE AS MACROSCOPIC PROPERTIES OF SIMPLE THERMODYNAMIC SYSTEMS INDEPENDENT OF, BUT MEASURED BY, HEAT

R. H. B. Exell

Asian Institute of Technology
Bangkok

SYNOPSIS

Equilibrium thermodynamics is developed macroscopically with entropy and temperature introduced solely in terms of the effects of adiabatic and diathermic walls through the relations of adiabatic accessibility between states of a system and thermal equilibrium between the system and its environment. Heat, defined separately in terms of internal energy and quasistatic work, is then used to establish measures of entropy and temperature through laws equivalent to the second law of thermodynamics. The method aims to provide an intuitive understanding of entropy and temperature that is difficult to acquire through historical approaches.

INTRODUCTION

Equilibrium thermodynamics has developed historically from a variety of contributions that to this day have not been sorted out into a theory generally accepted as satisfactory for teaching purposes. There is confusion in many expositions of the subject even though the principles are essentially understood and are no longer topics of fundamental research. Among the difficulties that exist in the traditional approach one may cite the dominance of heat engines and refrigerators in the Kelvin-Planck and Clausius statements of the Second Law leading to results unrelated to engineering, and the apparent dependence of the concept of entropy on heat and the Kelvin temperature scale.

This situation was criticized by Born (1949), who had earlier induced Carathéodory (1909) to reformulate the theory. Unfortunately, the mathematical difficulties in Carathéodory's treatment of the Second Law make it unattractive for teaching purposes.

Texts which try to escape from these problems through a statistical approach often complicate the issue by mixing the microscopic concept of disorder with macroscopic concepts in their explanations of the meaning of entropy. Statistical treatments of thermodynamics are, of course, valid in themselves, but it is improper to use statistical concepts to patch up holes in a macroscopic theory.

Detailed axiomatic studies, such as that of Boyling (1972), were clearly never intended for teaching purposes, while the postulational approach in the textbook by Callen (1960) demands considerable maturity in the reader for an appreciation of its contents.

The presentation of equilibrium thermodynamics outlined in this paper is designed to provide a macroscopic treatment of the subject in which the physical concepts and laws are clearly delineated and set in a simple mathematical structure. The theory is such that all the concepts and statements used could in principle be demonstrated directly by observation and experiment. It is also intended to be logically rigorous, although complete rigour cannot be attempted in this short exposition.

The following sections are concerned with presenting only the essentials of the theory. In practice it needs to be taught through illustrative examples.

Every logical theory begins with undefined terms representing the entities under discussion, which must be introduced by description and example. In the text below the undefined terms, and some important defined terms, are underlined.

We also introduce equivalence and order relations with the help of the axioms listed in the Appendix. Following Redlich (1968) we maintain that these axioms merely define the relations with which the theory is concerned and should not be regarded as laws of thermodynamics.

CLOSED THERMODYNAMIC SYSTEMS

A closed thermodynamic system is separated from its environment by a container. It may consist, for example, of a quantity of matter, or blackbody radiation. The system has a set of equilibrium states which can be recognized by observing the values of certain non-thermal variables associated with them, such as pressure and volume. A transition is a change from one equilibrium state to another.

A variable, such as volume, which can be freely manipulated, and whose change involves the performance of work on the system, is called an external parameter. A set of states in which all the external parameters are constant is called an isometric set.

ADIABATIC WALLS AND ENTROPY

A system may be thermally isolated from its environment by adiabatic walls. Any change of state of a thermally isolated system is called an adiabatic transition.

Experience shows that the states in certain sets, which we call isentropic sets, are mutually accessible from each other by adiabatic transitions, and that for any two different isentropic sets the states of one set are adiabatically accessible from the states of the other set but not vice versa. The isentropic sets are the equivalence classes of the relation of mutual adiabatic accessibility, and the relation of one-way adiabatic accessibility establishes a simple ordering of these classes.

Numerical entropies may be assigned to the isentropic sets in such a manner as to preserve their ordering. Thus entropy is a measure of the relative adiabatic accessibility between the equilibrium states of a thermodynamic system. This way of introducing entropy is due to Buchdahl (1958).

DIATHERMIC WALLS AND TEMPERATURE

A system may be placed in thermal contact with its environment through a diathermic wall. A state that remains unchanged when the system is in thermal contact with its environment is in thermal equilibrium with the environment.

A set of states all in thermal equilibrium with the same environment is called an isothermal set, and the states in the set are said to be in mutual thermal equilibrium with each other. Consider two isothermal sets, the first set in thermal equilibrium with the environment and the second not. If the system in a state of the second set when placed in thermal contact with the environment with its external parameters held fixed passes to a state of the first that is adiabatically inaccessible from the initial state, then the second set is said to be "hotter than" the first set. The isothermal sets are the equivalence classes of the relation of mutual thermal equilibrium between states, and the relation "hotter than" establishes a simple ordering of these classes.

The so-called Zeroth Law of thermodynamics is merely a detail in this part of the theory (axiom (A.3) in the Appendix).

Numerical temperatures may be assigned to the isothermal sets in such a manner as to preserve their ordering. Thus temperature is an indicator for the direction of spontaneous transitions with respect to adiabatic accessibility in isometric sets when there is disequilibrium between a system and its environment.

THE FIRST LAW OF THERMODYNAMICS

Our treatment of the First Law of thermodynamics follows that of Carathéodory. Experience shows that the amount of energy transferred to or from a thermodynamic system in an adiabatic transition from a given initial state to a given final state, measured directly or indirectly in terms of work in the system's environment, depends only on the two end states.

It follows that we can define an internal energy potential U for each state such that the energy transferred in any transition from state A to state B is $U_B - U_A$.

A path in the set of equilibrium states through a succession of neighbouring states is called a quasistatic path. Associated with it there is an amount of quasistatic work W done on the system defined in terms of the non-thermal variables (such as pressure and volume). The internal energy change between the end states A and B of the path can be written $U_B - U_A = W + Q$, where Q is the quasistatic heat associated with the path.

Note that W and Q are properties of quasistatic paths. They must not be confused with the work done and the heat transferred in transitions between states.

THE FIRST PART OF THE SECOND LAW

It is convenient to divide the content of the Second Law of thermodynamics into two parts. The first part provides universal measures of temperature and entropy in terms of quasistatic heat and a thermal unit.

Suppose that two different fixed environments are given. Now choose _any_ thermodynamic system and _any_ pair of isentropic sets of the system. Observations show that the quasistatic heats Q_1 and Q_2 associated with isothermal quasistatic paths in thermal equilibrium with the two environments from the first isentropic set to the second

(1) are always in the same ratio $Q_1:Q_2$, and
(2) always have the same sign (positive or negative).

Statement (1) above enables one to define _absolute temperatures_ T_1 and T_2 for the two environments by means of the formula $T_1/T_2 = Q_1/Q_2$, with a standard value 273.16 K at the triple point of water. Then it follows from statement (2) that all absolute temperatures are positive. The kelvin thus defined serves as the basic thermal unit of thermodynamics.

We can now define a _metric entropy_ potential S for each isentropic set such that for the two sets chosen $S_2 - S_1 = Q_1/T_1 = Q_2/T_2$. It is instructive to note that one could define an entropy unit 1 eu equal to the entropy change associated with quasistatic heat Q = 273.16 J at the triple point of water. Then the unit of absolute temperature would be 1 J/eu and the numerical measures of temperature and entropy would be the same as in S.I. units.

THE SECOND PART OF THE SECOND LAW

This part associates the relations of adiabatic accessibility and "hotter than" with the appropriate directions of the entropy and temperature scales.

Observations show

(1) that the states of an isentropic set with metric entropy S_2 are adiabatically accessible from the states of an isentropic set with entropy S_1 but not vice versa if and only if $S_2 > S_1$, and

(2) that an isothermal set with absolute temperature T_2 is "hotter than" an isothermal set with absolute temperature T_1 if and only if $T_2 > T_1$.

These laws imply that when a system loses heat an transition takes place to a state that is adiabatically inaccessible from the initial state, and that a system at a higher absolute temperature than its environment loses heat to the environment when placed in thermal contact with it.

SOME APPLICATIONS

Many of the applications of this theory are straightforward. However, there are certain problems that deserve special mention because they require new methods of solution.

Consider the operation of a cyclic engine converting heat from a reservoir into work. Since all absolute temperatures are positive, the system after increasing its entropy by absorbing heat from the reservoir must reject some heat to return to its initial state. The greatest amount of work is obtained when the heat rejected is least. This means that the heat should be rejected at the lowest possible temperature and with the minimum change in entropy. It can be shown from these considerations that the most efficient cycle operating between given high and low temperatures is the Carnot cycle.

When this theorem has been established the concept of exergy and the availability potential can be introduced in the usual way.

Imagine two thermally isolated thermodynamic systems at different pressures coupled by a movable piston initially held fixed. By the laws of mechanics the piston when released must move towards the low pressure side. In the absence of friction the piston would oscillate indefinitely. However, in reality friction occurs and the piston comes to rest with the pressures on each side equal. It can be shown that under these circumstances the sum of the entropies of the two systems must increase.

It follows from the second part of the Second Law that when two thermodynamic systems at different temperatures are placed in thermal contact heat flows from the system at the higher temperature until the temperatures are equal, and the sum of the entropies of the two systems increases in the process.

These results generalize to coupled systems the principle of increase of entropy that is established for a simple system by the second part of the Second Law.

DISCUSSION

To make it logically complete the theory sketched so far requires extension. In particular, one requires postulates describing the topology of quasistatic paths between states in isometric, isentropic and isothermal sets, and postulates for the manner in which these sets intersect.

The inclusion of the Third Law of thermodynamics and a treatment of negative temperatures are also needed in certain circumstances.

Several lectures on this theory were given recently to Master Degree students in the Asian Institute of Technology. All the students had studied undergraduate thermodynamics in their home universities before entering the Institute. Most of them understood the interpretations of entropy and temperature as indicators of adiabatic accessibility and thermal equilibrium, and found the ideas helpful. However, it was difficult for them to separate these concepts from the measures of entropy and temperature in terms of quasistatic heat. It was also difficult for them to deduce conclusions from the postulates. These difficulties were caused by the nature of their previous training, in which absolute temperature had been practically taken for granted and entropy had been defined at the outset in terms of heat, while the art of deductive reasoning had been neglected. The response of well prepared undergraduates would probably be much better.

REFERENCES

1. Born, M., "The Natural Philosophy of Cause and Chance", O.U.P., 1949, Chapter V.

2. Carathéodory, C., "Untersuchungen über die grundlagen der thermodynamik", Math. Ann., 1909, 67, 355-386; English translation in Kestin, J. (Editor), "The Second Law of Thermodynamics", Dowden, Hutchinson and Ross, Inc., 1976, 229-256.

3. Boyling, J. B., "An axiomatic approach to classical thermodynamics", Proc. Roy. Soc., 1972, A.329, 35-70.

4. Callen, H. B., "Thermodynamics", John Wiley and Sons, Inc., 1960.

5. Redlich, O., "Fundamental thermodynamics since Carathéodory", Rev. Mod. Phys., 1968, 40, 556-563.

6. Buchdahl, H. A., "A Formal Treatment of the Consequences of the Second Law of Thermodynamics in Carathéodory's Formulation". Zeitschrift für Physik, 1958, 152, 425-439.

APPENDIX

Let A, B and C denote states of a system, and let and < represent binary relations between the states that satisfy the following axioms.

(A.1) A ~ A.
(A.2) If A ~ B, then B ~ A.
(A.3) If A ~ B and B ~ C, then A ~ C.
(A.4) For any two states A and B, exactly one of the following relations holds: A < B, A ~ B, B < A.
(A.5) If A < B and B < C, then A < C.

Then ~ is an equivalence relation that divides the sets of states into equivalence classes, and < defines a simple ordering of these classes. As applied in this paper A ~ B may mean that A and B are mutually accessible from each other by adiabatic transitions, while A < B means that B is adiabatically accessible from A but not vice versa. Also A ~ B may mean that A and B are in mutual thermal equilibrium with each other, while A < B means that B is "hotter than" A.

THERMODYNAMIC TEMPERATURE - DIMENSIONAL ANALYSIS

J. D. Lewins

University of Cambridge

INTRODUCTION

The development of a thermodynamic scale of temperature, based on the physical arguments of Carnot and Kelvin for a reversible thermodynamic device operating between two identifiable thermal states, may be treated as an exercise in dimensional analysis. Such an exercise brings out the following points for study:

(1) The role of what Bridgman [1] calls the 'Absolute Significance of the Relative Magnitude' in establishing a cardinal scale.

(2) The Principle of Coherence for metric properties.

(3) The restrictions on inversion of functional relationships.

(4) The choice of scale still open at the end of dimensional arguments, to be determined on other grounds.

PHYSICAL BASIS

As usual, the physical basis of the problem must be established before applying dimensional arguments. Empirical scales of temperature, e.g. a mercury-in-glass thermometer scale, are ordinal scales serving only to place different states of thermal equilibrium in an order. There is no valid

*This paper is reproduced form the Int. J. Mech. Engr. Educ. (Vol. 12, No. 3) by permission of the copyright holders, Ellis Horwood, Ltd.

interpretation of lengths on such a device as a cardinal scale other than in its own terms (i.e. 2 cm mercury-in-glass is twice 1 cm) since there is no quantitative connection with any other thermal property or other empirically chosen thermometer (such as a platinum resistance thermometer).

The basis for a cardinal scale that is also free of the choice of thermometric medium, is offered by Carnot's theory of reversible heat engines operating between two fixed thermal reservoirs at temperatures θ_1 and θ_2 say in Fig. 1. These reservoirs are identifiable on physical grounds as being in determined states of thermal equilibrium and our task is to assign numerical temperatures to characterize these states.

The thermal efficiency, η, of such a cyclic machine is defined as the net work out, W, from the device as a ratio to the gross heat taken in, Q_{in}:

$$\eta = \frac{W}{Q_{in}} \tag{1}$$

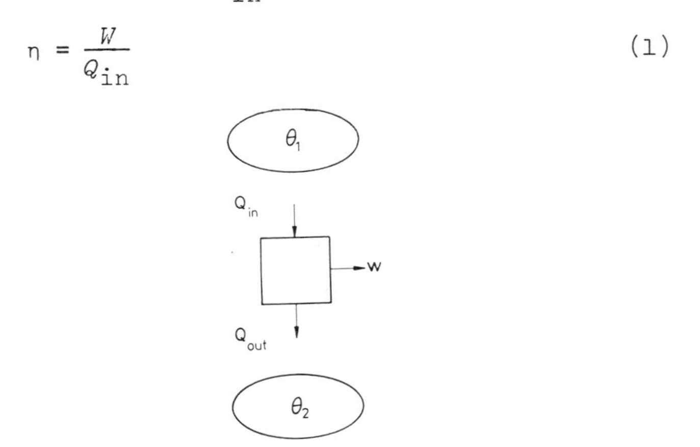

Fig. 1. Carnot engine.

The First Law of Thermodynamics provides a connection between the heat taken in, the heat rejected, Q_{out}, and the net work as

$$W = Q_{in} - Q_{out} \tag{2}$$

Carnot's theory for the reversible engine, i.e. a device that can be operated backwards to restore the heat to the hotter reservoir and all at the cost of the same work as produced in the forward running, is that the efficiency must be the same for any such reversible device operating between the fixed temperatures and is therefore independent of the working medium chosen. In this way, we have an 'absolute' device not dependent on any empirical choice of working medium or thermometric substance. This statement may be expressed in the form of a functional relation between the Carnot thermal efficiency, η, and the two temperatures, θ_1 and θ_2 say, either in the form

$$\eta = f(\theta_1, \theta_2) \tag{3}$$

or

$$\psi(\eta, \theta_1, \theta_2) = 0 \tag{4}$$

But we are not entitled to invert this function and write

$$\theta_1 = g(\eta_1, \theta_2) \tag{5}$$

without in fact showing that such an inverse should hold. Indeed it does not hold; we may have the same efficiency (e.g. 90%) for two distinct cases, $\theta_1 < \theta_2$ and $\theta_2 < \theta_1$. In that case, θ_1 cannot be given unequivocally by a function of η and θ_2.

Instead of the Carnot efficiency therefore we introduce R, the Carnot ratio, the ratio of the heat rejected, Q_2, to the heat received, Q_1, in the Carnot cycle producing work. It will be found that expressions based on R admit inversion. The physics of the problem is now expressed by the functional relation

$$R_{12} = F(\theta_1, \theta_2) \tag{6}$$

The Second Law of Thermodynamics provides that if two reservoirs have the same temperature, then no work can be produced by heat flow between them; $W = 0$ and $Q_{in} = Q_{out}$. The range of η is (0, 1) and R_t $(0, \infty)$.

COHERENCE AS A BASIS FOR A CARDINAL SCALE

For temperature or any other physical concept to have a measurable, extensive or metric property requires the existence of a general attribute we call 'coherence'. The inertial property of matter can be measured on a scale of mass because the process of weighing is coherent. That is, if we have a

$$F(\theta_1, \theta_2) = f(\theta_1)/f(\theta_2) \tag{13}$$

as claimed. The general two-dimensional temperature dependence is therefore reduced to a function of a single variable; the denominator $f(\theta_2)$ serves merely to normalize the function f so that $F(\theta_1, \theta_1) = 1$ as required of the Carnot ratio for reservoirs at the same temperature.

The effect of coherence then is to reduce the arbitrary, unknown function F of two variables to the choice of a function f of one variable.

CHOICE OF SCALES

As the Carnot ratio is determined by the temperatures of two established thermal reservoirs, two possibilities are open to us:

(1) The ratio depends only on the difference $\tau_{12} \equiv \theta_1 - \theta_2$. This leads to what may be called a first-type scale.

(2) The ratio depends on θ_1 and θ_2 separately. This leads to what may be called a zeroth-type scale.

These two forms are both subject to the coherency requirement.

CARDINAL SCALES AND THE DIMENSIONAL ANALYSIS

Any cardinal scale involves the selection of a man-chosen unit, sometimes called the fiducial unit (as in pound sterling being the U.K. fiducial unit of currency). The básis of dimensional analysis then is that no physical relation can depend directly on such an arbitrary selection of unit. It is preposterous, for example, to suppose that the choice of a lump of platinum-iridium and the accident that this has been ascribed a mass of 1000 gramme, can enter as a fundamental part of any expression for physical behaviour. It follows that any equation relating physical dimensions measured on a cardinal scale must be representable in terms of a ratio of the concept to the chosen fiducial unit. Bridgman [1] calls this succinctly the 'Absolute Supremacy of the Relative Magnitude' - ASRM.

Therefore in the present case we seek to use a ratio.

But in the light of the last discussion, this ratio is either the difference, $(\theta_1 - \theta_2)/(\theta_a - \theta_b)$ where θ_a and θ_b are the a unit mass and establish, by the operation of weighing, an identical unit mass, we can weigh both of these against a mass thus determined to be two-units. Then a four-unit mass will weigh as much as either two two-units or four one-units; in this sense, weighing is coherent. In the absence of coherency, a set of weights would be useless, having no unique relation to the operation of weighing.

Similarly the process of mensuration is coherent. Unit length, established as the distance between two defined scratch marks, can be used to build up a coherent scale of length where it will not matter if we express four metres as four times identical one-metre lengths or two times two-metre lengths. There is a physical property of coherency in mensuration.

The Carnot theory similarly provides coherency on which a cardinal scale of temperature can be based. Any succession of Carnot engines, as in the cascade of Fig. 2, is itself a Carnot engine operating between the end temperatures and thus provides coherency when such a cascade is used as a scale of temperature.

Corresponding to the non-uniqueness of the Carnot efficiency, there is no simple relationship between the thermal efficiencies of parts of the cascade and the overall cascade itself. There is a simple relation between the Carnot ratios. For two successive stages in the cascade

$$R_{13} = R_{12}R_{23} \tag{7}$$

which turns out to be the constructive form to exploit coherence. It is readily seen that this last relationship leads to a restriction on the unknown function $F(\theta_1, \theta_2)$ which must satisfy the coherent relation

$$F(\theta_1, \theta_3) = F(\theta_1, \theta_2)\ F(\theta_2, \theta_3) \tag{8}$$

and that this is satisfied by a product form

$$R_{12} = f(\theta_1)/f(\theta_2) \tag{9}$$

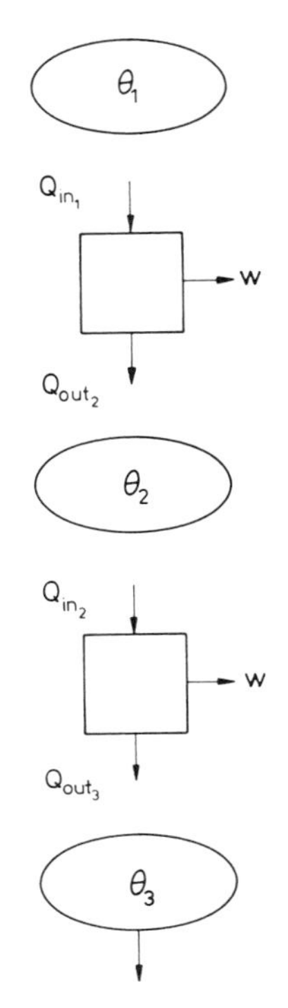

Fig. 2 Carnot cascade.

This is necessary as well as sufficient condition. To show this, we use an argument given by Kestin [4]. Let θ_2 and θ_3 be fixed in the above as θ_1 varies; $F(\theta_2, \theta_3)$ is constant in this investigation and

$$F(\theta_1, \theta_3) = CF(\theta_1, \theta_2) \tag{10}$$

That is, the two functions of θ_1 are proportional, the same within a parameter depending on θ_3 or θ_2 respectively. Then

$$F(\theta_1, \theta_3) = f(\theta_1)g(\theta_3) \text{ and } F(\theta_1, \theta_2) = f(\theta_1)g(\theta_2) \tag{11}$$

To show that $g = 1/f$, substitute the product forms in the original equation so that

$$f(\theta_1)g(\theta_3) = g(\theta_1)g(\theta_2)f(\theta_2)g(\theta_3) \tag{12}$$

and hence

two 'scratch marks' that determine a temperature interval analogous to the unit of length (first type-scale), or in terms of the individual temperatures as θ/θ_a where θ_b is a fiducial unit analogous to the unit of mass (zeroth-type scale). Consider each case in turn.

(1) For the first-type we have

$$R_{12} = f(\tau_{12}) \tag{14}$$

since this states the Carnot theory.

(2) Similarly for the zeroth-type we may write

$$R_{12} = F(\theta_1, \theta_2) = f(\theta_1/\theta_2) \tag{15}$$

in the light of ASRM.

First type scale. For the first-type scale equation (14), coherency requires that for any third temperature, θ_3, (not necessarily 'between' θ_1 and θ_2) we have

$$R_{13} = R_{12}\, R_{23} = f(\theta_1-\theta_3) = f(\theta_1-\theta_2)f(\theta_2-\theta_3) \tag{16}$$

If we differentiate this expression with respect to θ_2 and then let θ_3 approach θ_2, we obtain

$$\frac{f'(\theta_1 - \theta_2)}{f(\theta_1 - \theta_2)} = \frac{f'(\theta_2 - \theta_3)}{f\ \ \theta_2 - \theta_3)} = C \tag{17}$$

since $f(0) = 1$ and where C is the value $f'(0)$.

Therefore $R = f(\tau) = \exp(C\tau)$ and we can invert the expression to

$$\tau = \ell n\ R/C \tag{18}$$

That is, the first-type scale leads to a logarithmic dependence of temperature θ on the Carnot heat ratio R.

Coherence is seen to be satisfied because

$$R_{12} = e^{C\theta_1}/e^{C\theta_2} \tag{19}$$

Zeroth-type scale. Similarly for an arbitrary intermediate temperature, θ, we have (following Bridgman [1])

$$R_{13} = R_{12}R_{23} = f(\theta_1/\theta_2)f(\theta_2/\theta_3) \tag{20}$$

Again differentiate with respect to θ_2, and allow $\theta_3 \to \theta_2$ to obtain

$$\frac{f'(\theta_1/\theta_2)}{f(\theta_1/\theta_2)} = \frac{\theta_2^2}{\theta_1\theta_3} \frac{f'(\theta_2/\theta_3)}{f(\theta_2/\theta_3)} \to \frac{1}{n}\frac{\theta_2}{\theta_1} \tag{21}$$

where $f'(1)$ has been written as $1/n$. The solution is

$$\ell n\, f(\frac{\theta_1}{\theta_2}) - \frac{1}{n}\,\ell n\,(\frac{\theta_1}{\theta_2}) = K = 0 \to f = R^{1/n} \tag{22}$$

where the constant of integration is evaluated by reference to $\theta_1 = \theta_2$ as before.

The zeroth-type scale is then seen to be restricted to a power expression in R. The value of n must be further restricted if the expression is to have an inverse, allowing temperature to be expressed as a function of R. When it is, we have

$$\frac{\theta_1}{\theta_2} = (R)^n \tag{23}$$

CHOICE OF SCALE

If the logarithmic, first-type scale is chosen, two reference temperatures are needed to establish a fiducial interval. These might, for example, be the freezing point and boiling point of water at standard pressure to identify the states concerned. It is then open to associate these states with any two numerical temperatures and write $\Delta\theta^* = \theta_a - \theta_b$. If θ_2 is now regarded as fixed, temperature θ_1, is mapped in the range $(-\infty, \infty)$ as the domain corresponding to the range of $R_t(0, \infty)$. A 'natural' normalization might be to take the fiducial interval as ℓnR_{ab} so that

$$\tau = \ell nR \tag{24}$$

More generally, identifying θ_2 with θ_b

$$\theta_1 = \theta_b + \Delta\theta^* \frac{\ln R_{\ell b}}{\ln R_{ab}} \tag{25}$$

Although equation (25 does not apparently show coherence, coherence has already been shown to be satisfied in equation (19) and we are merely covering up the functional relation in writing the fiducial interval $\theta_a - \theta_b$ as $\Delta\theta^*$; and identifying θ_2 with θ_b.

The choice of a positive fiducial interval (boiling higher than freezing) secures the 'hotter' is 'higher' convention.

For a zeroth-type scale, only one fiducial value is required since the vanishing of R as its lower limit provides a natural 'absolute' zero of temperature. The choice of $\theta_0 > 0$ secures the 'hotter is higher' convention.

We must also seek a value of n that will allow an inversion of the function f. The choice $n = 0$ for example, would only serve to map all temperatures to the value 0. Complex or imaginary numbers will not serve nor even real numbers other than 1, unless artificial additions are made (such as choosing only the positive square root when $n = 2$). Only the choice $n = \pm 1$ will serve.

$n = -1$. This maps temperatures into the domain $(\infty, 0)$.
$n = +1$. This maps temperatures into the domain $(0, \infty)$.

INTERNATIONAL CHOSEN SCALE

The final choice is not a matter of dimensional analysis as such. There is international agreement that the zeroth-type scale, with $n = 1$, is chosen and that the fiducial value shall be to ascribe the number 273.16 kelvin to the triple point of water. This scale, in essence the second scale proposed by Kelvin, may be called Kelvin's scale. Conventionally, temperatures on this scale are represented by the symbol T rather than θ.

The choice of $T_o = 273.16$ K makes the current thermodynamic scale little different from earlier scales and thus saves having to recalibrate any but the most sensitive thermometers. We have the important relation

$$T/T_o = R = Q/Q_o)_{rev} \tag{26}$$

The choice of $n = -1$ gives a temperature mapped as $\theta = 1/T$ and this expression has the virtue of appearing naturally in such expressions as the Clausius-Clapeyron equation, statistical thermodynamics, Arrhenius rate equations, etc.

The choice of a first-type temperature, mapping θ to $\ell n(T)$, has the minor advantage suggesting the difficulty of approaching low temperatures, which are now mapped to $-\infty$. This property reflects the effort needed to approach absolute zero on the Kelvin scale.

But the major reason for the choice $n = 1$ is that only this choice secures the extensive, metric nature of entropy, a further property explored in thermodynamics.

HISTORICAL NOTE

Kelvin's first thermodynamic scale of temperature of 1848 [3] had a logarithmic form of the first-type discussed here. This was an accidental result, not obtained by the arguments given here since, at that time, Kelvin had not accepted the First Law of Thermodynamics. Using the caloric theory, he supposed that no heat was converted in the thermal engine but was passed on in unchanged amount in descending the cascade. Although Kelvin had the correct (Carnot) expression for the efficiency, this caloric viewpoint gave too high a work output in the lower stages of the cascade and hence the mapping of zero absolute temperature to minus infinity.

In his second scale of 1854 [3], Kelvin had adopted the thermodynamic theory of heat (First Law) and proposed the linear scale we now use. He credits 'Mr. Joule of Manchester' with the suggestion of measuring temperature from an absolute zero corresponding to a zero of the Carnot-R ratio.

The 'hotter is higher' convention, like taking the fiducial temperatures to be positive, is purely man-made. Celsius, credited with the first centigrade first-type scale having 100 divisions of the fiducial unit, proposed zero for the boiling point of water and 100 for its freezing point!

REFERENCES

1. Bridgman, P. W. (1922), "Dimensional Analysis", Yale University Press, New Haven.

2. Celsius, A. (1971), "Observationer om twanne bestandiga grader pa en thermometer", (1742), (see "Dictionary of Scientific Biography", Vol. III, Scribner, New York.

3. Kelvin, Lrd (W. Thomson) (1890), "Mathematical and Physical Papers", Cambridge University Press. (a)"On an absolute thermometric scale", Vol. I, p.100 (1848). (b)"On the dynamical theory of heat", Vol. I, p.233 (1854).

4. Kestin, J. (1968), "A Course in Thermodynamics", (2 vols.), Blaisdell, p.197.

CONSTRUCTING TEMPERATURE AND ENTROPY SCALES

J. D. Lewins

University of Cambridge

INTRODUCTION

Many students find difficulty with the concept of entropy. In a traditional mechanical engineering course of thermodynamics, based on cyclic heat engines, a thermodynamic scale temperature is constructed making use of such a device. At the end of a long argument, it is found that a certain integral has a change of value that is independent of the path taken between end states. The student is invited to call this 'magic' number the entropy. In a course for chemical engineers, a more axiomatic approach may be used in which a symbol S - or more precisely, dS - is proposed and from certain axioms the student is invited to derive the Second Law of Thermodynamics and associated results. Generations of students bear witness (not always mute) to the difficulties of either approach in a first course of thermodynamics.

In this teaching note, I suggest a course of development for the derivation of thermodynamic temperatures and entropy which I believe to have the following advantages over traditional methods:

(1) It starts from the easily accepted Kelvin-Planck form of the Second Law and does not employ an abstract axiomatic approach [5, 1].

---

* This paper is reproduced from the Int. J. Mech. Engr. Educ. (Vol. 12, No. 3) by permission of the copyright holders, Ellis Horwood, Ltd.

(2) The concepts of entropy and temperature are introduced in a qualitative fashion and then as invariants of restricted processes before attempting to extend the ideas to include scales for these concepts. In other words, the concept of entropy comes first, the scale after.

(3) The fundamental relation of the Second Law, 'Time's Arrow' $(\Delta S)_{Q \equiv 0} \geq 0$) is developed before a cardinal scale for temperature and entropy is proposed.

(4) Scales for temperature and entropy are constructed simultaneously and show the underlying asymmetry to be associated with heat flow.

The development is not without difficulties, chief to my mind being a use of the reversible process, which I take to be a most sophisticated concept dealing as it does with a limit which is approached but never achieved by real man-fashioned processes. However, this particular difficulty is inherent in any macroscopic account of thermodynamics and the present development attempts to represent the limiting process in the better-known form of limiting ratios as differential coefficients. These are concepts that the young engineer may be able to assimilate in relation to the differential calculus and may serve to align thermodynamic teaching with the main stream of engineering science. I have also included a number of small conventional points that would be necessary to make this a complete account suitable for teaching purposes.

ASSUMPTIONS

The following matters are assumed to have been developed already, some from school physics and some in previous teaching of the thermodynamics course:

(a) A system (or more explicitly a system closed to matter transfer) consists of a defined type and quantity of matter, which does not in itself rule out systems able to transmute themselves by a chemical change to other types of matter. 'Relativity' is ignored.

(b) Work transfer to a system leads to energy increase (using the work-out positive convention). The mechanical definition of work as arising when a force moves the point of its application is perhaps best presented in concrete terms as the idealized linear spring rather than the more traditional raising a weight against gravity since there are times when it is desirable to exclude the effects of gravity. Work transfer is taken to include any action, such as electrical work, that can in principle be considered entirely equivalent to mechanical work in all its effects.

(c) Heat transfer is recognized as a form of energy transfer that is similar but not identical to work. It is similar in that work may be freely turned into heat and hence, by measuring the equivalent work transfer (as in electrical calorimetry) heat transfer may be measured in quantity and direction. We use the convention of positive heat in, a gain in energy. The dissimilarity is expressed in the Second Law of Thermodynamics, that heat may not be freely turned into work. This dissimilarity leads to the essential asymmetry of processes involving heat transfer.

(d) Isolated systems brought into communication (e.g. by conduction or radiation) may evidence a transfer of energy that is not work and is therefore called heat. The transfer ceasing, such systems are said to be in thermal equilibrium. (The transient stage is necessary to assure us that there is no adiathermal barrier preventing communication or we could not be satisfied of thermal equilibrium. It may be dispenses with if we have previously established that there is no barrier and the communication is diathermal.)

(e) Reversible processes exist between any pair of end states of a system, such that the original state of the system may be restored after a change together with all effects of the original process external to the system having been effaced. This is known not to be true in real life except as an idealization or as fluctuations in finite natural systems. (The moving finger write, and having writ, moves on). This idealization is given precise mathematical basis in the form of limiting ratios or differential coefficients in the following developments.

This final assumption, (e), is interesting as in effect defining the domain of classical thermodynamics. There are cases where it is not valid and in those cases (for example the negative spin temperatures described by statistical thermodynamics), classical thermodynamics has a negligible role to play.

VIGOUR

Energy transferred as work in Newtonian, i.e. conservative non-heat reversible processes, is quantified in joules with no distinction necessary as to the type of work, mechanical, electrical and so on. However, when the analysis is extended to include heat, we recognize that the quantification of the transfer in joules is only part of the description. The transfer now has quality as well as quantity, the quality being related to the extent to which the transfer can ultimately do work. We can later identify this quality as the temperature at which the energy is transferred; work can be associated with an infinite temperature of transfer so that its quality is unbounded whereas heat has a finite quality dependent on its temperature, indicating the restrictions on its subsequent transformation to work.

We lack as yet words that describe this situation of the two-dimensional nature of the transfer (quality and quantity) and we are driven to the awkwardness of using the same word, energy, for both the general concept of what is being transferred and for its particular quantifiable aspect. In Newtonian mechanics, however, sufficient time has passed for words to be accepted that make the analogous distinction. We may speak of the transfer of material in terms of the quality of the matter being transferred (lead, water etc.) and the quantity, expressed as mass with units of kilograms and based on the common inertial property of matter. A general name corresponding to matter in this context would be helpful in avoiding the neologisms necessary at the moment in say discussing a power station and speaking of it as either 1000 MW electrical or 1000 MW thermal.

I propose therefore the word 'vigour' for the general name for what is transferred between thermodynamic (closed) systems. Vigour would then be quantifiable in energy, as joules and in quality as temperature, in kelvins.

The alternative expression of the quality of the transfer is of course to use the term exergy. This is not inconsistent with our suggestion for the general description 'vigour' but the role of exergy should be considered carefully. Exergy too is measured in joules and has the disadvantage of not being a property of the system. Rather, exergy depends not only on the system but the condition, temperature and pressure in particular, of the environment the system occupies.

Certainly the available work from the system depends in this way on the environment but at least a statement as to the temperature at which heat is passed avoids the full specification of the environment and allows the nature of the system and its transfer to be isolated in the analysis.

INVARIANTS

It is helpful to express the principal physical concepts in the form of invariants. We can associate a numerical value with such an invariant and begin to quantify a concept. We already have:

- Mass, symbol M, is the inertial property of matter. When numbers are assigned to M.

  MASS IS A NUMERICAL INVARIANT OF A (CLOSED) SYSTEM

- Energy, symbol E, is a property of a system associated with heat and work transfer between systems. When numbers are assigned to E.

ENERGY IS A NUMERICAL INVARIANT OF AN ISOLATED SYSTEM

Of course mass is also invariant in a closed and isolated system but it is easily shown that both concepts are needed when work or heat transfer occur.

Scales for both these concepts are established in a way that allows different systems to have the same number, M or E, without inconsistency since systems are not necessarily able to transmute themselves one to another. Saying that system A is 1 kg of water, system B 1 kg of lead does not imply that water may be changed to lead. But if the matter may transmute by chemical change, then mass is an invariant of the closed system reaction. Because there is no inconsistency in so doing, both mass and energy may be taken to be extensive properties, the sum of the property values of their component systems:

$$M = \sum_i M_i \qquad U = \sum_i U_i \tag{1}$$

We define temperature, symbol T, as the invariant of a set of systems in thermal equilibrium.

TEMPERATURE IS A NUMERICAL INVARIANT OF SYSTEMS IN THERMAL EQUILIBRIUM

Temperature is seen to be an intensive property not an extensive property, by reference to subdivision of a system in thermal equilibrium - the sub-systems have the same temperature.

We will seek to devise a scale of temperatures for different systems not in thermal equilibrium with each other but we might note that T must be a single number if assigned a numerical value (real and not complex say) and that a useful convention might be that 'hotter' corresponds to a higher temperature number, where hotter bodies communicate heat to colder bodies (those receiving heat). There is nothing necessary about this convention; indeed Celsius, in proposing a centigrade scale, suggested zero for the boiling point of water and one hundred for its freezing point.

ACCESSIBLE STATES

Consider a system in state 1 which can therefore be transformed by a reversible process into any other state in which the system may exist. Such a generality is too broad to be useful. We restrict ourselves therefore to reversible processes that are adiathermal, i.e. with no transfer of heat. (The word adiathermal is more precise than adiabatic which only means 'no through path' without reference to heat; furthermore, many distinguished writers in thermodynamics have used adiabatic to mean both reversible and adiathermal which we certainly do not intend.)

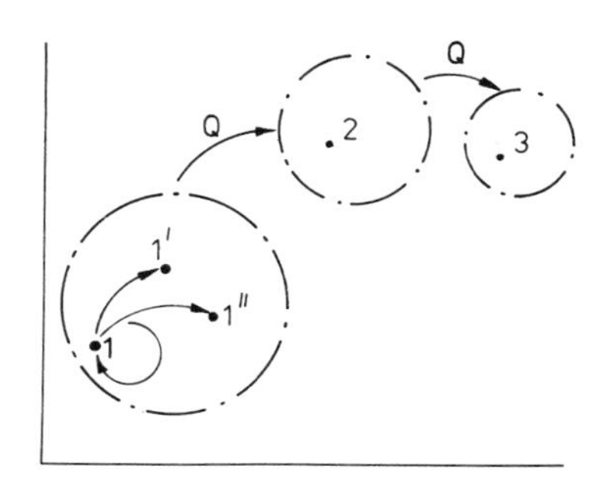

Fig. 1 - Accessible and non-accessible states

Then from state 1 some states 1', 1" etc., say, may be reached by a reversible and adiathermal process. Clearly the set of such accessible states, which includes 1 itself in view of the reversible nature of the process, has something in common, some common property. We call this common property their accessibility. Anticipating at least an ordinal scale, we would assign a number to this concept as first step in metrication. Such a number is called the entropy, symbol S, of a set of accessible states of a given system.

ENTROPY IS A NUMERICAL INVARIANT OF A SYSTEM UNDERGOING A REVERSIBLE AND ADIATHERMAL PROCESS BETWEEN ACCESSIBLE STATES

For this concept to be useful, we have to show that entropy is different to the other end-state properties of a system. In the first place, mass is also an invariant of this process since it is an invariant of any closed process; we must show that there are some states, 2, 3 etc., that are inaccessible from state 1 so that although they have the same mass they have different entropies. Entropy has been introduced by studying a Newtonian or conservative process within the context of systems admitting heat transfer and irreversible processes. There is no claim however that if the entropy of a process is constant, the process is necessarily both adiathermal and reversible (Newtonian); it is readily shown that two of the three epithets 'reversible', 'adiathermal' and 'isentropic' are necessary to prove the third for a process in a closed system. It is this that allows the modelling to be extended to non-reversible processes where entropy increases coupled with heat removing processes where entropy may decrease.

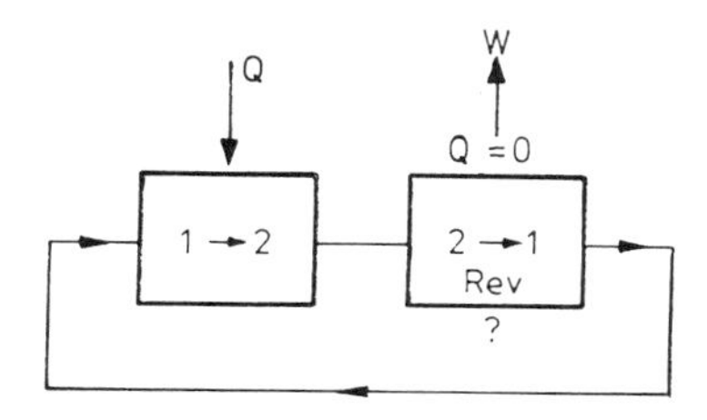

Fig. 2 - Inaccessible state 1 → 2

Consider therefore any process from state 1 that accepts heat, reversibly or irreversibly. Suppose that there was a reversible and adiathermal process from 1 to 2. Then it could be reversed. But if heat is not given out in the reversed process, then the equivalent amount of work must be given out to restore the original energy of state 1. The two processes may be coupled to provide a device that accepts heat and gives out work in contradiction to the Second Law of Thermodynamics in the form:

> Heat may not be turned solely, entirely and continuously into work.

Given that some reversible path exists (assumption (e)), our supposition is false and there is no reversible and adiathermal process from 1 to 2. Any heat receipt involves a change of entropy.

This argument has not attempted to prove that any rejection of heat also changes the entropy and indeed that would not generally be true. We can say that any reversible processes rejecting heat changes the entropy since if the process is indeed reversed, the entropy must return to its original (property) value on a process accepting heat and we know accepting heat changes entropy.

Clearly as states accepting further heat transfer are considered, there is a sequence of states and entropies 1, 2, 3 etc. which suggests we can assign a scale to entropy. We might expect to employ a convention that entropy increases as heat is received. Such a convention is certainly followed but this and the convention for thermodynamic temperature are established simultaneously by devising a scale for both.

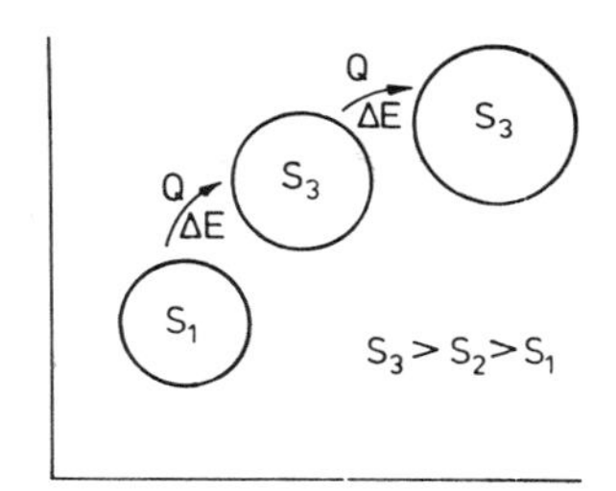

Fig. 3 - Entropy convention.

But it remains to show that entropy is not one of the other thermodynamic properties such as pressure, volume or even temperature itself. Briefly, if a reversible and adiathermal process is carried out (conceptually, or as a limiting process) then certainly work may be transferred. Thus energy is not invariant and from our experience we know that pressure and volume may also be changed by work transfer. Similarly temperatures may be changed. We may reasonably take entropy to be a new invariant property. But we note that in Newtonian and conservative system, where no heat is transferred, all processes are reversible and adiathermal so no distinction is made between mass and entropy; the latter is not necessary to the discussion and is not employed.

We now make the major assumption that entropy is an extensive property; this assumption is subsequently justified constructively, by our success in finding such a form of entropy.

TIME'S ARROW

With this assumption, with the definition of entropy and the sign convention, increasing heat means increasing entropy, we may derive the statement called by Sir Arthur Eddington 'Time's Arrow'

$$\Delta S)Q \equiv 0 \geq 0$$

Suppose that any adiathermal process, 1 to 2, at $Q = 0$, in a system reduced the entropy. Couple such a process with a reversible process to restore the initial properties of the system and thus produce a cyclic device. In the second part of this cycle, 2 to 1, entropy must increase. Thus heat must be taken in. And since no heat is given out 1 to 2, the restoration of the energy at the end of the cycle to its initial value demands that work is done, positive out. Such a device is a Perpetual Motion Machine of the Second Kind contradicting the Second Law.

Two assumptions were made: assumption (e) and the counter-assumption to Time's Arrow. One of these is false. In the realm of classical thermodynamics therefore, the counter-assumption is false and Time's Arrow has been proved. The role of time, t, in the result may be put more prominently by writing

$$\frac{\partial S}{\partial t}\Big)_Q \equiv 0 \geq 0 \qquad (2)$$

HEAT TRANSFER IN THE LIMIT OF REVERSIBILITY

If a system is composed of sub-systems, while the system entropy is constant in the adiathermal, reversible process, the entropy of the sub-systems individually may change as heat is exchanged reversibly between them. Since our immediate need is to prescribe a way of measuring temperature, we use this observation to obtain a ratio that is the same in all systems in thermal equilibrium, a ratio that can be used therefore to measure their common temperature. We shall find that in establishing a thermodynamic scale for measuring temperature, independent of the properties of any specific medium or empirical choice of thermometer, we have also established a scale to measure changes of entropy.

Heat transfer as a result of thermal disequilibrium is irreversible. The irreversibility is decreased as the two bodies in communication approach thermal equilibrium. However, at this equilibrium, the rate of heat transfer vanishes. To discuss the limit, one could suppose that the limiting process included either an increasing time available for heat transfer or an increasing size of system (or both) such that these aspects became infinitely large as the rate became infinitesimal. Since we shall also want the change in thermal equilibrium to be negligible (i.e. no change in temperature of the test systems despite heat transfer), a more attractive way to secure all these together with proper limiting values is to consider differentials of heat and entropy changes as ratios.

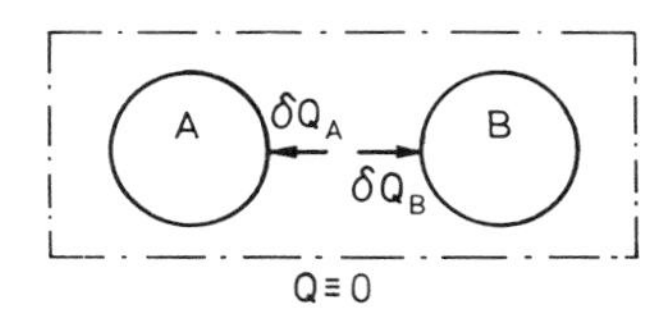

Fig. 4 - Heat exchange: thermal equilibrium.

The limit of reversible heat transfer in thermal equilibrium may be approached in two ways. First consider an adiathermal system made up of two sub-systems, A and B. Each of A and B is in thermal equilibrium with itself but initially

A and B are not in thermal equilibrium. In these circumstances, exchange of heat between them is subject to the overall system being adiathermal:

$$\delta Q_A + \delta Q_B = 0$$

and

$$\frac{\delta Q_A}{\delta S_B} = -1 \text{ (isothermal)} \tag{3}$$

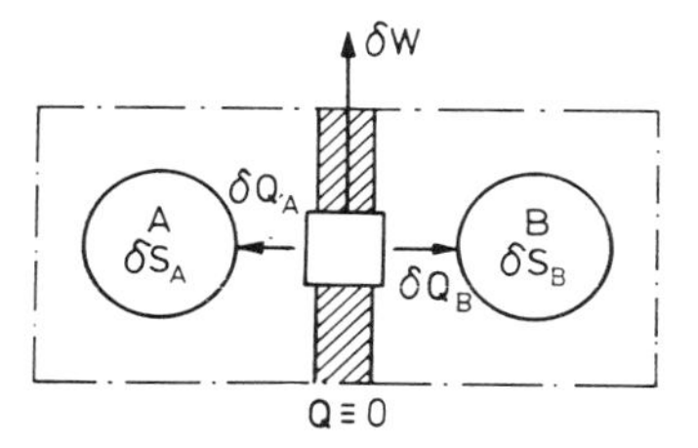

Fig. 5 - Heat exchange: reversible non-equilibrium.

Secondly, let the two sub-systems be separated by an adiathermal barrier but allowed to communicate through a reversible device which will produce work from the heat exchange. The heat exchanges are now not subject to (3) but rather the new arrangement is overall isentropic, being adiathermal and reversible: $\delta S_A + \delta S_B = 0$ so that

$$\frac{\delta S_A}{\delta S_B} = -1 \text{ (rev)} \tag{4}$$

In the limit of thermal equilibrium, both ratios, equations (3) and (4), remain the same so that we obtain proper differentials and

$$\mathcal{L}_{\text{equilibrium}} \quad \left.\frac{dQ_A}{dQ_B}\right)_{\text{rev}} = \frac{dS_A}{dS_B} = -1 \tag{5}$$

Combining these two differentials gives

$$\left.\frac{dQ}{dS}\right)_{\text{rev A}} = \left.\frac{dQ}{dS}\right)_{\text{rev B}} \tag{6}$$

and observe that the term $dQ/dS)$rev is the same in both systems in thermal equilibrium. We therefore chose to use this common value as the thermodynamic temperature T and have chosen a thermodynamic thermometric device such that

$$\int T\,dS = \int dQ)\text{rev} \quad \text{or} \quad \int dS = \int dQ/T)\text{rev} \tag{7}$$

SCALING ENTROPY AND TEMPERATURE

To construct a scale of temperature, we return to the general case of reversible changes between sub-systems at different temperatures. By definition wer have

$$T_A = \frac{dQ}{dS})\text{rev A}; \quad T_B = \frac{dQ}{dS})\text{rev B} \tag{8}$$

and employing the entropy differential

$$\frac{T}{T_o} = -\frac{dQ}{dQ_o)\text{rev}} \tag{9}$$

This result is of course the Kelvin proposal [3] except that, in his words, 'heat used is to the heat rejected in the proportion of the temperature of the source to the temperature of the source to the temperature of the refrigerator' we find no mention of the necessary qualification reversible'. We offer instead:

> Ratios of temperatures on the scale are as ratios of heat transferred reversibly between them with due regard to sign.

(Note carefully that one of the heats is positive, one negative, whether A is hotter than B or vice versa.) The scale is properly called thermodynamic since the assumption of the reversibility of heat transfer it requires at different temperatures implies the conversion of heat to work. It is also of course independent of any empirical choice of thermometric medium such as mercury-in-glass.

It may also be noted that both temperatures will have the same sign; if we select a fiducial value for one temperature to be positive, there will be no negative temperatures. It is the province of the Third Law to discuss the unobtainibility of the limit of the absolute zero of temperature.

The scale is completed by assigning a fiducial number to T_o, the temperature of some reference state. By international agreement, the reference state is chosen to be the (principal) Triple Point State of water substance, where ice, water and steam can coexist in equilibrium. This state is fairly readily reproduced in different laboratories. The agreement extends to assigning the fiducial number 273.16 to T_o. 1/273.16 of the interval between zero and T_o is then called a ^{o}kelvin, symbol K.

This choice of number is no more arbitrary than the definition of the metre involving 1 650763.73 wavelengths of light. In both cases, the number was chosen with regard to custom and practice so that most existing thermometers and their practical scales did not have to be discarded.

Entropy has now been scaled at least as far as a change of entropy is concerned since

$$S - S_o = \int_o \mathrm{d}Q/T)\mathrm{rev} \tag{10}$$

and its units are therefore joules/kelvin (J/K or often kJ/K). Since T is now positive, the desired convention for an increase in entropy has been achieved. Only the difference on entropies between two states is fundamental so that an international convention for the zero of entropy is less necessary. With water, for example, it is customary to make the liquid state have zero entropy at the Triple Point. On other occasions, it is supposed that all substances have the same (zero) entropy at or approaching absolute zero (Third Law).

ENTROPY AS AN EXTENSIVE PROPERTY

The particular choice of ratio, $\mathrm{d}Q/\mathrm{d}S$)rev to be T, the thermodynamic temperature, is crucial in the development. It leads, and is the only form that leads, to the extensive property of entropy. To show this we first establish that indeed entropy as defined is extensive. This is proved constructively, by carrying out the construction in three parts:

(a) For exchanges between two parts of the adiabatic system at the same temperature, $\mathrm{d}S = \mathrm{d}S_A + \mathrm{d}S_B = 0$, a result turning on the extensive property of energy exchanged as heat between the two sub-systems.

$$\frac{dS_A}{dS_B} = \frac{dQ_A}{dQ_B}\Big)_{rev} = -1 \text{ isothermal, reversible}$$

(b) For exchanges at different temperatures, the ratio choice $dQ/dS)\text{rev} = T$ ensures that the entropy changes remain constant

$$dS_A + dS_B = \frac{dQ_A}{T_A} + \frac{dQ_B}{T_B}{}_B = 0 \text{ reversible}$$

(c) We may now add any reference value making $S_o = S_{AO} + S_{BO}$ so that the entropy overall as well as changes in entropy are extensive.

Now suppose we sought some other temperature θ related to T as $\theta = f(T)$. Such a temperature could still satisfy the requirements of thermal equilibrium and still be found to be an integrating factor in the equations such as $dU = \theta d\Sigma - pdV$ if we introduce the generalized entropy

$$\Delta\Sigma = \int \frac{dQ)\text{rev}}{f(T)}$$

But now consider the sum of component generalized entropies in relation to the elements of dS:

$$d\Sigma_A + d\Sigma_B = \frac{dQ_A}{f(T_A)}\Big)\text{rev} + \frac{dQ_B}{f(T_B)}$$

We see that $d\Sigma$ vanishes only for the cases

(i) $T_A = T_B$ thermal equilibrium

and for general T

(ii) $\theta = f(T) = T/T_o$

where T_o is a constant, i.e. the arbitrary fiducial value. Thus only the linear form of temperature scale provides the corresponding entropy with the property of being extensive in nature.

NOMENCLATURE

In this development we have emphasized the separate steps in relation to each concept and it might be useful to categorize the steps with a suggested nomenclature. For matter, we have a general concept for which inertia is an invariant. Mass, symbol M, is then a metrical form for the invariant to which we have ascribed numbers. Inertia itself is subject to an ordinal scale while mass may be ascribed (through suitable operations) a number on a cardinal scale.

For energy, the convoluted history of the development of thermodynamics leaves us with no separate words to describe the sequence of 'concept', 'invariant' and 'metric'. Perhaps regarding the symbol E itself as the metricated energy is the best we can do. We then suggested the general term 'vigour' for the overall exchange involving quantity (E) and quality (T).

Thermal equilibrium plays the role of an invariant while temperature is its metricated form, capable of being assigned numbers operationally on a cardinal scale. Accessibility is our suggestion for the invariant for which entropy, S, is a metric form. Really heat should be regarded as the common concept underlying both these invariants and their metrication.

CONCLUSIONS

It will be seen that our present development at least meets the stated aims of providing concepts for entropy and temperature before entering the details of thermodynamic scales. It rests on the ideas of accessibility and non-accessibility, the later being inherently associated with the occurrence of heat transfer to a system. The subsequent development emphasizes the asymmetry involved.

One should make a careful distinction between removing the source of irreversibility - the new process will then proceed to a different end point while giving out more work - as opposed to finding an equivalent reversible process in which more work is obtained and more heat accepted, the additional heat being taken from some available reservoir at temperature T say. The only way to determine the increase of entropy due to irreversibility is indeed to devise some equivalent reversible path and, in conjunction with T, say that the lost work is $T_0 \, \Delta S$.

In summary, the important ideas are:

- Heat and work are manifestions of the transfer of energy between systems; work may be freely converted to heat but heat may not be freely converted to work, the origin of the asymmetry of the subsequent development.

- Entropy is an invariant of reversible and adiathermal processes in a closed system.

- Temperature is an invariant of isolated systems in thermal equilibrium.

Entropy increases inevitably with the receipt of heat whether this is received reversibly or irreversibly. It is 'that bourne from which no thermodynamic traveller returns'. In the absence of heat transfer, entropy increases with irreversible processes and at best is constant for reversible processes.

The thermodynamic temperature device and scale are selected and defined by making use of the common property of the heat to entropy ratio for reversible processes between systems in equilibrium.

ACKNOWLEDGEMENT

Pippard in his monograph gives a similar interpretation to entropy as the invariant of an 'adiabatic' process but without following the present route to an ordinal and then cardinal scale of entropy and temperature.

I would wish to acknowledge my debt to colleagues in Cambridge who have discussed these matters with me: H. Daneshyar, A. Organ, the late Professor J. A. Shercliff (whose own teaching suggested the need for discussion of the metric entropy) and M. D. Wood. The last words however might go to Kelvin [4], lecturing exactly one hundred years ago on electrical measurement: 'I often say that when you can measure what you are speaking about, and express it in numbers, you know something about it'.

REFERENCES

1. Carathéodory, C. (1909), "Investigation in the foundations of thermodynamics", Math. Ann., 67, 335.

2. Clausius, R. (transl. W. R. Browne), (1879), "The Mechanical Theory of Heat", Macmillan.

3. Kelvin, Lord (Thomson, W.) (1882), "Mathematical and Physical Papers", Vol. 1, Cambridge University Press.

4. Kelvin, Lord (1891), "Popular Lectures and Addresses", Vol. 1, Macmillan.

5. Planck, M., (transl. A. Ogg) (1903), "Treatise on Thermodynamics", Longmans.

6. Pippard, A. B., "Elements of Classical Thermodynamics", Cambridge University Press (1957).

"THERMODYNAMICS, A GOD WITH FEET OF CLAY?"

S. P. S. Andrew

Imperial Chemical Industries PLC

INTRODUCTION

This short paper is based on the author's personal acquaintance with a number of practical problems requiring some grasp of thermodynamics - considered in its broader sense - and the failure of those engineers concerned with these problems to understand them correctly. In all cases the engineers had been university trained in thermodynamics and therefore a knoweldge of their misconceptions may be of interest to those who teach thermodynamics.

EXAMPLE 1 - THE CARTESIAN FALLACY

I met this example in my first year as a student in engineering. Two professors of engineering with whom I have discussed this case also met the same difficulty in their time as students when similarly taught.

My lecturer explained Bernoulli's equation for flow of incompressible fluid along a streamline as: 'in the absence of friction, the total energy of the fluid at any point along the streamline is, per unit mass (or volume) of fluid, constant - being the sum of the kinetic plus the pressure plus the potential energy components'. As the fluid (being incompressible) can have no 'pressure energy' I was unconvinced, and by likening the whole process to a horse - the shafts - and a cart, came to the correct conclusion that energy, like (it is said) angels, can travel from place to place without traversing the intervening space (the shafts!).

This example shows how an early intrusion of the concept of energy into mechanics can readily wreck the good work done by Newton three centuries ago in ejecting energy in favour of what we now call momentum from such calculations. How many students of mechanics are over concerned with energy from an unfortunate presentation of its importance?

EXAMPLE 2 - THE UNREAL VISION

An engineer unacquainted with the theory of gas centrifuges but of an inventive turn of mind decided that their performance could be improved by incorporating within them a multiplicity of conical discs in a stack such as is to be found in centrifuges for removing small metal particles from oil. The fall of the heavier molecules would (he hoped) occur in a lesser time as the distance of fall would be diminished, as occurred with the particles of metal in the oil, the throughput of the centrifuge, when operating at a given g-force would thereby be much increased.

His error was a result of a lack of appreciation of the magnitudes of the relevant physical phenomena together with an inability to recognise that if his views were correct he would have been living in an atmosphere which near ground level would have been virtually pure carbon dioxide.

EXAMPLE 3 - THERMODYNAMICS OUT OF CONTEXT

Some remnants of the concept of minimum energy are left in most engineers' minds after thermodynamics lectures. These concepts tend to spread and blur as time passes so that even in the most impossible circumstances they may be found misapplied.

This error is frequently committed by chemical engineers concerned with relating the steady-state interfacial area in a vessel in which a gas liquid mixture is continuously and violently agitated to the properties of the liquid. Having noted that the effect of adding a surface active agent to water results in a great increase in the area, they "explain" this on the basis of the reduction in the gas-liquid interfacial tension favouring a high dispersion of the gas. In fact the interfacial tension has only a relatively small effect compared with influence of the presence of the surface agent on bubble coalescence which is greatly hindered.

EXAMPLE 4 - SLIPPING UP ON SOLIDS

The novice in the thermodynamic field may be forgiven for being unaware of the slippery nature of the thermodynamic properties of what may at first appear to be simple single component solids such as carbon. He is typically to be found calculating the carbon formation equilibrium $2CO \rightleftharpoons CO_2 + C$ leading to solid "soot". The product analyses as relatively pure carbon and is black and clearly not diamonds. He therefore assumes that it must be graphite, the only other form to be found in the tables of thermodynamic properties and on this basis calculates the soot formation equilibrium. Little does he realise that "soot" has a wide range of fugacities, from those of graphite increasing to very much higher values as the carbon atom lattice increases in disorder.

EXAMPLE 5 - THE UNFORESEEN SPECIES

The previous example shows the nature of one hazard encountered in chemical thermodynamics when applying measured constants to be found in thermodynamic property tables. A different form of this hazard is the result of that gross chemical ignorance which is not uncommon in engineers. The unsuspecting engineer, who has in this example the task of designing a cooler for carbonated ammoniacal liquor needs to know the crystallisation temperature of the liquor so as to avoid deposition on the walls of the cooler tubes.

He therefore makes up a solution of the correct NH_3 - CO_2 - H_2O content by, let us say, dissolving appropriate quantities of solid $NH_4\ HCO_3$ in aqueous ammonia liquor of a known strength. He then cools a sample of this solution slowly noting the temperature at which crystals first appear during cooling and then redissolve during reheating. With a little care he thereby measures with sufficient accuracy the crystallisation point of this solution. His results however are worse than useless as he had not made up the correct solution! His chemical education was sufficiently extensive to acquaint him with ammonia, the bicarbonate and the carbonate ions, but he had never heard of the carbamate ions ($NH_2\ COO^-$) nor the fact that in strong partially carbonated ammonia liquors this is the predominant reservoir holding CO_2 (not HCO_3^- or $CO_3^=$), nor the even more important fact that the rate of formation of this ion from aqueous NH_3 and aqueous HCO_3^- is very slow below 0 $^\circ$C. In consequence the solution he made-up, though containing the correct total quantities of H_2O, NH_3 and CO_2, contained a totally inadequate amount of NH_2COO^- and a grossly excess quantity of HCO_3^- compared with a true equilibrium mixture. In consequence the crystallisation point (of NH_4HCO_3) was many degrees higher than it should have been.

CONCLUSION

Though the God may not have feet of clay, his would be followers often have - particularly when they attempt to blindly follow his teachings being ill-informed on the nature of physical reality.

THERMODYNAMICS AS ENGINEERING SCIENCE

Alan Cottrell

Jesus College
Cambridge

The limitations of twentieth-century science are obvious. We still play with pebbles on Newton's seashores, surrounded by oceans of undiscovered truth. The physical sciences are confined to an island, cut off from the biological ones and even more from the psychological ones by the lack of a place in physics and chemistry for concepts such as organisation and purpose. Open any textbook on the properties of matter. You will find no chapters on these concepts. Indeed, they seem so utterly different from the simple, dull, properties of simple, dull, matter that we can hardly imagine that they could ever come within the scope of physical science.

It might be different in a hundred years' time. The textbooks then may be dealing with much more exciting properties of matter. If this becomes so, much of the credit is likely to belong to the pure science of engineering. But how can this be? What pure science is there in this? The key insight for this was in fact realised as long ago as the seventeenth century, when Sir Thomas Browne said that 'all things are artificial, for nature is the art of God' (Religio Medici, 1643). This puts all things on the same footing as objects of scientific interest, from crystals to cells and cars. They are all constructions and have properties, from simple and dull to complex and interesting. If the works of engineers are artifices then so is a thunderstorm, trilobite or tendril. If the works of nature are natural, then so is a thermostat, trireme or telephone.

As objects of scientific interest, the works of engineers have one quality which is not unique to them but is very obvious in them. They have purpose. This gives them two features of great scientific significance. First, in designing and constructing their machines, engineers have usually tried to obtain the maximum benefit from the forces of nature put to work in them and in so doing have thereby uncovered limits, set by nature, to the magnitudes of those benefits. In this way have some of the laws of nature been discovered. Second, although these machines work according to the rules of physics and chemistry, they nevertheless perform like biological organisms. They behave as integral bodies which act in complex ways to fulfil purposes. Viewed as a heap of gears and cogs a machine is a mere illustration of elementary and well-worn principles of mechanics. But such a view misses the point. Seen as an organism, working purposively, a machine is a spectacularly strange object of nature, a most improbable structure with some extraordinary properties, something far more exciting and challenging, as a scientific object, than any pebble on the seashore.

The quintessential work of the engineer was the steam engine. From the purely scientific point of view the steam engine still has, as we shall see, some interest for us today, although it is more remembered for inspiring in the nineteenth century the discovery of the laws of thermodynamics: the (first) law of energy conservation and the (second) law of entropy increase, i.e. increase in the disorderliness of structures and processes. The second law, discovered in fact before the first, arose directly out of Watt's efforts to make steam engines deliver more work, for a given input of heat, and of Carnot's perception that the temperatures of the boiler and condenser impose a fundamental limit on this efficiency. Thc concepts of energy and entropy are of endless scientific interest. They allow us to regard the whole universe as an engine, with energy as its working 'fluid' and entropy as the measure of its state of evolution or old age. Even the direction of time might be a result of the law of entropy increase. Thus, one version of the second law of thermodynamics, i.e., 'heat cannot pass of itself from a colder to a hotter body', introduces the direction of time through the verb 'to pass' and could thus be turned around into a law about time itself, i.e., 'later is when entropy is greater'.

For a more modern lesson from the steam engine we turn to biology. Another popular version of the second law is that 'everything decays; order inevitably gives way to disorder'. Applied to a plant or animal this seems obviously wrong. Living things are fruitful and multiply. The order embodied in them increases. There is no contradiction however, but instead an interesting loophole is revealed in the second law; and Watt's engine shows us where it is. An engine is an extractor of energy. It feeds on high-temperature heat (medium entropy), extracts the pure form of orderly mechanical or electrical energy (zero entropy) from it, and ejects the remaining low temperature heat (high entropy) as large-scale waste. But this is in effect just what a living system does, as the first step towards using such extracted orderly energy to sustain itself and multiply. The steam engine shows that the extraction is achieved by means of instabilities. In this particular case the instabilities are provided by the smooth, oiled, surfaces of the cylinder, piston, crank-shaft, etc., which enable parts of the engine to slide and rotate, so giving way to the forces from the steam and delivering mechanical work. This giving way is a deliberately contrived instability. The Belgian physical chemist, Prigogine, recently received a Nobel Prize for proving that practically anything will develop an instability when overloaded with ordered energy. The effect is a general one. There can be many kinds of chemical, thermal, electrical or nuclear instabilities, as well as the simple mechanical ones of the steam engine. A living organism, through its metabolical apparatus, uses biochemical instabilities to extract orderly energy - embodied in molecules - from sunlight or foodstuffs. This engine principle, the ability to extract orderly energy through an instability, is the bridge leading from physics and chemistry to the living world, and the engineer's engine is in this respect a true, if very simple and incomplete, organism.

The twentieth-century contribution of engineering to pure science was the discovery, through efforts to make telephone wires and radio channels carry more messages, that information can be viewed as a scientific commodity, much like energy, entropy, heat and work. Information narrows the range of possibilities in situations where alternatives exist. For example, in chess if you say on which square the White Queen sits you give more information than by merely saying that she is somewhere on the board. The most economical way of giving it is a series of yes/no answers to well-chosen questions. First, is she in the left half of the board? The answer, yes or no, gives one bit of information. Next, is she in the upper half of the board? Again, one bit for the answer. And so on. Six yes/no answers are needed to locate the particular square, so that the information necessary for this is six bits more than that she is merely on one square of the board.

What has this to do with the natural world? The link came through the realisation that when any change occurs, e.g., the expansion of a gas into a vacuum chamber, the information (I) about the positions of the gas molecules decreases by precisely the amount that the entropy (S) of the gas increases, i.e.,

$$S + I = \text{constant}.$$

This was an extraordinary discovery since it relates a physical quantity, entropy, which can be measured with calorimeters and thermometers, to something which represents a simple form of human knowledge.

The theory of information has an obvious place in modern biology. A strand of the long-chain DNA in the nucleus of a biological cell functions rather like a computer tape carrying a message in morse code. Each tri-molecular unit of the strand spells out one letter of a genetic alphabet and thereby selects one molecular protein constituent, out of a choice of 20, for the replenishment and further construction of the organism. A choice of 1 in 20 involves about five bits of information. The DNA in a human egg cell carries about a billion (10^9) letter units and so about five billion bits are required to write down the genetic instructions to make a human being. Since the 20-letter genetic alphabet is numerically close to our 26-letter alphabet, we would similarly need a few billion of our own alphabetic letters to express this amount of information; about four sets of the Encyclopaedia Britannica, in fact. If written out instead in ordinary English words, which convey much less information per letter, the same message would require a library of about a thousand books.

Thermodynamics and information theory deal with order, but the organisation of a machine is something much more than this. Suppose that you are trying to mend an old mechanical watch and then, in the way that all too often happens, it suddenly jumps apart into a heap of disorganised cogs and springs. If you had performed this depressing act in a calorimeter, you would have concluded that, thermodynamically very little had happened, that the entropy of the watch was not appreciably altered by its disintegration. But you no longer have a working machine. A quality has been lost, which we immediately recognise as organisation but which lies outside present physics and chemistry, a quality which machines share with organisms. Does this mean that we could look yet again to engineering, to inspire still another chapter of theoretical physics, to open the way to a 'super-thermodynamics' that could link organisation and functional value to the traditional thermodynamical entities and information?

We might try to define this special quality of machines in terms of their structures. After all, most machines considered strictly as natural objects have spectacularly improbable structures. If Neil Armstrong in 1969 had found a motor cycle on the Moon, he would surely have concluded that someone had been there before him, that no freak action of blind natural forces could have wrought such an improbable result. But structure is nevertheless not suitable for our definition, because a working machine is not significantly different, considered purely structurally, from a non-functional imitation, such as a wrongly-assembled clockwork, a misconnected radio circuit, or a piece of modern sculpture. Performance, not structure, is what matters, even though an elaborately improbable structure is generally necessary for a virtuoso performance. This raises a difficulty for thermodynamics, for this science has concerned itself traditionally with structure as the primary feature of systems and taken performance to be a merely secondary consequence of structure. But there is a more profound point which goes back to Darwin: performance, through the competitive advantages which it can give, has enabled the forces of nature, slowly working through aeons of evolutionary history, to construct far more complex and improbable structures than motor cycles. Biological and engineers' machines both owe their existence and survival to the advantages that accrue from their performances. Could thermodynamics then be expanded to take this into account? Possibly. We can temporarily bypass the really hard problem - that of treating organisation, function and purpose as scientific entities - by simply supposing that a machine has a certain property of somehow causing copies of itself to be made; this could be done either by von Neumann's automatic self-reproducing fantasy, or more commonly through the machine proving so useful to us that we are inclined to look after it and make copies of it. By-passing all these facinating possibilities, the property itself is then reduced to a kind of autocatalysis, something which is familiar in chemistry and can be handled by conventional thermodynamics.

Such a simplistic reduction of course fails to catch that scientifically elusive quality of a machine, its functional value. The overall effect of this is summarised by an auto-catalysis factor which represents how much we are prepared to give up in the way of other opportunities in order to retain and have more of that functional value. A far-reaching consequence of the engine instability principle, working its way through ever more sophisticated and competitive constructions, is thus that opportunity cost joins the traditional scientific qualities of energy, entropy, heat, work, and information. In other words, that thermo-dynamics takes over economics, a delicious prospect for any natural scientist.

This paper is based on an article that first appeared in the Cambridge Review, February, 1983, (Taylor and Francis, Ltd.).

INTRODUCING BASIC CONCEPTS OF THERMODYNAMICS

H. V. Rao

Mechanical Engineering Department
Huddersfield Polytechnic

Engineering students are usually introduced to the basic concepts of Thermodynamics through the conventional classical approach, which is also the general theme followed in most of the popular text books. Many of the recently published text books include reference to Statistical Thermodynamics and Quantum Mechanical principles, especially as an aid to achieve an "insight" of the property, Entropy. However, this is done as an appendix to the main theme and it is doubtful to what extent students are benefitted.

In this paper, an alternative method of introducing the basic concepts is suggested, starting from the elementary Quantum Mechanical principles and the concept of thermodynamic probability of an equilibrium state of a system.

Work and Heat transfers are defined by considering the energy transfer due to the movement of atoms/molecules at a surface. Energy transfer other than due to macroscopic mass transfer across the surface may be divided into (i) work transfer, due to macroscopic displacement of a microscopic force and (ii) heat transfer, due to the microscopic displacements and associated forces. This method avoids the possible formation of the misconception of "Heat Storage" or "Work Storage" of a system and logically leads to First Law of Thermodynamics as a principle of energy conservation.

The fundamental postulate of Quantum Mechanics and the concept of Equilibrium state lead to the relationship between Thermodynamic Probability (Z) and the distribution of the number of particles (n_i) amongst the energy levels (E_i):

$$\frac{\partial \ln Z}{\partial n_i} = A + \frac{1}{B} E_i \tag{1}$$

The Lagrangian multiplier (1/B) which is introduced in the maximisation process of Z, is established as a function of temperature only, by considering the thermal equilibrium of a composite system. This result is independent of the type of particles of the system and whether these follow M-B, B-E or F-D statistics. Hence, without loss of generality, by considering the case of mono-atomic gas, the absolute temperature of any system is defined as:

$$T = B/K \tag{2}$$

where k is Boltzmann Constant.

The concept of reversible change is introduced as a sequence of equilibrium states. It follows that heat and work transfers to a system during an infinitesimal reversible change are given by:

$$dQ_r = \sum E_i \, dn_i \tag{3}$$

$$dW_r = \sum n_i \, dE_i \tag{4}$$

Entropy of a system is defined by

$$S = k \ln Z \tag{5}$$

From equations (1), (2) and (3) it follows that:

$$dS = k \, d(\ln Z) = \frac{k}{B} \sum E_i \, dn_i = dQ_r/t \tag{6}$$

The related concept of "reversible heat engine" is clarified. It is suggested that the terminology, "Heat source" and "Heat sink" are to be replaced by "Energy Source for Heat Transfer" and "Energy Sink for Heat Transfer", respectively, as the former terminology implies the erroneous concept of "Heat Storage", originating from Calorific Theory.

ON THE RATIONAL DEFINITION OF AN IDEAL GAS

Y. R. Mayhew

Department of Mechanical Engineering
University of Bristol

SYNOPSIS

It is common to introduce an ideal gas as one which obeys the equation of state $pv = RT$. This is often represented as a 'semi-empirical' fact: the behaviour of the gas approaches this equation as the pressure p tends to zero.

The Joule law $u = u(T)$ or the Joule-Thomson law $h = h(T)$, whether they are introduced later or simultaneously as part of the definition of an ideal gas, beg the question because the instrument used to measure T, the absolute thermodynamic temperature, relies on the result $pv = RT\big|_{p \to 0}$. This can lead to confusion or the impression that a tautology is involved.

Let us first consider the two parts of the definition of an ideal gas.

(1) Boyle's law states that, as $p \to 0$, pv becomes a function of temperature only, i.e. pv = constant, when Θ, which can be any empirical temperature, is kept constant. Hence we can define an ideal-gas temperature Θ_G by

$$pv = A\Theta_G \tag{1}$$

(2) The Joule-Thomson law states that, as $p \to 0$, h becomes a function of temperature only, i.e. is constant when Θ, and therefore Θ_G is kept constant. Hence we can write

$$h = f(\Theta_G) \tag{2}$$

(N.B. Joule's law $u = \Phi(\Theta_G)$ could be used, but it is less securely founded in experiment than the Joule-Thomson law.)

From the First Law and elementary ideas of reversibility we can write

$$dQ_{rev} = dh - v\, dp \tag{3}$$

and hence

$$\frac{dQ_{rev}}{\Theta_G} = \frac{1}{\Theta_G} \left[\frac{\partial f(\Theta_G)}{\partial \Theta_G}\right] d\Theta_G - \left(\frac{v}{\Theta_G}\right) dp \tag{4}$$

If it can be shown that

$$\frac{\partial}{\partial p}\left[\frac{1}{\Theta_G}\,\frac{\partial f(\Theta_G)}{\partial \Theta_G}\right]_{\Theta_G} = -\left[\frac{\partial (v/\Theta_G)}{\partial \Theta_G}\right]_p \tag{5}$$

then dQ_{rev}/Θ_G is a perfect differential. Now LHS must be zero because the term in brackets is function only of Θ_G. For RHS

$$\left[\frac{\partial (v/\Theta_G)}{\partial \Theta_G}\right]_p = \left[\frac{\partial (A/p)}{\partial \Theta_G}\right]_p = 0 \tag{6}$$

This proves that Θ_G is an integrating factor for equation (3). Thus we arrive at the limited concept of an ideal-gas entropy, s_G, where

$$s_G = \frac{dQ_{rev}}{\Theta_G} \tag{7}$$

with recourse to the First Law alone!

Indeed it is known from analysis that, for functions with only two independent variables, there must exist an integrating factor, and we have shown that Θ_G defined by equation (1) is one such factor. What does the Second Law then contribute?

(a) The Second Law, which leads to the equation

$$ds = \frac{dh}{T} - \frac{v}{T}\, dp \tag{8}$$

shows that the thermodynamic temperature T is one possible integrating factor.

(b) By a mathematical argument analogous to that developed above, it can be shown that for a suitable choice of scale factor (A)

$$T \equiv \Theta_G \tag{9}$$

(c) For systems with more than two independent variables, the existence of an integrating factor is not a foregone conclusion. The Second Law stipulates in effect, that integrating factors definitely exist even for systems with more than two independent variables, and that T is one such universal factor, whatever the substances involved.

(d) The ideal-gas thermometer leads to a practical procedure of measure Θ_G and hence T.

It is suggested that a rational sequence of introducing the ideal gas is via equations (1) and (2), and proof of the identity (9).

SOME THOUGHTS ON BORN-CARATHEODORY

Y. R. Mayhew
Department of Mechanical Engineering
University of Bristol

There is some doubt to what extent concepts in mathematics can be defined a priori without reference to experience, but this author subscribes to R. B. Braithwaite's view that in physical science, concepts must develop as the understanding of a subject is reinforced by experience. There have been several two- and even one-axiom structures of thermodynamics proposed over the years, unparalleled in other branches of physics, and the limitation of such attempts must be recognised. The reason for the need of such structures probably arises from (i) the elusive nature of 'heat', (ii) the negative form of the Second Law, and (iii) the amazing generality or power of the two Laws of Thermodynamics.

Of the two-axiom structures of thermodynamics, that of Born-Carathéodory (B-C) is the most rigorous, although many teachers of thermodynamics are oblivious of its existence. The structure probably still used most widely in the U.S.A. and the U.K. is essentially that of Keenan (K). Much of the German tradition is still to avoid the issue and state the two Laws in a number of equivalent forms. In view of the opening remarks, is rigour to be preferred to ease of physical appreciation however imprecise, at least in a first-round engineering course?

Looking initially at the First Law, K starts on the basis of an intuitive understanding of heat which turns out to be a measurable quantity and definable as a mode of energy transfer. The First Law can then be stated as an axiom in terms of a cycle undergone by a closed system, and this leads to the concept of internal energy. The difficulties lie in the concept of heat.

B-C start with adiabatic processes which leads to a statement of the First Law which does not involve heat. Heat then emerges as a derived concept and quantity occurring in non-adiabatic processes. The difficulty arises that adiabatic is difficult to define; it cannot be defined with the aid of the concept of heat but must fall back on a definition based on properties of a system.

K follows Planck's negative statement of the Second Law, involving ideas of work and heat and cyclic processes, leading via the Helmholtz inequality to the concept of entropy. B-C rely on an ingenious negative statement in terms of properties: that there exist thermodynamic states in the neighbourhood of any state which cannot be reached via adiabatic processes. Following difficult arguments in analysis about Pfaffian differential expressions involving more than two independent variables (the arguments are unnecessary for simple systms with only two independent variables), B-C arrive at the existence of a universal integrating factor for all systems and at the concept of entropy.

It would be interesting to discover whether the undoubted extra rigour achieved by B-C is rewarded in a first course of engineering thermodynamics. The author's view is that K's approach is still the most easily assimilated by students, although those who need to be exposed to a second or even third round of teaching might well benefit from B-C.

DISCUSSION IN SESSION 4 "PRINCIPLES OF THERMODYNAMICS"

N. Hay - University of Nottingham

H. K. Zienkiewicz - University of Exeter

On the papers by R. W. Haywood, "Teaching by the Single Axiom Approach", and by P. H. Brazier, "The Thermodynamic Laws from the Law of Stable Equilibrium"

Professor C. Gurney noted that these papers use the definition of work popularised by Keenan. He has found this unsatisfactory. How, on this basis, does one reply to a student, who asks if the energy transmission from a laser is work or heat? It is much more satisfactory to define heat as energy transmission whose direction depends on temperature difference, and work whose direction does not depend on temperature difference. On this basis, a student could soon discover that laser action involves work.

In a written reply to Professor Gurney, Mr. R. W. Haywood noted that his own paper was directed to first-year students and he would not expect such a student to be so equipped with knowledge of laser technology that he would be disappointed if his lecturer did not give him an instant answer to Professor Gurney's query. It was not dissimilar from the older question about the nature of radiation and radiant heat transfer. However, the answer which Professor Gurney gave presupposed that the three concepts mentioned by him were introduced in the sequence - temperature difference - heat - work. In the Single-Axiom Approach, and in his own paper, the sequence was - work - heat - temperature difference, the exact opposite of that presupposed by Professor Gurney. Nevertheless, once he himself had presented those concepts in the latter sequence, he would be perfectly happy to answer the laser query by asking in return - "Well, does the energy transmission from a laser depend on a temperature difference?"

That, of course, could fairly be described as evading the issue, but detailed answers to difficult questions were sometimes better left until later. In the first year, the student had enough with which to cope in taking on board the fundamental concepts and theorems of thermodynamics.

Dr. R. I. Crane asked whether Mr. Haywood, or any other participant, has had much experience of teaching by the single-axiom approach, particularly to first-year engineering students? Do students find one "axiom" and its corollaries more readily acceptable than two "axioms"?

Mr. Brazier replied that he has not as yet used this approach but he would be doing so in a minor way this session.

Mr. Haywood stated in a written reply that he did have several years' experience of teaching by the Single-Axiom Approach, but not to first-year studetns. He was nevertheless convinced that first-year students would readily absorb it if it were presented in appropriately simple language. It was for that reason that he had written his paper. He added that he would like to outline how he came to be interested in this relatively new Approach.

In the Sixties, two rather challenging books came on to the American market, that by Hatsopoulos and Keenan presenting the so-called Single-Axiom Approach to classical thermodynamics, and that by Tribus presenting the Information-Theory Approach to statistical thermodynamics. He had found the ideas presented in both those books challenging and exciting, but difficult to absorb from the way in which they were presented in the books. That was particularly so with the book by H. and K., and he decided that he would not fully understand and appreciate their new Approach unless he first taught it to students. He had less difficulty with the book by Tribus, since he had had the pleasure of attending a week's course of lectures by Professor Tribus in Edinburgh in 1965. He nevertheless felt similarly that the only way fully to understand the latter's Approach was to teach it. He therefore suggested to his colleagues in the Thermodynamics Group at Cambridge that he should give, to third-year students specialising in thermodynamics, a short course of lectures on both topics, with three lectures on each Approach. His colleagues had demurred, claiming that the material would be too difficult for the students. He had nevertheless gone ahead and the lectures had proved rewarding and exciting to students and teacher alike, with no great difficulty of understanding being evident on the part of the students. After a few years perfecting his own understanding in this way, with great help from the students, he had decided to write his own book on the Single-Axiom Approach, with the aim of making its presentation much more acceptable than in the book by H. and K. That task took him several years. He would have liked to conclude with a chapter on the Information-Theory Approach to statistical thermodynamics, but that would have made the book too long and expensive. He was surprised that that latter topic had not even been mentioned in any of the Papers presented at the Workshop.

With regard to Dr. Crane's question as to whether students found one "axiom" and its corollaries more readily acceptable than two "axioms", he said that that was certainly so. All those students with whom he had discussed this had shared his own view that it was much more satisfactory to start from the single Law of Stable Equilibrium, which had a clearly physical basis and from which the other so-called Laws could be _proved_, than to be required to accept those 'Laws' as _simple acts of faith_, particularly when they were expressed in their cyclic form. Thermodynamics was a science, not a religion.

Professor M. J. French expressed the opinion that the derivation from a single axiom is deceptive, because some assumptions are made which are not acknowledged. For instance, in Figure 4 of Haywood's paper, it is not obvious that the effect of moving the two weights separately is the same as that of moving them together. Because we know some mechanics and the First Law, we can say that they are, but it should not be assumed.

In a written reply, Mr. Haywood noted that, in teaching thermodynamics, one of course started with an already acquired body of knowledge, particularly in mechanics, and it was not always easy to decide precisely if and where that degree of pre-knowledge may have entered unawares into the argument. Because he considered the so-called 'Single-Axion Approach' to be a method of presenting classical thermodynamics that was so much more logical than, and so greatly superior to, the more conventional 'Keenan' approach of 1941, he was not prepared to throw away the baby with the bathwater simply because some raised doubts as to whether it was, in fact, based on a single axiom. He was happy to let the philosophers argue at length about that point. Even if they came down in favour of denying it the benefit of being based on a single axiom, that would not stop his advocacy of the Approach as being a very good way of presenting thermodynamics logically to students. He would simply regret that it had got lumbered with that title, which was, however, a good, simple title which had caught on. He would be quite happy to use another description of it if anyone came up with a better title, but he saw no reason for rejecting the method, even if the single-axiom argument continued to provide endless disputation for the philosophers.

The arguments about whether the Approach was in fact based on a single axiom started right from the date of its publication by Hatsopoulos and Keenan; even, and perhaps particularly, amongst those who had scarcely had time to study it properly. It was because he was aware of these arguments that he had come up with the idea of constructing his 'Thermodynamic Family Tree'. He had hoped that that would help to abate some of the argument, but it had evidently been too much to expect that this device could stem the tide of endless philosophical disputation. He had deliberately avoided the inclusion of the sub-title - "Single-Axiom Approach" - in his book because of his awareness of this question.

Dr. D. Jones pointed out that since physics involves two fundamental laws namely the law of conservation of energy and the law of equal probability of all accessible quantum states, thermodynamics cannot therefore itself be derivable from a single axiom.

A response to this point was included in Mr. Haywood's written reply to Professor French.

Dr. Y. R. Mayhew commented that he was happy to accept the approach but he did not regard the LSE as intuitively obvious.

In a written comment on the LSE approach, Dr. T. H. Frost pointed out the necessity to distinguish precisely between a system in a stable state and one in an equilibrium state, and gave as an illustrative example a system which he regarded to be in mechanical equilibrium but not in a stable equilibrium state according to corollary 2 in Section 4.3 of Mr. Brazier's paper. Mr. Haywood replied that the essence of the matter was the necessity to distinguish between a constrained and an unconstrained system. Thus, Mr. Brazier's corollary 2 was perfectly acceptable once the word "constrained" was inserted before the word "system".

On the paper by H. V. Rao, "Introducing basic concepts of thermodynamics"

Dr. B. M. Burnside commented that a good British compromise on the teaching of Thermodynamics to Engineers is best. At Heriot-Watt the traditional approach is used, illustrated partly by molecular and statistical-mechanical ideas; also a form of Haywood's Lost Work Theorems is used where and when it is felt that this helps students. He did not think that greater rigour than is necessary need be employed if there is a danger of obscuring the presentation.

Dr. N. Hay thought that the major benefit of the Workshop to most delegates will probably be the new ideas and approaches and compromises that were aired in the papers and discussions and which delegates will take home to incorporate in their respective courses.

On the paper by R. H. D. Exell, "Entropy and temperature independent but measured by heat" (presented in Professor Exell's absence by Miss Rosalind Armson)

In the absence of the author discussion was limited to a few comments. In reply to a query regarding the student level at which the paper was aimed, Dr. Hay pointed out that Professor Excell mentions in the paper that he had used this approach in his lectures to Master Degree students who found it difficult. Dr. Hay added that what he liked particularly about the paper was the rigorous logical way in which everything was defined and then talked about. The approach used by the author of underlining every new concept or quantity when it was first mentioned to indicate that it is a new departure was a very useful one to employ in this or any other approach to thermodynamics.

On the papers by Y. R. Mayhew, "On the rational definition of an ideal gas" and "Some thoughts on Born-Carathéodory"

Dr. T. H. Frost said that he appreciated this presentation of the ideal gas relations and of the integrating factor. He wondered whether there was an advantage in relating the ideal gas temperature to kinetic theory in order to elucidate what is happening at microscopic level.

Dr. Mayhew replied that he felt no real need for the microscopic approach and that he preferred to keep the macroscopic and microscopic approaches separate.

Dr. J. D. Lewins thought that Joule's Law was worth restating if only for historical reasons. He also submitted the following written contribution on Boyle's Law:

1. Observational. There are two crucial experiments

a. Establishing a region in which Boyle's law is satisfied: $pv = \theta$ const. The θ is an ordinal temperature. It is only possible to say that $\theta = \theta(T)$ where T is the thermodynamic temperature.

b. Establishing a region in which the Joule-Thomson coefficient is zero (or equivalent Joule coefficient), i.e. enthalpy is a function of temperature only. If both J-T and J coefficients are zero, we have a proof that pv is a function of temperature only.

In principle these are separate regions. Where the Boyle and J-T regions overlap, Maxwell relations lead to the result that $pv = RT$, i.e. that the empirical temperature can be identified with T.

2. Molecular Model. If the simple 'point' model of molecules is used, we find that both the Boyle behaviour and the vanishing J-T behaviour are predicted. Starting from this point we should still distinguish between this common behaviour and separate behaviour since more complicated molecular dynamics might lead to one but not the other.

3. Mollier Chart. In describing the structure of the Mollier chart it is tempting to say that the ideal gas region is indicated by horizontal isotherms and thus the ideal gas can extend at low pressures to the saturation line. But logically we are only in the J-T region. (And one goes on to say the perfect gas is only a locally uniform region where the vertical spacings of the isotherms are uniform). Is there any obvious feature to show we are in the Boyle region?

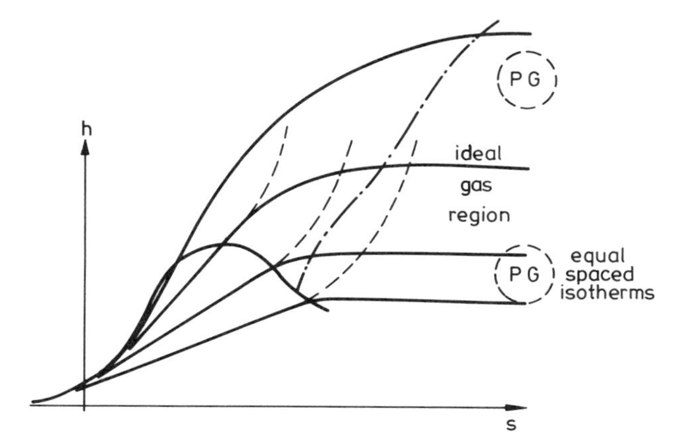

Fig. 1 The 'ideal' and 'perfect' gas regions.

Professor P. T. Landsberg made the following points on the "Born-Caratheodory" approach:

1. The separation between heat and work by adiabatic partitions is achieved by allowing transmission of mechanical energy through the partition. As in a case of a balloon one may dent it. If long-range forces are kept fixed, the additional energy which can pass on making the partition transparent is heat. That, very roughly, is how 'heat' is introduced.

2. The adiabatically inaccessible states from a given state S_o are readily recognised after thermodynamics has been constructed. They are the states of lower entropy than tne entropy of S_o. They occur in any neighbourhood of S_o, however small.

3. The first geometrical treatment based on Carathéodory (and also using information theory as a basis of statistical mechanics) was, he believed, his 1961 book referred to in his paper at this Conference. A relevant paper is also one he published in the American Journal of Physics (1984) on the occasion of the Born Centenary.

The Born Centenary: Remarks about classical thermodynamics. Am. J. Phys. 51 (9), September 1983, pp.842-5.

Mr. B. R. Wakeford wished to emphasize the importance of building upon a student's current knowledge. In a school physics course the student will hopefully have met the term "heat" and will hopefully know that "heat" is a form of energy in transition and is due to a differential temperature. To then pretend that "heat" has not been heard of and to introduce it as a name for something new and resulting from obscure perambulations will confuse the majority of first year students.

Dr. Mayhew in reply doubted whether those students who have taken Nuffield Physics would have heard of heat; they are more likely to have a wrong view of internal energy.

On the papers by J. D. Lewins "Thermodynamic temperature-dimensional analysis" and "Constructing temperature and entropy scales"

Presenting his papers, Dr. Lewins drew attention to a link between his establishment of entropy and the so-called Single-Axiom approach in the form of his Figure 1.

When considering the accessibility concept, different states 1, 1', 1" were shown as accessible from state 1 by a reversible and adiabatic process. Equally, in discussing the states of an isolated system, one could use the same diagram with the direction reversed. The first use of the diagram established the concept of entropy before quantifying it; the second establishes the concept of energy before quantifying it.

Energy is the invariant of all those states of a system that terminate in the same (stable) state by isolated processes.

Entropy is the invariant of all those states that a system can reach with adiabatic (no heat) and reversible processes.

Mr. Haywood expressed his concern at the frequency with which thermodynamic temperature is defined in terms of a scale.

Dr. Lewins stressed that although the word scale appears in the title of the paper, the paper in fact was concerned with defining temperature in terms of functions rather than scales.

On the paper by S. P. S. Andrew "Thermodynamics; a God with feet of clay?"

Mr. D. G. Walton felt that Industry does not take effective steps to disseminate within itself knowledge and experience relevant to it. Universities educate students for 3 years after leaving school at the age of 18. If the graduate then spends, say, 15 years in Industry, Industry generally takes no steps to reinforce his knowledge base through retraining and post experience course. In this respect Industry is at fault not Universities.

In reply Dr. Andrew said that what one must understand is that the first thing you discover when you get into industry is that the problems you are faced with are exceedingly diverse both from the technical point of view and for that matter from the management point of view. Limiting ourselves now to the technical side of the job, the problem is diverse in the sense that it invariably becomes an interdisciplinary problem of a great width; therefore you have to be able to understand or get to understand a considerable range of reality. The main problem one is faced with in educating people is that courses come cut and dried in little packages, whereas reality does not come in that form.

Following on from Mr. Walton's comment, Professor J. C. R. Turner suggested that too many students regarded their education as finished when they leave university. They were, he regretted to say, intellectually idle.

He also pointed out that the last example of Dr. Andrew's paper involved rates and thermodynamics. These are different subjects but are tangled together in the problem.

In reply to the comment by Mr. Walton regarding training after graduation, Dr. B. N. Furber stated that his own company is involved in many aspects of training. All fresh engineering graduates receive two years general training (to satisfy IMechE requirements). Subsequently in-service training, attendance at conferences all contribute. He would however recommend stronger links between industry and universities to their mutual benefit.

Professor R. S. Silver referred to his long experience in Industry and made the point that where data are available problems are easily tackled. Otherwise recourse has to be made to scientific knowledge including thermodynamics to generate such data and sometimes one has to resort to intelligent guesses.

In reply Dr. Andrew said that he thoroughly agreed with these points and wished to emphasise that the greatest error is not to realise that an important phenomen exists.

Dr. Hay thought that a lot of what was discussed in this context actually can be defined as experience and it is very difficult to get experience across to undergraduates in the time that is available and because they have not got the background against which to reflect findings and gather experience. If there is a way in which experience could be gelled and got across, then we would solve this problem, but he thought that time is the only element that will get experience across. To get people back to do refresher courses after say 10 years would be extremely useful because they are then better able to assimilate the information than they could when fresh from school.

Mr. P. Tucker remarked that experience in real engineering may be obtained by students on sandwich courses, which may help to reduce the kind of errors described in the paper. Difficulties in placement could be alleviated by cooperation between industry and academic institutions.

In reply Dr. Andrew said that he accepted this point, but drew attention to the commercial and organisational difficulties involved.

To a question as to whether there were enough engineers in Industry, Dr. Andrew replied that more does not necessarily mean better. If there are too many then each one gets less experience. If there are fewer then each one gets more problems thrown at him so on occasions Dr. Andrew argued for fewer rather than for more.

On the paper by Sir Alan Cotterell, "Thermodynamics as engineering science"

The paper was introduced by the Session Chairman in the absence of Sir Alan abroad. Dr. Hay said that the paper gave wider meanings to the ideas of thermodynamics and made some challenging future projections which should give the gathering of thermodynamics teachers present heart to keep teaching and researching the subject.

SESSION 5

APPLICATIONS OF THERMODYNAMICS TO DESIGN ASSESSMENT

Chairman - Professor G. D. S. MacLellan

Secretary - K. W. Ramsden.

NEW CONCEPTS IN THERMODYNAMICS FOR BETTER CHEMICAL PROCESS DESIGN*

B. Linnhoff

Department of Chemical Engineering
University of Manchester
Institute of Science and Technology

Most chemical processes are networks of different pieces of equipment. Usually, even the best pieces of equipment will give a poor overall process if linked up inappropriately in the network. This paper describes principles and procedures for better process network design. Development of the procedures began in 1972. In the years since, industrial applications have led to significant improvements in even the most modern processes.

The paper begins with a fresh look at thermodynamic Second Law analysis. This classical analysis highlights inefficient parts of complex systems, drawing the engineer's attention to excessive losses of potential. Unfortunately, the analysis is both difficult to produce and difficult to interpret. To tackle the first problem, the paper describes how Second Law information can be obtained from conventional heat and mass balances. There is no need for additional data. To tackle the second problem, the paper introduces a general distinction between 'avoidable' and 'inevitable' inefficiencies. This makes an interpretation of the analysis practically more meaningful.

Next, the paper describes thermodynamic procedures and principles for specialized sub-tasks in process design. Emphasis is placed on heat recovery networks. Here, the problem is to recover as much heat as is economically justified within a process before externally supplied heat is used. The concept of 'inevitable' inefficiencies leads

* The Esso Energy Award Lecture, 1981. Republished with permission from the Proc. Roy. Soc. London, A386, 1-33 (1983).

to techniques for the prediction of the 'inevitable' amount of external heating. This amount is called the energy target. The target either stimulates the engineer into achieving it or gives him confidence that his design is optimal.

The paper continues by describing the concept of the heat recovery 'pinch'. The pinch leads to the design of, first, heat exchanger networks, which achieve the energy targets, and, second, overall processes, which keep the targets low.

Two common threads in all these procedures are the attempt to keep the engineer involved (they do *not* constitute 'automatic' design) and the attempt to make best practical use of inefficiencies that are 'inevitable' anyway. Owing to these features, the procedures usually help the engineer to find processes that are elegant in a general sense. Many designs found in practice were not only energy efficienct but easily operated and maintained, safe, had relatively simple network structures and, most surprisingly, were cheap to build as well as cheap to run.

INTRODUCTION

(a) A personal word

This paper would never have been written had it not been for the 1981 Esso Energy Award. Admittedly, I was planning to write a review paper anyway, but, without the Award, that review would have been conventional. It would have attempted to give a balanced description of the entire work and to give a clear history of previous research, and it would have been aimed at a somewhat specialist readership. The present paper is different. It reviews not so much an area but one particular development in that area. In a way, it describes one man's approach. As such it mixes facts with preferences, experience with speculation, and science with judgement. In short, it lacks discipline when compared with a conventional review. Also, it is written for the non-specialist reader. I admit I thoroughly enjoyed letting go of discipline and chatting to the non-specialist and I feel indebted to the Royal Society for this opportunity as well as for the Award itself.

(b) Process integration

Figure 1 shows an outline diagram of a typical process for the production of ethylene from naphtha. The diagram is much simplified with only the most essential process steps shown. Even so it is a complex network of reactors, distillation columns, compressors, heat exchangers, driers, furnaces, quench units, etc.

When designing such processes, the chemical engineer faces two tasks. First, there is the conventional engineering task of designing the individual pieces of equipment. Second, there is the task of designing the overall process. This involves decisions of where and how to integrate the various pieces of equipment or, quite frequently, whether to integrate at all.

In many design offices, 'integration' has a bad name. Too many promising schemes have not fulfilled their promise in the past. Generally, integrated processes cannot readily accommodate changes in operating conditions, local equipment failure can more easily affect the entire plant, maintenance procedures become more difficult, etc. In addition, integration can be costly in terms of capital.

The procedures described in this paper have repeatedly led to practical integration schemes and have proven that the common scepticism regarding integration is not always justified. Table 1 shows a summary of applications in I.C.I. (1977-1981). The experience gained in these studies is as follows: while it is correct that inappropriate integration can do great harm it is also correct that elegant integration can yield great benefits. Integration is simply an aspect of process design that should not be approached casually.

(c) A complex problem

A 'casual approach' may appear understandable when considering what an enormous task a thorough approach would be. Finding the 'best' integrated process network is difficult. For example, there are impurities in the feedstock. Should we separate them from the feed or should we carry them through the process and separate them from the products? Should we compress the combined products before we separate the by-product or should we introduce the compression afterwards? Which (by-) product should we separate first? Not

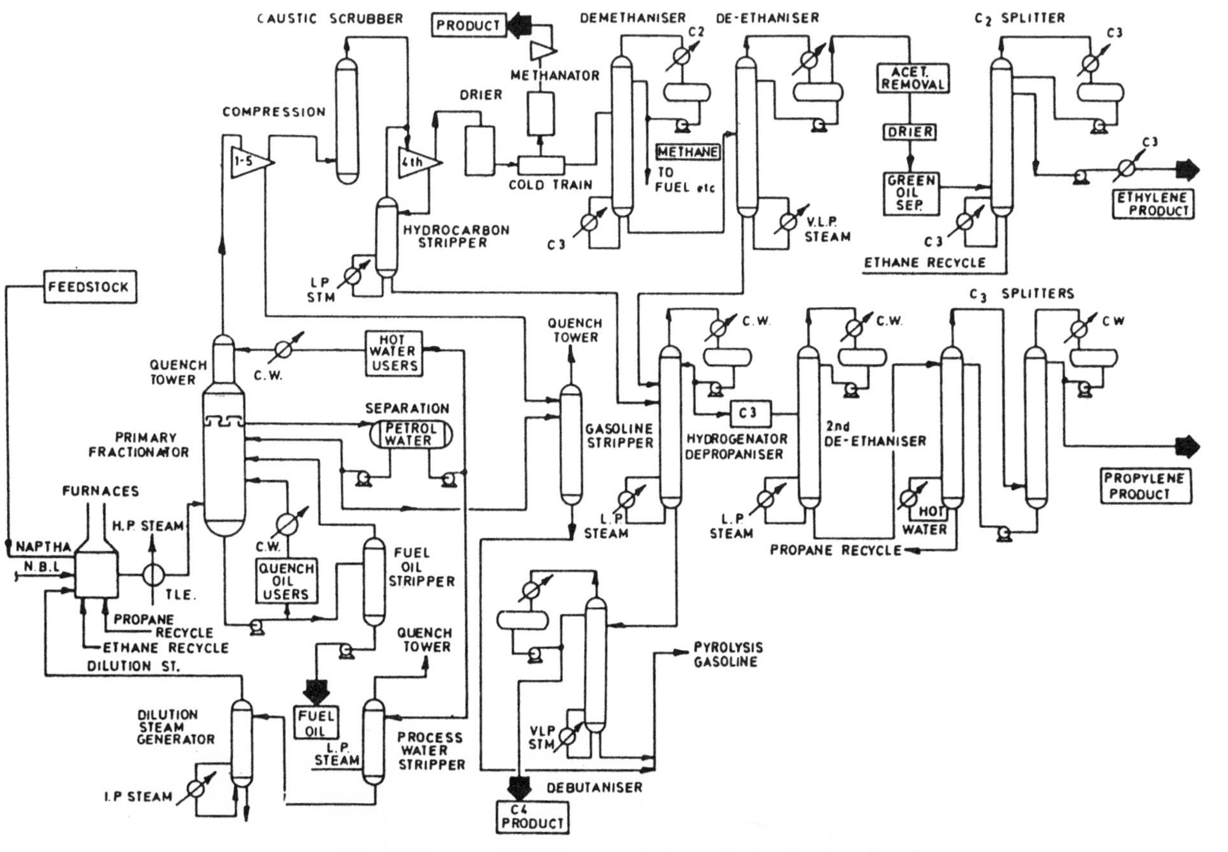

Fig. 1. Simplified diagram of an integrated chemical process (By courtesy of Stone and Webster).

counting ancillary items such as pumps and valves there may be 50 or more major pieces of equipment in many chemical processes. If there are, say, three possible positions for each of these in an overall network then there are 3^{50} or

$$\text{approx. } 7 \times 10^{23}$$

different networks! Evidently this estimate is based on some gross simplifications. It does, however, make the point that 'best integration' is a very complex problem indeed.

In practice 'best' networks are hardly ever found. It takes many years, and many different consecutive installations, before the design of a chemical process evolves towards the 'best'. This progress is documented for quite a few processes in the industry in so-called 'learning curves'. Figure 2 is an example for the ammonia process technology (Marsden 1980).

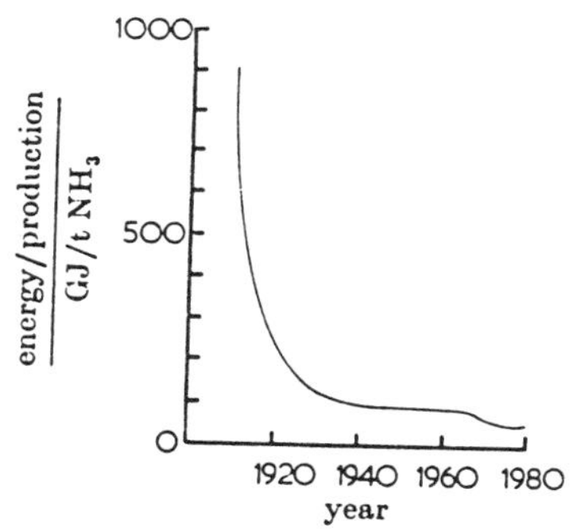

Figure 2. Learning curve for improvements in the ammonia process technology (Marsden 1980).

(d) Early research

Research in process network design is aimed at generalized procedures which, for any process, will lead to a 'best' network. Generally, such procedures are called 'Process Synthesis' procedures. An extensive review in Process Synthesis has recently been published by Nishida et al. (1981). Early research in Process Synthesis accepted the argument of there being many different possible designs. Thus, a large number of networks were generated and evaluated on the computer. The research then focused on two main endeavours: first, the development of mathematically efficient search techniques to scan quickly as many designs as possible; second, the development of 'heuristics', or rules-of-thumb. These were used to prefer or reject families of solutions on the basis of common sense. Examples of such heuristics are: (i) the most plentiful product must be separated first (this avoids repeated processing of the biggest mass flow; (ii) the most difficult separation must be done last (this ensures that the most difficult operation is carried out on the smallest mass flow).

There were two main problems with this approach. First, even the fastest mathematics were too slow for processes of realistic size. Second, heuristics make sense in isolation but what do we do if the most plentiful product is also the most difficult one to separate?

(e) Using thermodynamics

Why then use thermodynamics? Today, I could argue 'it's obvious'. However, I have to admit that, about eight years ago, I did not know much about chemical process design. Had I known more, I might have followed the conventional research route. The problem was that I was much more interested in thermodynamics than in process design. Like so many young researchers, I was not really driven by wanting to solve a problem but by wanting to try a problem-solving technique. In the event, the technique was appropriate. It led to a deeper understanding of the characteristics of process networks. This understanding in turn allows the engineer to identify better processes.

SECOND LAW ANALYSIS

(a) Entropy, exergy, and practical design

Entropy, available energy, or exergy analysis (i.e. Second Law analysis) is often referred to as an obvious tool for studying the quality of engineering designs (see for example Denbigh 1971). Different from a First Law analysis (or heat balance), a Second Law analysis compares actual performance with best possible performance.

Take a combustion engine. The actual engine may deliver 0.24 units of work per unit of fuel consumed, i.e.

$$\eta_{1st\ Law} = 0.24 .$$

If an ideal engine operating under the same conditions could deliver 0.3 units of work per unit of fuel consumed, then the Second Law efficiency of the actual engine is

$$\eta_{2nd\ Law} = 0.24/0.3 = 0.8 .$$

Next, a domestic central heating system may produce 0.7 units of useful heat per unit of fuel consumed. Thus:

$$\eta_{1st\ Law} = 0.7 .$$

If an ideal system in the form of, say, a heat pump could produce 2.7 units of useful heat per unit of fuel, then

$$\eta_{2nd\ Law} = 0.7/2.7 = 0.26 .$$

For the engine, the First Law efficiency seems quite low whereas the Second Law efficiency is high. In the second example, the First Law efficiency gives us a false sense of security. Generally, the Second Law assessment emphasizes what would be possible.

Apart from efficiency, Second Law analysis evaluates the amount of useful energy that is lost in a process. This loss is usually measured in terms of entropy gain or exergy loss.

Entropy, as a concept, is intuitively less meaningful. The entropy of a substance increases with temperature but decreases with pressure. Thus, the entropy of steam at p = 1 bar ($\approx 10^5$ Pa) and T = 100 ^oC is the same as that of steam at p = 100 bar and T = 690 ^oC! To make matters worse, no process can be assessed without considering the entropy gain of the universe... ! Little wonder that generations of engineering students have wondered what entropy is all about.

Keenan (1932) is often said to have improved on this situation by introducing the concept of available energy. However, the concept goes back at least as far as Gouy (1889) and his essay 'Sur l'energy utilisable'. Rant (1956) introduced the term 'exergy' as a word which has a meaningful similarity to 'energy' and is easily pronounced in many languages. Generally, exergy is regarded as a much simpler concept than entropy. A system's exergy is its ability to deliver work. If exergy is 'lost', then the system can deliver less work. Exergy decreases with both temperature and pressure and, when a process is analysed, the universe can be ignored.

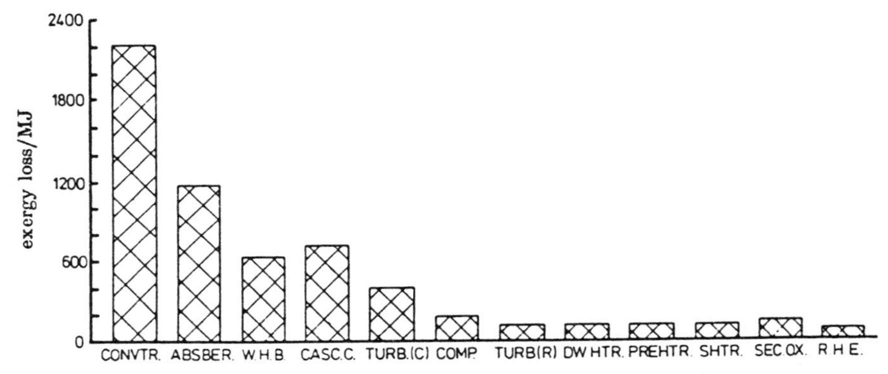

Figure 3. Exergy losses in a typical nitric acid process (Linnhoff 1979).

Analysing a nitric acid process in terms of exergy losses gives an assessment such as that shown in Figure 3. This is said to give the engineer clues as to where improvements might be possible.

So far, I have summarized conventional textbook knowledge. Eight years ago I knew as much as this about Second Law analysis. What I have found out since is that there are two basic problems when it comes to applying the analysis in earnest in the context of chemical process design. First, generating the analysis is not easy. The data and correlations needed are not usually well documented. It takes a long time to produce results such as those shown in Figure 3. Second, it is not at all clear how the analysis should be interpreted. Usually there are some good practical reasons for entropy gains and exergy losses. For example, rapid cooling might cause a large exergy loss. Unfortunately, without rapid cooling, the wrong chemical reactions would take place. In complete contrast to the science of thermodynamics, the interpretation of Second Law analysis is anything but scientific. To find out which exergy loss matters is more an art than a science. The following is a description of how both problems were tackled, i.e. that of producing and that of interpreting Second Law analysis.

(b) Producing an analysis

Figure 4 summarizes the conventional way of producing both a First Law and a Second Law analysis for a steady-state chemical process. In both cases, we know temperature, pressure and composition of the process flows. When producing a First Law analysis, enthalpy correlations allow us to calculate enthalpy changes ΔH as a function of temperature, T, pressure, p, and composition, c. When producing a Second Law analysis, entropy correlations allow us to calculate entropy changes ΔS.

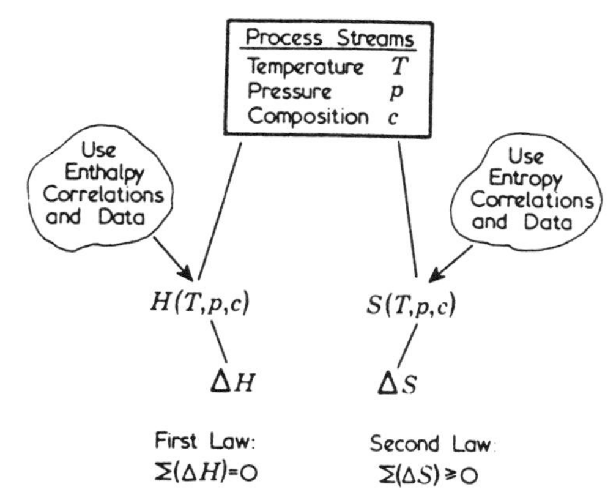

Figure 4. Traditional way of producing First and Second Law analyses.

This seems straightforward enough. But what is the situation in industrial practice when a Second Law analysis is required? Quite often, a heat and mass balance, i.e. a First Law analysis, is available. In other words, the enthalpies of process streams are known. By contrast, exact stream compositions, correlations and data may not be available. (Design contractors often keep their data and correlations confidential and only communicate the overall results of calculations). To produce a Second Law analysis from such information is impossible by the approach described in Figure 4.

But do we need these data and correlations if we have a completed heat and mass balance? Consider heating a liquid at constant pressure. Enthalpy increases with temperature according to

$$\Delta H = m \int_{T_1}^{T_2} c_p \, dT. \qquad (1)$$

Entropy increases according to

$$\Delta S = m \int_{T_1}^{T_2} \frac{c_p \, dT}{T} \, . \qquad (2)$$

Assuming c_p = constant, we can write

$$\frac{\Delta S}{\Delta H} = \int_{T_1}^{T_2} \frac{dT}{T} \Big/ \int_{T_1}^{T_2} dT = \frac{\ln(T_2/T_1)}{T_2 - T_1} \, . \qquad (3)$$

Further, we can write

$$\frac{\ln(T_2/T_1)}{T_2 - T_1} = \frac{\ln(T_1/T_2)}{T_1 - T_2} = [(T_{LM})_{1,2}]^{-1} , \qquad (4)$$

where $(T_{LM})_{1,2}$ is the logarithmic mean of T_1, T_2, and obtain by combining equations (3) and (4)

$$\Delta S = \Delta H [(T_{LM})_{1,2}]^{-1} \, . \qquad (5)$$

To simplify further, we may write

$$(T_{AM})_{1,2} \approx (T_{LM})_{1,2} , \qquad (6)$$

where $(T_{AM})_{1,2}$ is the arithmetic mean of T_1, T_2, and obtain

$$\Delta S \approx \Delta H \, [(T_{AM})_{1,2}]^{-1} \; . \qquad (7)$$

Note that temperatures in these equations are absolute.

Equations (5) and (7) demonstrate a general approach. They allow us to calculate entropy changes as a function of enthalpy changes and operating conditions. There is no need for explicit entropy data or correlations. Better still, there is no need to know which chemicals, and in what composition, are making up the stream! This is demonstrated by way of a simple example in the Appendix.

A summary of the approach is given in Figure 5. Provided enthalpy changes ΔH are known, we can define functions of operating conditions, X, and calculate approximate entropy changes:

$$\Delta S = f(\Delta H, X). \qquad (8)$$

Linnhoff and Carpenter (1981) describe a small number of transfer functions X_1, X_2, X_3, ..., which allow the engineer to relate any entropy change to its corresponding enthalpy change. *This includes the treatment of chemical reactions, obliterating the need for reaction-free energy data.* Approximate exergy changes are obtained analagously through transfer functions Y:

$$\Delta Ex = f(\Delta H, Y). \qquad (9)$$

The simplicity of the approach becomes apparent in the Appendix. Typically, the analysis of a chemical process of average size will take a day or two.

What about accuracy? There was a gross simplification in the derivation of equation (5) and another in that of equation (7). Let us remember the purpose of a Second Law analysis. We do not need the results to size equipment or to predict the behaviour of a process. All we need the results for is to establish a 'batting order' in our mind as to where we think that process improvements will be possible or likely. The approach exemplified in equations (8) and (9) has been used in many project studies to date and accuracy has not presented a problem. Sometimes, there is a choice between transfer functions of different accuracy. An example is the choice between equations (5) and (7). Initially, engineers tend to be cautious. As they become more experienced, they tend to use the simplest function available.

Lastly, the approach has another unexpected benefit. It tends to enhance the engineer's understanding of the way in which First Law and Second Law principles interact and complement each other. This may be apparent in the Appendix. Since the ultimate aim of the analysis is to improve the engineer's understanding, this benefit is of considerable importance.

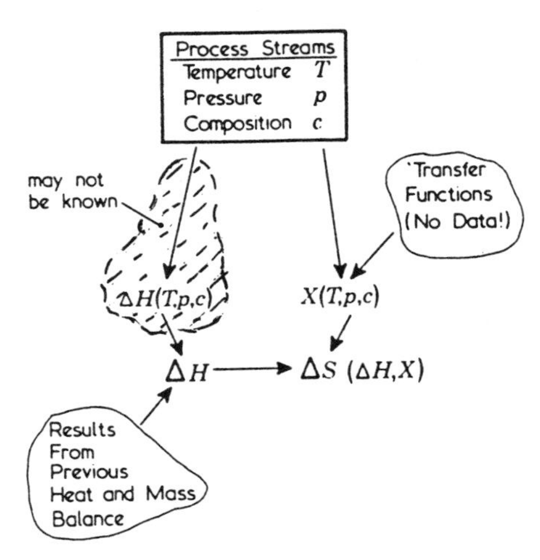

Figure 5. Producing Second Law analysis from First Law analysis through transfer functions.

(c) Interpreting an analysis

Over the plast ten years or so many papers have been written on the subject of entropy or exergy studies of chemical processes. Examples are papers by Riekert (1974), and Gaggioli and Petit (1977). A tongue-in-cheek summary of most such papers is as follows. First, there is an explanation of thermodynamic concepts, i.e. of entropy, T, S-diagrams, exergy, free energy, etc. Second, there is a 'case study'. Exergy balances and efficiencies are calculated for an example process with a table of losses resulting similar to Figure 3. Third and last, the reader finds a paragraph or two saying that this exercise will enable the design engineer to improve his system. There is a marked lack of explanation of *how* information such as that gathered in Figure 3 should lead to improved designs. Sadly, most papers do not discuss the key question: how do explicit design changes follow from the study of exergy (or entropy)?

Consequently, the engineer's search for the exergy loss that can be avoided remains not unlike the search of the sleuth for the criminal, working from a mixture of honest as well as misleading clues. Let us look, then, at how the most famous sleuth of all tackles the problem.

Once, when discussing a murder case, Dr. Watson gave to Sherlock Holmes his opinion on who seemed most likely to be the murderer. Holmes interrupted 'This is not a scientist's approach, Watson. A scientist does not work by guesses but by elimination. There may be fifty suspects. Draw up a list, considering everyone. And rather than speculating how likely a suspect looks, find reasons why he or she can be eliminated from the list. For example, an old lady might be listed. If the murder weapon was a heavy club, eliminate her from the list. Finally, there will be one suspect left on the list and you cannot find a reason why he should be eliminated. He must be the murderer.'

Following Holmes' reasoning, Linnhoff (1981 a) lists a number of 'themes', based on experience, which explain why exergy losses may have to be eliminated from the list of 'suspects'. The engineer is asked to examine each loss against each theme. If one or more of the themes excuse the loss, or part of the loss, then the loss, or part of it, is called 'inevitable'. We cannot improve the process by eliminating this particular loss. If no theme excuses the loss, then it is called 'avoidable'. We should be able to improve the process by eliminating this avoidable loss, or part of it. In some cases, the extent to which losses are avoidable can be calculated by following defined rules. Most importantly, this applies to chemical reactors (Flower and Linnhoff 1979). In other cases, the engineer has to make estimates. Two of the non-quantitative themes follow to serve as examples.

(d) Losses breed losses

Figure 6 shows a chemical process requiring two services, steam and refrigerant. There are two energy conservation schemes, 1 and 2. Scheme 1 would save ten units of Second Law weighted energy (i.e. exergy), in refrigerant. Both schemes have the same capital cost. On first inspection, the schemes seem equally promising.

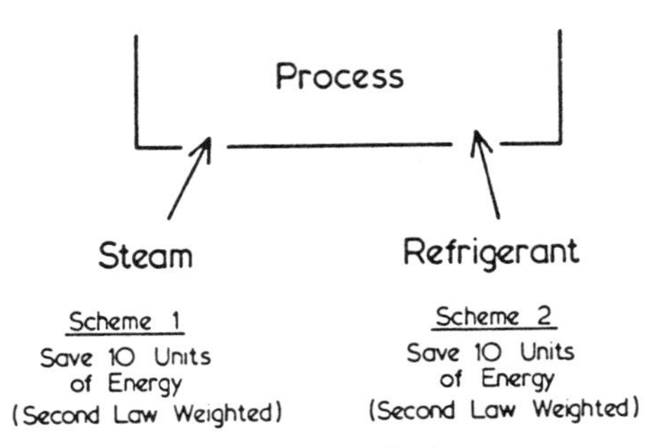

Figure 6. A process requiring two services.

Now consider Figure 7. A simple diagram shows how both steam and refrigerant are produced from primary fuel. Fuel is burned in the boiler to generate steam at an efficiency of $\eta = 0.8$. Some of that steam is consumed by the process. The remaining steam is fed through turbines ($\eta = 0.66$) to provide shaft work for the refrigerant system. The refrigeration system ($\eta = 0.25$) provides the refrigerant. The efficiency values given in Figure 7 are typical of many industrial systems. Consider now the implications of saving ten units of steam exergy or ten units of refrigerant exergy. In terms of primary fuel supply the steam savings will lead to $10/0.8 = 12.5$ units of fuel exergy saved. The savings in refrigerant will lead to $10/(0.25 \times 0.66 \times 0.8) = 75$ units of fuel exergy saved. We recognize that there is a multiplication chain of inefficiencies. The further away we are from the original source of energy, the more pronounced the overall effect will be of our schemes on fuel supply. In conclusion, scheme 2 seems much more promising than scheme 1.

In a general sense, the theme 'losses breed losses' excuses losses, or parts of losses that happen close to the original source of energy. It places emphasis on losses that take place far away. In practice, the theme has been found by many designers to yield worth-while insights.

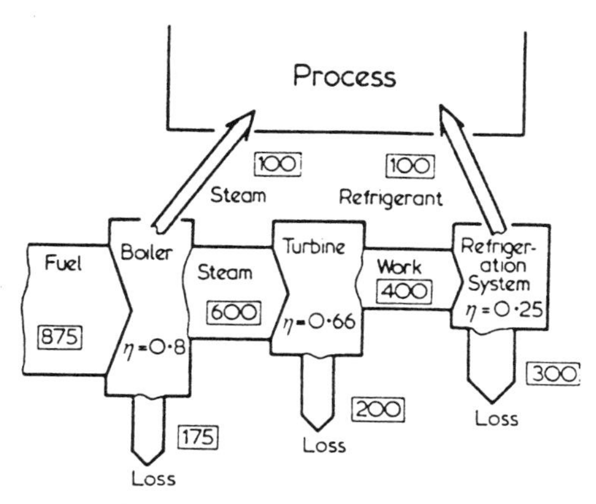

Figure 7. The provision of refrigerant involves more losses than the provision of steam.

Having discussed losses breed losses, let us reflect on industrial energy conservation campaigns. After the oil price rice in 1974, many large companies introduced systematic energy conservation campaigns (see, for example, Robertson 1974).

Most of these campaigns seemed to place heavy emphasis on measures in the boiler house and the utility systems. No doubt, such measures have in their favour economy of scale. On the other hand, losses breed losses makes it clear that the leverage on overall fuel supply is low.

(e) Load against level

Figure 8 shows two quantities of heat, a large amount at low temperature and a small amount at high temperature. Amounts and temperatures have been chosen such that the Second Law weighted energy (exergy) is the same for both. (The ambient temperature assumed is 298 K.) This then has the following implication: with ideal machinery (Carnot engines and heat pumps), it is possible to convert each quantity of heat into the other. Heat loads and temperatures are interchangeable at the 'rate of exchange' fixed by ideal machinery. This is the thermodynamic assessment. The design engineer's assessment of this situation is different. If he requires heat at 500 K, then high temperature heat could be used by simple heat transfer but low temperature heat could not. If he requires heat at 350 K then both sources of heat can be used. We recognize that the temperature level of heat is of prime importance to the engineer. It determines whether or not heat can be used without resorting to heat pumps. This is in contrast to the thermodynamic assessment, which considers temperatures and heat loads interchangeable.

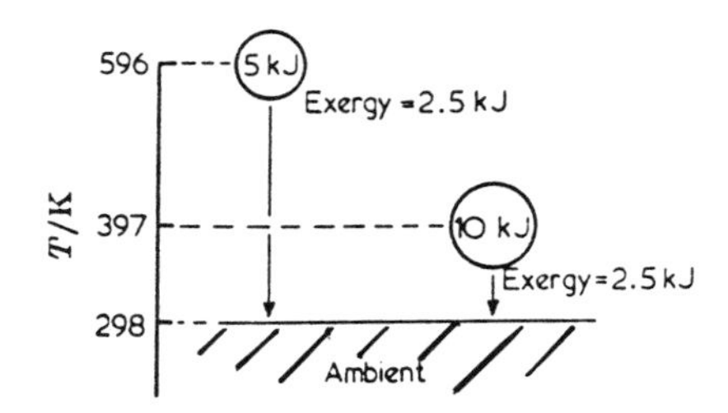

Figure 8. Relative roles of heat load and temperature in Second Law analysis: more heat at lower temperature has the same exergy as less heat at higher temperature.

In general terms 'load against level' excuses Second Law losses which take place at levels of potential that are too low to be useful in simple operations. These losses are considered as practically 'inevitable'. On the other hand, losses that take place at high levels of relative potential are emphasized as being potentially 'avoidable'.

Having discussed load against level, it is interesting to reflect that Second Law analysis is quite commonly used by mechanical engineers. Mechanical engineering systems (such as combustion engines, refrigeration cycles and power stations) are generally designed to fulfil energy functions. Heat loads and temperature levels are usually chosen to suit these energy functions. By contrast, chemical engineering systems are designed to fulfil chemical functions. Temperature levels and heat loads have to be compatible with the requirements of the chemical reactions and separations. Energy transfers and transformations are considered of secondary importance. In this situation, load against level makes Second Law analysis a much less meaningful tool than it is for the mechanical engineer.

(f) Summary

Figure 9 shows the distribution of exergy losses in the nitric acid process from Figure 3 after allowance has been made for 'inevitable' losses. The procedure has only been applied to the chemical reactions but, even so, different priorities from those in Figure 3 suggest themselves to the engineer as to where in the process improvements might be possible.

Both the concept of producing Second Law analyses from heat balances by using equations (8) and (9) and the concept of inevitable and avoidable losses have been applied to date in many industrial studies in I.C.I. In almost all cases, improved designs followed; see Table 1. Two studies have been described in detail by Townsend (1981) and another has been described by Linnhoff (1982).

Why does the approach seem to work? I feel that both the use of the transfer functions for numerical work and the use of the 'themes' for conceptual reflexion have the same effect: they enhance the engineer's understanding of the process under examination *and* of thermodynamic principles in general. Quite possibly, the themes do not really represent a firm method of interpreting Second Law analysis.

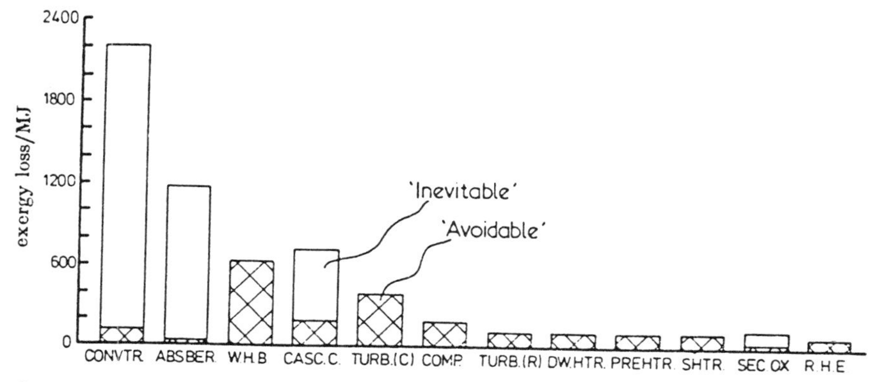

Figure 9. Energy losses in nitric acid process after exclusion of 'inevitable' losses in chemical reactions.

If this is so, then the improvements obtained followed from people's better understanding. I would be inclined to think that this would be an even more encouraging results.

HEAT RECOVERY NETWORKS

(a) Description of the problem

As has become apparent then, a Second Law analysis is an evolutionary tool. A process design is required as a starting point. This left me some years ago with the desire to tackle the problem of designing integrated networks from 'scratch'.

Like so many researchers before me, I started with heat recovery networks. The chemical engineering content of this problem is simple, making it easy to focus on the problem of networking. In brief, the problem can be described as follows. The major pieces of equipment such as reactors, distillation columns, compressors, etc. are given. The process streams from and to these pieces of equipment are known. Some streams have to be heated while others have to be cooled. Mass flow rates, heat loads, and temperature specifications are given. The simplest design of a network for heat transfer would use utility heaters on all streams that have to be heated and utility coolers on all streams that have to be cooled, see Figure 10a. This type of design is usually believed to minimize capital cost. A more complex design would 'match' streams that have to be cooled against others that have to be heated in heat exchangers, avoiding the use of utilities but incurring capital cost, see Figure 10 b. The objective in heat recovery network design is to select streams for matching such that the network is cheapest in terms of utility and capital cost required.

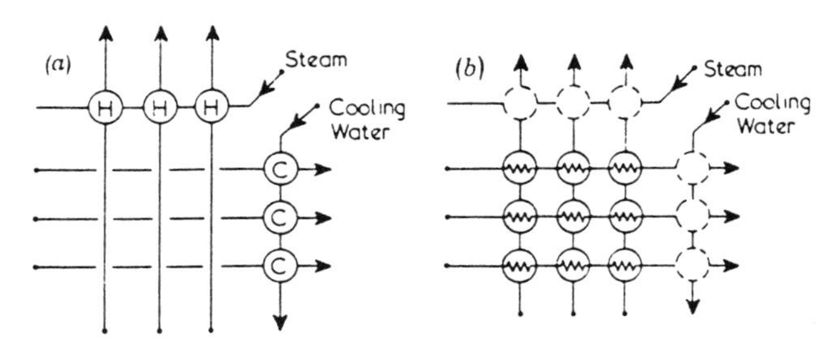

Figure 10.(a) A heat recovery network without recovery.
(b) A heat recovery network with maximum recovery.

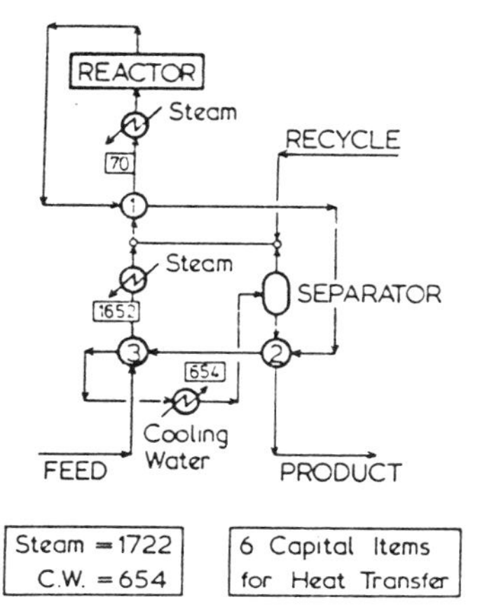

Figure 11. Front end of speciality chemical process.

A practical example of a heat recovery network is shown in Figure 11. The system represents the front end of a speciality chemicals process (Boland and Linnhoff 1979). Feed is heated before reaction in two heat exchangers and two utility heaters. The reactor products are cooled in three heat exchangers and a cooler and are fed to a separator. The bottom product from the separator is heated by the reactor products. The top product is mixed with a recycle flow from the main process and fed back into the reactor. The task in the present context is to find another heat recovery network that would achieve the same objectives at lower total cost of energy and capital.

(b) Early research

In many systems, most streams could potentially be available for heat interchange with most others. This gives rise to many possible networks. Ponton and Donaldson (1974) give an estimate of the size of the problem: a process having five streams to cool and five streams to heat (this is small by normal industrial standards) could have as many as

$$\text{approx. } 1.6 \times 10^{25}$$

different solutions! Consequently, research in heat exchanger networks followed the usual Process Synthesis pattern: large computer programs used efficient mathematical search techniques supported by heuristics. The problems were much the same as in general Process Synthesis.

(c) Performance targets

The economics of heat exchanger networks mainly depend on the cost of energy and capital. Energy costs tend to be dominated by the utility heat load. Capital costs for a process of a given production rate (i.e. size) tend to be dominated by the number of different pieces of equipment or, in short 'capital items'. Thus, it is relevant to ask two questions about the system in Figure 11. First, 'Do we need 1,722 units of heating?' Second, 'Do we need six capital items for heat transfer?'. Clearly, there must be a minimum figure for both the energy and the number of capital items. Can we determine what these minimum figures are? In other words, can we set 'targets'?

(i) Energy targets. To establish the energy target, the Second Law analysis theme of load against level is relevant. It tells us that it is chiefly temperature that is important for heat recovery. In the present context the situation is complicated by the fact that, as heat is exchanged with a process stream, the stream's temperature changes. This consideration leads to Figure 12.

Figure 12, in turn, leads directly to Figure 13. In Figure 13a, the T, H-graphs of process streams between key items of equipment such as reactors, distillation columns, compressors etc. are shown. One graph represents a stream to be cooled from 550 K to 350 K. There are two other streams to be heated. Then, the overall amount of heat available or required within ranges of temperature is calculated from the sum of all heat capacities within each range (Figure 13b). The resulting graph in Figure 13b is called the 'Grand Composite' T, H-graph for a process. Let us examine its physical significance.

Where the slope of the Grand Composite is positive, more streams (in terms of total heat capacity, not number) have to be heated than cooled. Where its slope is negative, more cooling is required than heating. Thus, sections G-F and D-B represent overall heating requirements whereas sections A-B, D-F and G-H represent overall cooling requirements. We can see that the temperatures allow us in the limit to recover heat from section A-B to section C-B, and from section E-F to section G-F. This leaves us with the external heating requirement D-C and cooling requirements D-E and G-H. We have established energy targets for the overall process!

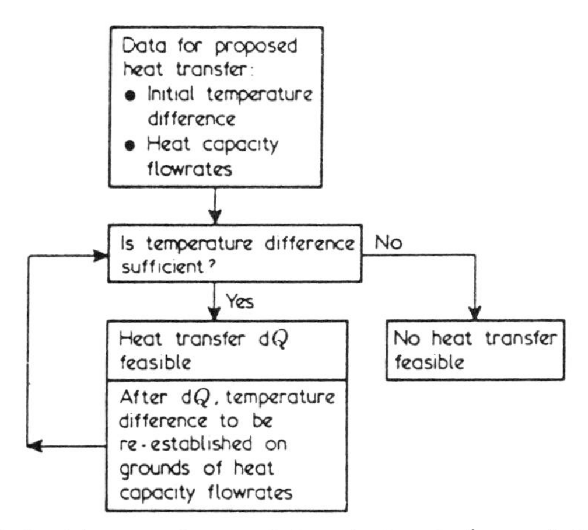

Figure 12. Relative roles of heat load (specific heat) and temperature in simple heat transfer.

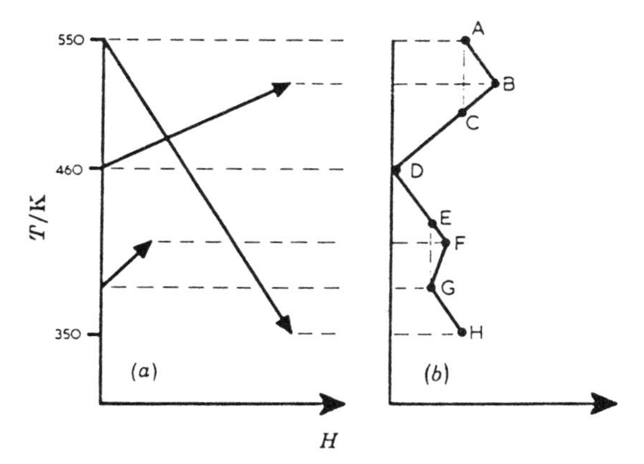

Figure 13. Setting the energy target:
(a) T, H-graphs for individual process streams;
(b) Grand Composite T, H-graph for overall process.

It is clearly possible to construct Grand Composites for any number of process streams. The task is ideally suited for simple computer programs. With little effort, it is possible to include allowances for minimum temperature differences (Figure 13 is based on the assumption of zero temperature differences: this is obvious at points B and F) and for the use of discrete utility levels (Linnhoff and Hindmarsh 1982). Also, it is possible to consider constraints such as: there must be no heat transfer between a given hot stream and a given cold stream, see Cerda et al. (1982). Figure 13 is based on the assumption that heat exchange is possible between any two streams). With these provisions for practicality, a simple and powerful tool emerges for the prediction of heating and cooling targets. *Note that only process stream data are required. We do not need to have a heat recovery network to set the target.*

(ii) Capital targets. The capital target does not follow directly from thermodynamics. However, it is derived from the same philosophy of trying to understand a problem.

In essence a heat recovery network connects separate heat sources (hot process streams and utility heat sources) and sinks (cold process streams and utility heat sinks). Once the energy targets are known, and therefore the utilities, we know all these sources and sinks. A heat exchanger then simply transfers heat from one source to one sink. Figure 14 illustrates this situation for the system in Figure 11. The hot process streams and the hot utility (predicted by means of the Grand Composite) are shown at the top of the diagram and the cold streams at the bottom. A possible arrangement of heat exchangers is represented by lines. (The arrangement is different from that shown in Figure 11). The question is: What is the minimum number of lines necessary to connect two hot streams with three cold streams?

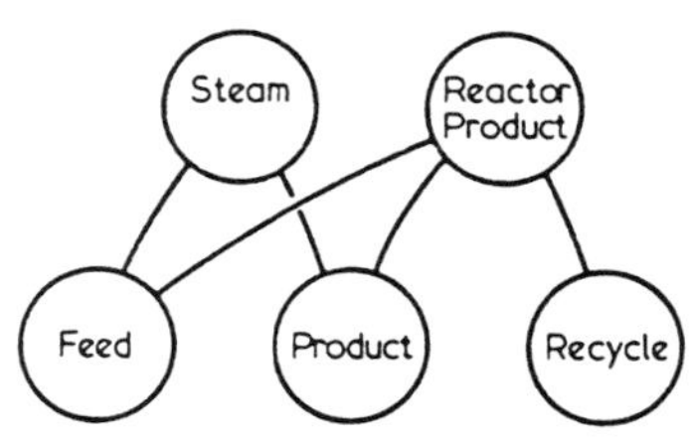

Figure 14. Heat exchanger network as a graph. Application of Graph Theory yields equation (14).

TABLE 1. APPLICATIONS IN I.C.I. 1977-81 (see ELLIS 1981)

Study	Process	Year of study	New process or modification of existing plant	Energy savings available (£/a)	Capital cost expenditure (£) or savings
1	organic bulk chemical	1977	new	400000	same
2	unspecified	1978	new	> 3 million	savings
3	speciality chemical	1978	new	800000	savings
4	bulk acid	1979	new	40000	70000
5	petrochemical	1979	mod	1 million	2 million
6	crude unit	1979	mod	> 500000	savings
7	inorganic bulk chemical	1979	new	160000	savings
8	speciality chemical	1979	mod	> 100000	80000
			new	> 100000	savings
9	unspecified	1979	mod	2 million	6 million
			new	2 million	6 million
10	general bulk chemical	1979	new	6% scope	unclear
11	inorganic bulk chemical	1979	new	100000	
					unclear
12	future plant	1979	new	30% - 40%	30% savings
13	speciality chemical	1979	new	50000	75000
14	unspecified	1979	mod	150000	500000
			new	150000	savings
15	general chemical	1980	new	180000	unclear
16	power house	1980	mod	phase 1, 34000	12-month payback
				phase 2,150000	24-month payback
17	petrochemical	1980	mod	phase 1,100000	100000
				phase 2,170000	300000
18	petrochemical	1981	mod	phase 1,600000	300000
				phase 2,600000	600000

For every study completed at the time the table shows the type of process, the year of the study, whether it was new design or a plant modification and the energy and capital cost implications. In all cases energy savings were identified (quoted as percentages compared with existing usage or as revenue) and where a number is given in the last column this represents the corresponding capital cost expenditure. However, in many studies the capital cost is either similar to or less than that for the existing 'state of the art' design.

The answer is given by Euler's general network theorem from graph theory (see Linnhoff et al. 1979):

$$E = N + L - S \ . \tag{10}$$

In this equation, E is the number of edges in a graph, N is the number of nodes, L is the number of loops, and S is the number of separate components into which a graph can be divided. Clearly, edges are heat exchangers in Figure 14 and nodes are streams and utilities. We shall normally have fully 'connected' networks with $S = 1$.

$$E = N + L - 1 \ . \tag{11}$$

Trying to minimize the number of heat transfer units (edges), we write

$$E_{min} = N_{min} + L_{min} - 1. \tag{12}$$

We cannot change the number of streams (they constitute the problem we are trying to solve)

$$E_{min} = N + L_{min} - 1 \tag{13}$$

but we can hope to reduce the number of loops to zero (Linnhoff 1981 b):

$$E_{min} = N - 1. \tag{14}$$

We conclude that, subject to a network being connected, the minimum number of heat transfer units is one less than the number of streams and utilities. The minimum number is obtained by eliminating loops from the network.

Experience has shown that equation (14) represents a meaningful and practical target. However, its application to real systems warrants some discussion. The necessary comments are given by Linnhoff and Turner (1981).

TABLE 2. COMPARISON OF ACTUAL PERFORMANCE WITH TARGETS FOR THE SYSTEM IN FIGURE 11.

	actual	target
utility heat in	1722	1068
utility heat out	654	0
number of heat transfer units	6	4

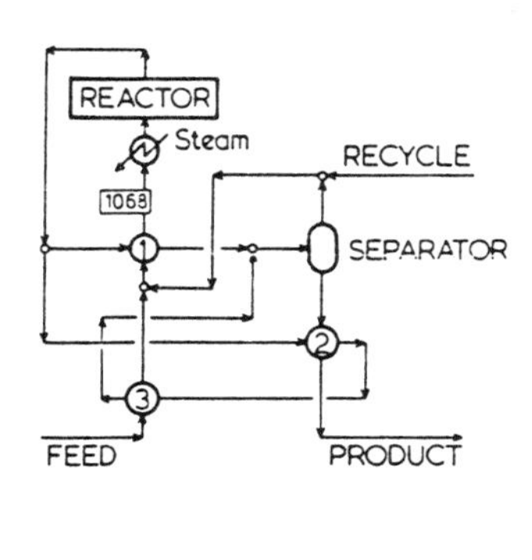

Figure 15. Improved front end of speciality chemicals process: approx. 40 % less energy with fewer capital items.

(iii) Applying the targets: an example. Table 2 shows a comparison between the actual performance achieved by the system in Figure 11 and the targets evaluated by using the principles just described. Evidently there is a great stimulus for improved performance in terms of both energy and capital. Design in this case is easy, yielding the system shown in Figure 15 (Boland and Linnhoff 1979). The system is cheaper in terms of energy and capital. It is safe and operable. It simply is a better system than that in Figure 11.

(iv) Targets, learning curves, and technology prompting. It is interesting to note that the concept of setting targets, as such, is well established. In many companies, management pressure is brought to bear on design departments by 'technology prompting'. The manager does not know what improvement is possible. However, instead of saying to his engineer 'improve' he says 'improve by 5 %'. The latter approach is said to exert stronger pressure on the engineer.

A special way of technology prompting is by extrapolation of learning curves of the type shown in Figure 2. In the early stages of a technology, good progress is expected. As the technology becomes more mature, there is less scope for improvement. Thus, the manager would initially prompt for large improvements but gradually reduce his challeneges.

In a way, the energy target described in this paper represents the asymptote to energy learning curves (if there is no development in chemistry and basic equipment technology). The implication is that the area under the learning curve could now be avoided. We know where the asymptote (target) lies and can prompt technology accordingly. The act of prompting is the same as in past practice. The prompt itself has changed. It is no longer based on guesses but on objective analysis.

(d) Network design

(i) Design with targets. The simplest form of design is by inspection keeping the targets in mind. The engineer uses his usual methods and techniques, but he is stimulated by knowing the targets. For the system in Figure 15, for example, this approach is adequate (Linnhoff 1981 b).

(ii) The network 'pinch'. We have not yet exhausted all implications of Figure 13. The most important feature of Grand Composites is probably the least obvious one: point D in Figure 13b represents a bottleneck for heat recovery. We call it the network 'pinch'.

Owing to the nature of their construction, Grand Composites have marked points of inflexion. One of these points will be at a position of lower enthalpy than any other. In Figure 13b this is point D. Above and below this point the Grand Composite extends to higher enthalpy. Above, this implies an overall positive slope (i.e. heat is required). Below, this implies an overall negative slope (i.e. there is heat surplus). It follows that the overall process is divided into two distinct parts: a heat sink (requiring heat input) above the pinch point and a heat source (requiring cooling) below the pinch point, see Figure 16.

Further, and this is an important observation, we can state the following by enthalpy balance over the sink and the source. An optimized system (in terms of energy) requires heat input A_{min} and heat output B_{min} with no heat flowing across the pinch, see Figure 17a. A sub-optimal design for the same system will feature heat input A, heat output B and heat flow across the pinch α, see Figure 17b. There is an exact correlation between the pinch heat flow and the excess utility consumption:

$$A - A_{min} = \alpha = B - B_{min}, \tag{15}$$

where A is the heat input to a system,
B is the heat output from a system, and
α is the heat flow across the pinch.

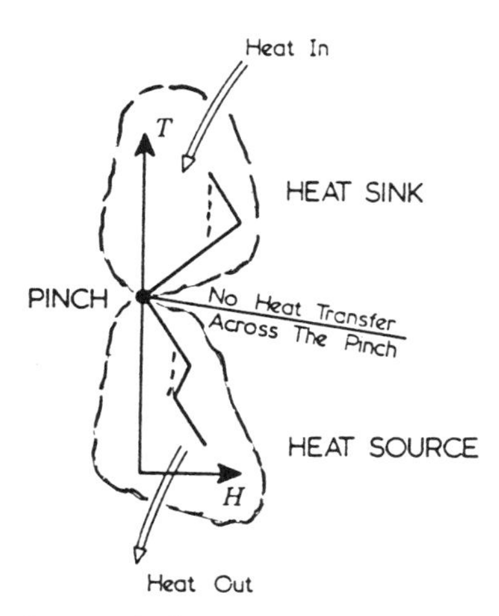

Figure 16. The pinch divides the heat recovery network problem into a sink and a source.

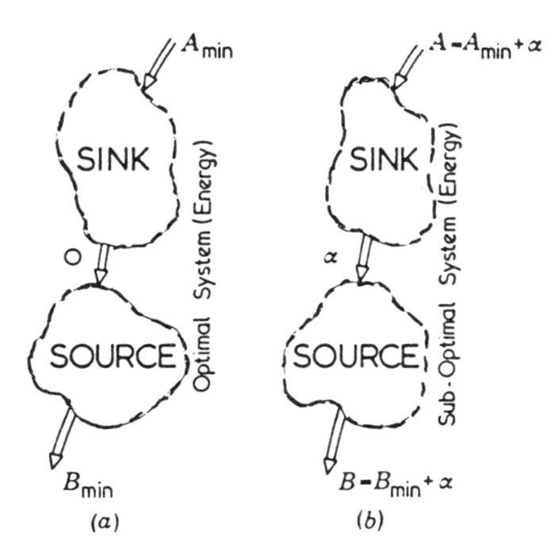

Figure 17. There is a fundamental correlation between the pinch heat flow and the excess utility requirements. (a) Optimal system (in terms of energy); (b) sub-optimal system.

In words, the pinch heat flow must always exactly correspond to the difference between (i) heat input and minimum input, and (ii) heat output and minimum output! Thus in Figure 13, any heat transfer from temperatures above 460 K to temperatures below 460 K must result in excess utility requirements, bot hot and cold.

This observation leads to two design principles, which are as simple as they are powerful.

(i) As long as heat transfer across the pinch is avoided the overall system is *guaranteed* to be optimal (in terms of energy).

(ii) If heat transfer across the pinch is introduced, the overall system is *guaranteed* to be less than optimal by as much as the pinch heat flow.

In a nutshell, the traditional problem of process design could be described as the engineer's inability to predict later consequences of early decisions. Who would start the design of a network such as that shown in Figure 1 and be confident that the early steps are 'right'? With the two simple principles just discussed, the designer can now have confidence in his early decisions. He can take them in full knowledge of their later implications. Heat transfer matches wholly above or wholly below the pinch are compatible with optimal overall performance (principle (i)). Heat transfer matches across the pinch are guaranteed not to be (principle (ii)).

(iii) Systematic design methods. In connection with the simple concepts of the pinch and the Grand Composite, more complex methods were developed for the systematic design of networks. It is interesting to note that this development did not evolve from the simple concepts towards the complex methods but that the reverse was the case. Five or six years ago, I felt like so many others that the most worth-while contribution in the area of Process Synthesis would be a systematic design method. Accordingly, I concentrated on this goal (Linnhoff and Flower, 1978a, b). Later, when trying to apply these methods in practical contexts, I realized that the basic understanding of important principles such as targets and the pinch can be a far more powerful asset in design than systematic methods. Systematic methods exclude the engineer from the design task. Above all, design is a creative process and to exclude the engineer from it must be a mistake. In recognition of this fact, I have shifted my emphasis over the years from the 'invention of methods' to the 'demonstration of principles in use'. A recent paper on heat exchanger network design (Linnhoff and Hindmarsh, 1982) emphasizes the need for the engineer to take his own decisions whenever possible and simply concentrate on the explanation of basic principles.

PRACTICAL ASPECTS

The approach to design discussed in this paper has proved practical. As far as I.C.I. is concerned, the list of applications in Table 1 bears witness to this fact. In more general terms, the experience gained from the studies documented in Table 1 can be summarized as follows:

- energy: 6 → 60 % savings,
- capital: expense → 25 % savings,
- return on investment: improved by factors of up to 4,
- often energy could be saved *alongside* capital,
- sometimes process flexibility was improved,
- design time was saved,
- concepts are valid for design of new processes and for modification of existing processes,
- concepts are applicable over a wide range of technology.

There must be many different reasons why a development bears such practical fruit. The single most important reason has been the enthusiasm of many of the people involved (see the acknowledgements). But how does enthusiasm develop? It tends to increase when things prove practical. The following is an attempt to pinpoint the most important reasons for the practicality of the approach. Some have been discussed before by Linnhoff and Turner (1980).

The pinch. The pinch was known before (see following) but its implications, i.e. the distinction between the process sink and source and equation (15), were not. Consider these concepts in the context of Figure 1. The separation of sink and source introduces a totally unsuspected division in a process. The divided parts are 'overlain'. Streams, and parts of streams, from everywhere in the process may belong to either the sink or the source. Without knowing about the concept of keeping sink and source separate the designer will make 'convenient' matches and the chances of obtaining an optimal network must be minimal.

Objective targets. Setting targets is quick and simple and does not require complex design calculations. The stimulating effect of targets undershooting a given design is great. The satisfaction that a designer obtains from seeing the targets confirming his design is great, too. Gone are the niggling doubts: have I really exhausted all possibilities? In short, people enjoy setting targets.

The concepts are simple. In my experience, techniques, procedures and computer programs are used in practice not because they save money (that is the theory) but because they are simple (that is the practice). A complicated procedure will usually require a strong management 'push' to be applied. A simple procedure will spread almost by itself. I believe the procedures and principles outlined in this paper find ready application in practice essentially because they are simple to grasp and easy to apply.

The approach is interactive. The approach represents a complete turnaround from today's usual black box computer design procedures. Black boxes sometimes insult the engineer's intelligence. The procedures and principles described here enhance it.

The philosophy is design oriented. This applies especially to §2, i.e. Second Law analysis. Traditionally, the emphasis in thermodynamic texts lies on the preciseness of concepts and definitions and on the correctness of reference states, datum points, system boundaries, etc. Most such subtleties are totally unnecessary in the context of process network evaluation. Indeed, they can be counterproductive. The average design engineer tends to respond by thinking that this is a task for the specialist. The approach described in §2 redresses the balance. There is little concern with the science of thermodynamics. The focus is on how the science is *used*.

Design time savings. One of the biggest hurdles for the practical use of the principles initially was the fact that industrial designs are usually carried out under time pressure. It is often more profitable to finish a design in time than to have it fully optimized. Who wants to start learning about exergy in this situation, and network pinches? However, it gradually became apparent that application of the principles saves time rather than costs time. The ability to screen alternatives at an early stage by comparative targeting, the tidyness of the pinch concept, they all help the designer to turn out designs faster than before and with greater confidence.

Integrated systems can be controlled. Another hurdle was the widespread scepticism regarding the control of integrated systems (see §1(b)). However, a modern process may be integrated anyway by, say 80 % in heat terms. Improving its energy load consumption by 30 % then means reducing the non-integrated load from 20 to 14 %, increasing integration from 80 to 86 %. This is a small change in heat load and, in any case, heat loads are not as important for control as is network connectivity. Usually the change in heat load is accompanied by changes in network topology, with the engineer's better understanding of loops and network connectivity, see equations (10)-(14). On balance, improved controllability can result (Linnhoff et al. 1982).

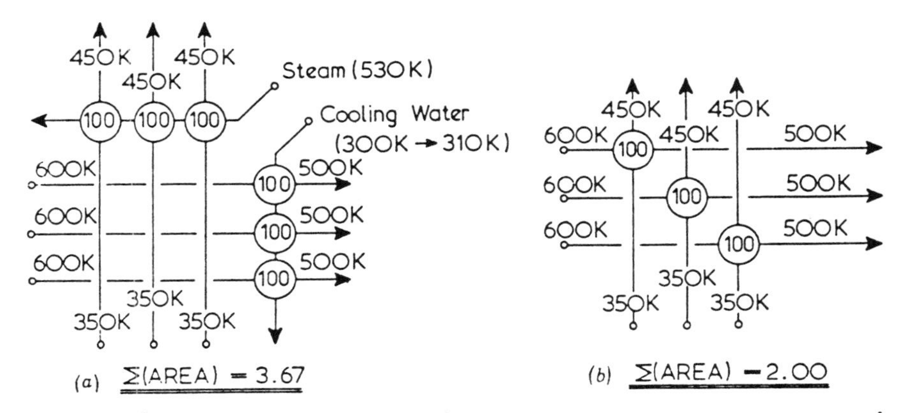

Figure 18. Surface area requirements for networks such as in Figure 10. The network that avoids recovery does *not* have the smallest surface area! This is due to the inordinate increase in the amount of heat transferred. (Mass flowrates, specific heats and heat transfer coefficients are assumed to be unity.)

Save capital by saving energy! The final hurdle for practical acceptance was the common belief that energy conservation requires investment. Surprisingly, many of the applications listed in Table 1 have led to energy savings *alongside* capital savings. In other words, compared with the 'old' design (which might be one of the most modern of processes) the new design was cheaper to run *and* cheaper to build. Initially, my colleagues and I simply thought this was a string of good fortune. As time went on, we could no longer fail to understand. It became apparent that proper recognition of the pinch led to reduced overall heat loads with only small increases in driving forces. Reduced heat loads led to reduced equipment sizes, which in turn led to lower capital costs. The message here is a general one. There are two effects that link energy and capital costs in processes: first, driving forces and second, load. Higher efficiency means smaller driving forces but also lower load. Smaller driving forces cost capital, but lower loads save capital. The overall effect might well be capital savings. This is exemplified in Figure 18. Our usual belief that energy conservation costs capital is based on a consideration of driving forces only. We tend to forget about the head load effect.

DISCUSSION

(a) Comparison with previous work

So far, I have only given generalized, and tongue-in-cheek, discussions of previous work. I felt this was justified as many of the principles and procedures discussed here represent a complete departure from most previous work. However, there are four previous workers I know of who have done similar work. First, there is Denbigh (1956), who distinguished avoidable and inevitable irreversibilities. Second, Frankenberger (1967) used a formula equivalent to equation (5) for exergy. Third, Hohmann (1970) developed a technique for setting energy targets in heat-exchanger networks and guessed equation (14). Fourth, Umeda et al. (1978) observed a 'bottleneck on heat recovery' in heat exchanger networks and called it the 'pinch'.

Initially, all these sources were unknown to me and with hindsight I am sure that this was a great blessing. Usually, different people make the same discovery using different approaches and each approach is carried by its own momentum beyond the discovery in a different direction. In the event, Denbigh did not generalize the distinction of 'avoidable' and 'inevitable' losses beyond the discussion of the Gibbs free-energy change in chemical reactors. Frankenberger used his transfer function solely to calculate energy changes in heat transfer. Hohmann used equation (14) without recognizing the pinch. This left him without design clues, and consequently, he found limited application for the energy target. Umeda et al. observed the existence of the pinch but did not recognize the principle of avoiding heat transfer across it.

It often only takes a little step beyond existing knowledge to greatly enhance practical usefulness. The chances of making such little steps are often better when existing knowledge is arrived at independently and from a different direction. The lesson to learn is that 'reading the literature' might be a bad way for a novice to enter a research field.

(b) On-going research

About 1979 when working with the heat recovery network techniques in I.C.I., my team and I came to recognize that the pinch is significant not only for heat recovery networks but also for thermal systems in general. Useful observations were made especially in the areas of integrated heat and power systems (Townsend and Linnhoff 1982) and of distillation processes (Dunford and Linnhoff 1981).

The power system most familiar to the chemical engineer is the steam turbine. It accepts high pressure steam for the generation of shaft work and rejects steam at lower pressure, either as waste or at a usable level (see Figure 19). If X is the heat input, Y is the heat output, and W the work generated, then by simple energy balance

$$X = W + Y. \tag{16}$$

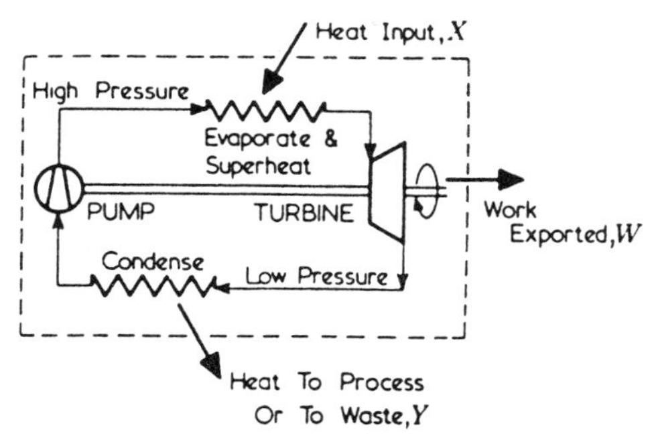

Figure 19. Simple model of a steam turbine.

Steam turbines are frequently employed in chemical processes with the exit steam used for process heating. An important question in design is where to use the steam.

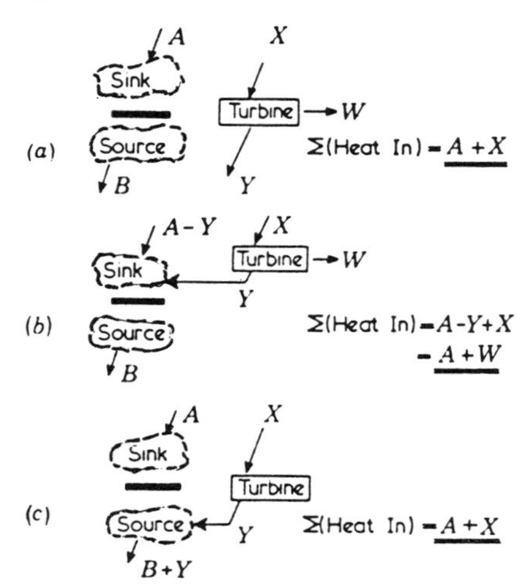

Figure 20. Integrating a steam turbine and a process in terms of heat recovery. (a) No integration. (b) Integration above the pinch is very worth while. Marginal fuel supply over stand-alone process is equal to work generated, W. (c) Integration below the pinch cannot confer any improvement compared with no integration.

The pinch offers an almost trivially simple answer, see Figure 20. If the steam is above the pinch (Figure 20b) the integration is highly worth while. The total heat input of the combined system (process and turbine) is

$$\Sigma(\text{heat in}) = (A - Y) + X, \tag{17}$$

which, with equation (16), reduces to

$$\begin{aligned}\Sigma(\text{heat in}) &= \{A - (X - W)\} + X \\ &= A + W.\end{aligned} \tag{18}$$

In other words, the integrated system requires an amount W of heat over and above the requirement of the process on its own. One could say, that, owing to the integration, heat has been converted to shaft work at 100 % efficiency! For those trying to reconcile this statement with Carnot's teachings, there has not been so much a conversion of heat but a reduction in the wastage of heat.

Figure 20c describes the situation if the steam is used below the pinch. The total heat input is no less than that for the separate systems. There is no benefit to be obtained from the integration. It follows that good designs should utilize turbine exit steam above the process pinch.

Surprising as it may seem, from this simple insight follows an entire approach for the design of integrated heat and power systems including the use of heat pumps. A detailed description has been given by Townsend and Linnhoff (1982).

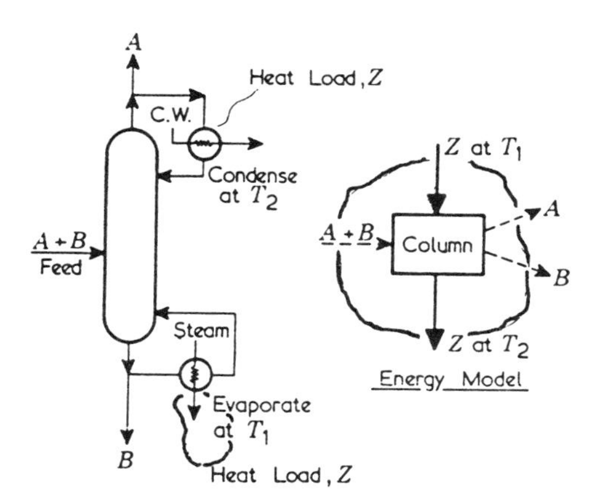

Figure 21. Distillation columns consume temperature not heat.

Another piece of equipment familiar to the chemical engineer is the distillation column. Not so well known to other engineers, a distillation column is often described as a separator 'running on heat'. This is misleading. Distillation columns tend to absorb heat at a given temperature and then to reject it (i.e. roughly the same amount) at a slightly lower temperature, see Figure 21. Thus, they do not really run on heat but on temperature.

This difference is important when considering the integration of distillation columns in processes, see Figure 22. Integration across the pinch (Figure 22c) gives no benefit compared with the separate systems. Integration above or below the pinch (Figure 22b) leads to the columns *and* the process requiring no more energy than the process would require on its own. The reason is that the columns do not consume heat but temperature. They simply make use of spare temperature differences available from the process. This simple insight, too, has far-reaching implications. A fuller description is given by Dunford and Linnhoff (1981).

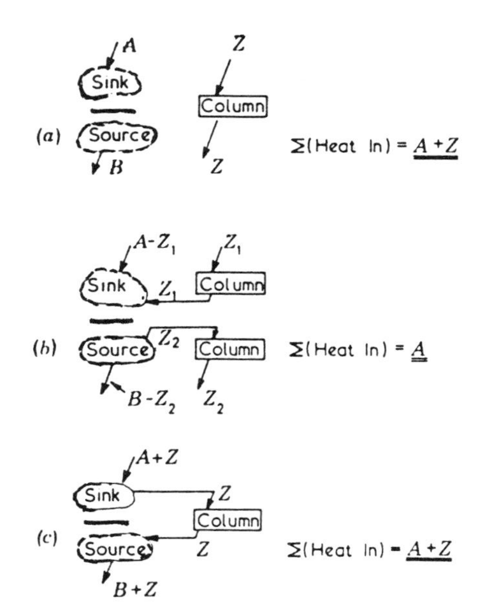

Figure 22. Integrating a distillation column and a process in terms of heat recovery. (a) No integration. (b) Integration above or below the pinch enables the column to run for 'free'. (c) Integration across the pinch cannot confer any improvement compared with no integration.

These principles for heat and power systems and distillation column integration appear to be just as practical as those for heat recovery networks. The reason is that they are similar in nature. They avoid hard and fast rules. As such, they represent a worth-while development in their own right. However, they are exciting for another reason, too.

Figure 23, the 'onion diagram', describes the hierarchical build-up of chemical processes. First, we need to know the chemistry and the network of reactors and chemical species. Once we know the reactors, we can define separators. Once we know the separators, we know pressure levels and vapour-liquid splits and we can define compression and expansion duties. Once we know compressors and expanders, we can design the heat recovery network. The traditional approach to process design is to build the process 'from the inside out'. There does not seem to be any other way. How can we design separators if we do not know yet what to separate?

Yet, the principles for heat and power for distillation column integration represent a departure from this approach. Rather than accepting the inner layers of the onion and saying: 'Here is the inner process, how do we design a heat exchanger network for it', we say: 'Here is the inner process, how do we change it'? We change the process by placing engines and distillation columns on the appropriate side of the pinch. This leads to different Grand Composite curves with lower energy targets.

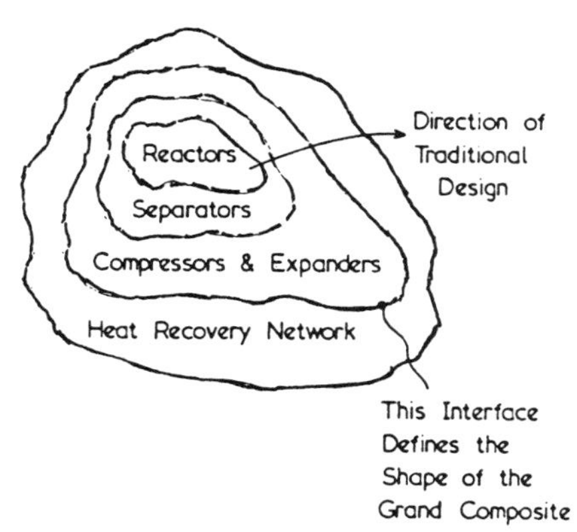

Figure 23. The 'onion' diagram represents the hierarchical nature of process design.

This departure from traditional practice is fundamental. By working 'from the outside in' in Figure 23 we learn how to improve the whole process in an evolutionary manner in such a way that the ultimate heating and cooling requirements become minimal. In my opinion, the research prospects for studying the implications on reactor design and the basic process mass balance are exciting.

(c) Implications outside the chemical industry

This paper has been discussing chemical processes. However, many of the principles discussed apply to the study of thermal systems of other kinds. I have restricted the discussion to the chemical industry in order to feel confident in terms of practical experience. Yet there is one temptation I cannot resist.

The overall economy, national and worldwide, could be described as a giant thermal network. Energy is supplied in the form of gas, coal, hydro and nuclear electricity, and crude oil, see Figure 24. Then, the energy industries generate coke from coal, electricity from coal, gas, and oil, refined oil from crude, etc. Next, the manufacturing industries consume this energy as well as one another's products. Finally, the consumers receive energy, goods, and services.

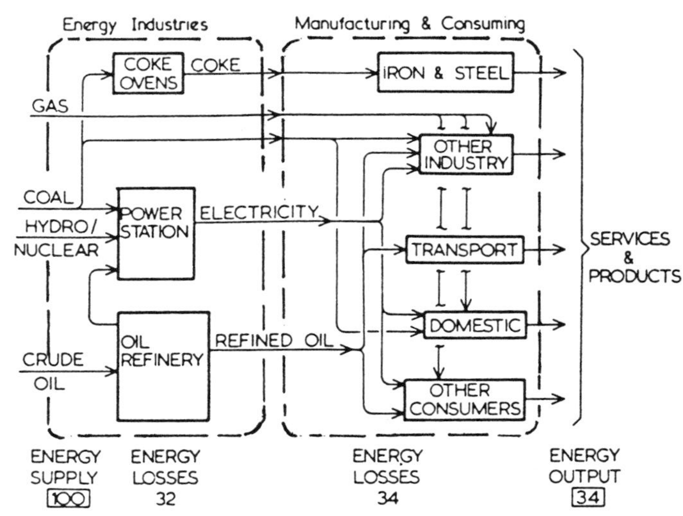

Figure 24. Simplified network of the national economy. Only one third of the primary energy supply is ultimately used. Two thirds are required to offset losses. (Approximated from information given in DoE 1981).

As shown in Figure 24, the direct energy value in Britain of the energy, goods, and services delivered to the consumer is only about one third of the primary energy supply to the economy. Similar figures apply in other countries.

Evidently, losses breed losses here. We conclude that one unit of energy saved in the manufacturing industries may save up to three units of energy on the oil tanker! In my opinion, this consideration makes it imperative that we devote efforts, as a nation, to the conservation of energy at the consumer end. I cannot help feeling an analogy between government policies and the corporate energy campaigns described earlier. Just as the companies concentrated on their boiler-houses, the national government concentrates on its power stations and the energy supply system. The overall returns are lower here than at the consumer end.

Sir Andrew Huxley, P.R.S., on making the Award, said that the 'bridging of gaps' between research and industrial practice was an important aspect of this work. My response is that without some good solid rock on either side of a gap there can be no bridge. I am deeply grateful to those people, in academia and in industry, that provided the rock. M. Berchtold, J. R. Flower, A. W. Westerberg, J. M. Douglas, G. Stephanopoulos, G. Sachs, T. A. Kantyka, R. L. Day, G. Hunter and C. W. Suckling have all had a profound influence on me and on the developments described in this paper. In addition, I have received much stimulation through the enthusiastic cooperation of many design engineers and the members of my team in I.C.I. Special thanks are due to P. Middleton, D. Boland, I. Wardle, D. R. Mason and, above all, D. W. Townsend.

APPENDIX
CALCULATING THE ENTROPY GAIN IN A SIMPLE HEAT EXCHANGER BY MEANS OF TRANSFER FUNCTIONS

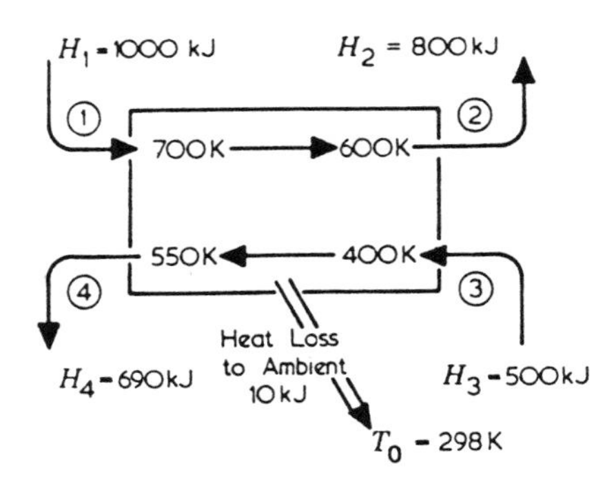

Figure 25. A simple counter-current heat exchanger with a small heat loss to ambient. Stream enthalpies and temperatures are known. Stream compositions are not known.

With equation (7):

$$\Delta S_{1,2} = (H_2 - H_1)(T_{AM})^{-1}_{1,2}$$

$$= (800\text{kJ} - 1000\text{kJ})\left(\frac{2}{600\text{ K} + 700\text{ K}}\right) = -0.308\text{ kJ/K}$$

and:

$$\Delta S_{3,4} = (H_4 - H_3)(T_{AM})^{-1}_{3,4}$$

$$= (690\text{kJ} - 500\text{ kJ})\left(\frac{2}{400\text{ K} + 550\text{ K}}\right) = 0.400\text{ kJ/K}.$$

The entropy gain of the surroundings is

$$Q/T_o = 0.034\text{ kJ/K}.$$

Thus, the overall entropy gain is

$$\Delta S_{tot} = -0.308\text{ kJ/K} + 0.400\text{ kJ/K} + 0.034\text{ kJ/K}$$
$$= 0.126\text{ kJ/K}.$$

Note that the use of equation (5) instead of equation (7) would have yielded the result

$$\Delta S_{tot} = 0.129 \text{ kJ/K}.$$

REFERENCES

1. Boland, D. and Linnhoff, B., 1979, "The Chemical Engineer", April, 9.

2. Cerda, J., Westerberg, A. W., Mason, D. R. and Linnhoff, B., 1983, Chem. Engng. Sci. 38, No. 3.

3. Denbigh, K. G., 1956, Chem. Engng. Sci., 6, 1.

4. Denbigh, K. G., 1971, "The principles of chemical equilibrium", 3rd edn. Cambridge University Press.

5. D.o.E. (Department of Energy), 1981, "Digest of United Kingdom Fnergy Statistics", 1981, London, H.M.S.O.

6. Dunford, H. A. and Linnhoff, B., 1981, I. Chem. E. Symposium Series No. 61, paper 10.

7. Ellis, G. V. (ed.) 1981, I.C.E. "Energy News", July, issue No. 6.

8. Flower, J. R. and Linnhoff, B., 1979, "Computers and Chemical Engng", 3, 283.

9. Frankenberger, R., 1967, "Zem. -Kalk-Gips" No. 1, 24.

10. Gaggioli, R. A. and Petit, P. J., 1977, CHEMTECH, August, p. 496.

11. Gouy, M., 1889, "J. Phys. Théor. Appl. II 8, 501.

12. Hohmann, E. C., 1970, Ph.D. Thesis, University of Southern California.

13. Keenan, J. H., 1932, "Mech. Engng", 54, 195.

14. Linnhoff, B. and Flower, J. R., 1978a, A.I.Ch.E. Jl. 24, 633.

15. Linnhoff, B. and Flower, J. R., 1978b, A.I.C.E. Jl.24, 642.

16. Linnhoff, B., 1979, Ph.D. thesis, University of Leeds.

17. Linnhoff, B., Mason, D. R. and Wardle, I., 1979, "Computers and Chemical Engng", 3, 295.

18. Linnhoff, B. and Turner, J. A., 1980, "The Chemical Engineer", December, 742.

19. Linnhoff, B., 1981a, "Exergy Analysis of Chemical Process Networks - a Short Course", Department of Chemical Engineering, UMIST.

20. Linnhoff, B. 1981b. "Design of Heat Exchanger Networks - a Short Course", Department of Chemical Engineering, UMIST.

21. Linhoff, B. and Carpenter, K. J., 1981, Paper presented at 2nd World Congress of Chemical Engineering, Montreal, Canada, 4-9 October.

22. Linnhoff, B. and Turner, J. A., 1981, "Chemical Engineering", November 2, 56.

23. Linnhoff, B., 1982, Paper presented at I.Chem.E. Jubilee Symposium, London, 5-7 April.

24. Linnhoff, B. and Hindmarsh, E., 1983, Chem. Engng. Sci. 38, No. 5.

25. Linnhoff, B., Townsend, D. W., Boland, D., Hewitt, G. F., Thomas, B. E. A., Guy, A. and Marsland, R. H., 1982, "Process Integration for the Efficient Use of Energy", I.Chem.E. User Guide.

26. Marsden, P. A., 1980, "ICI Catalyst Technical Paper", No. 8.

27. Nishida, N., Stephanopoulos, G. and Westerberg, A. W., 1981, A.I.Ch.E. Jl. 27, 321.

28. Ponton, J. W. and Donaldson, R. A. B., 1974, "Chem. Engng. Sci.", 29, 2375.

29. Rant, Z., 1956, "Forsch. Ing. Wes." 22, 36.

30. Riekert, L., 1974, "Chem. Engng. Sci." 29, 1613.

31. Robertson, J. C., 1974, "Chem. Engng.", 21 January, 104.

32. Townsend, D. W., 1980, "Chemical Engineer", October, 628.

33. Townsend, D. W. and Linnhoff, B., 1982, "Chemical Engineer", March, 91.

34. Umeda, T., Itoh, J. and Shiroko, K., 1978, "Chem. Engng. Prog.", July, 70.

RELEVANCE OF 2ND LAW ANALYSIS IN MECHANICAL ENGINEERING THERMODYNAMICS SYLLABUS

B. M. Burnside

Mechanical Engineering Department
Heriot-Watt University
Edinburgh

ABSTRACT

Second Law analysis is a valuable tool in locating and evaluating irreversibilities in thermal power plant. It is increasingly used in design of power and process plants. On this basis it is argued that mechanical engineering Applied Thermodynamics course should contain a sound background of second law methods with practical examples of applications including Thermoeconomic costing methods. An example of relative costing of cogenerated power and heat is given and references include a number of other applications. Design of heat exchangers for minimum irreversibility or for predetermined irreversibility and minimum heat transfer surface area is mentioned. It is suggested that students be introduced to these methods for the insight they give into not only the meaning of optimisation of the design of heat exchangers, but also into the effect of heat exchangers on the performance of power and process plant.

INTRODUCTION

It may seem odd to argue about the relevance of 2nd Law Analysis to mechanical engineers a century after Maxwell (1) referred to available energy and Gibbs (2) developed the concept. Early applications by Darrieus (3) and Keenan (4) to steam turbine performance evaluation and by Keenan (5, 6) to the vapour power cycle, absorption refrigeration, the Otto cycle and the combustion process are noteworthy for laying the foundations of the treatment of flow and non-flow processes. Since then the engineering literature has seen a great expansion of papers on availability analysis of process, power and extraction industrial processes particularly after the rapid increase in energy prices started in the 1970s. A recent thesis (7) lists 70 references on the subject and most current engineering thermodynamics texts at least refer to available energy analysis. The contributions of Obert and his colleagues in Wisconsin (8 - 18) illustrates application to heating and air conditioning systems (14, 15) to determining the true value of localised modifications to power plant (11, 12, 17) and to resolution of problems of pricing electricity and process heat from cogeneration plant (10, 13, 16, 18). These analyses and others of their type (21, 22) are thermo-economic in nature affected as much by market conditions and forecasts and by taxation and other government regulations as by thermodynamic factors. The importance of 2nd Law analysis in these evaluations is its inherent ability to grade energy and to express irreversibilities, both numerically. Moreover the concept of available energy is much closer to the layman's picture of "energy" - "the go of things" - than the thermodynamicists's (13). Since most major decisions on energy policy are approved ultimately by laymen this is a significant improvement in communication.

Recent applications of 2nd Law analysis to industrial processes (7, 21) have proved useful not only in identifying their rank order on an irreversibility scale but also in setting a theoretical standard process performance. These studies show for example that approximately 25% of all the available energy loss in process and allied industry in the United States is attributable to steam raising (7). To the untutored eye this may seem surprising because the conversion rate of chemical energy of the fuel into enthalpy flux of steam is relatively good. However, 2nd Law analysis correctly identifies the large combustion and heat transfer irreversibilities in the boilers. Boiler efficiency may be improved considerably by limiting air/fuel ratio and preheating combustion air by regeneration. 2nd Law analysis may be used to calculate the reduced loss of available energy due to raising steam pressure but it cautions that there will nevertheless be high irreversibility associated with the combustion. Typically 2nd Law analysis identifies and quantifies irreversibilities and sets the limits for improvement. Notwithstanding these advantages the author has the impression that 2nd Law analysis is not widely used in industry. For example, the Institution of Chemical

Engineers' User Guide on Process Integration for the Efficient Use of Energy (22) which is intimately concerned with reducing fuel and energy feedstocks does not even mention available energy. Closer examination of the guide reveals why. Intuitively the conditions for minimum irreversibility, apart from considerations of coupled pumping power or pressure loss, are being met. Heat transfer temperature differences are kept to a minimum and unnecessary (on a 1st Law basis) utility heat supply, which is then degraded to cooling utility temperature, is avoided. Thus minimum available energy loss is achieved.

A similar situation exists in power plant design. Summation over infinitesimally small Carnot engines and noting that heat transfer temperature differences must be minimised to reduce irreversibility indicates that the temperatures of heat addition and rejection must be as high and as low as possible respectively. Obviously dissipation of working fluid kinetic energy must be minimised. Again, although isentropic efficiency does not in itself make proper allowance for the increase in available energy of working fluid leaving a stage attributable to irreversibility therein (3) designers are well aware of this advantage as the reheat effect. The same can be said about most thermal equipment design. Designers rely on experience and an intuitive 2nd Law approach. However, among these designers who have a negative view of 2nd Law analysis some (23) admit to the benefits of having an available energy loss "checklist" for their plant to complement design experience.

Notwithstanding these remarks it seems that 2nd Law analysis is used increasingly in industry. Together with the insight it gives to inexperienced students of the true source and value of dissipations in thermal plant this convinces the author of the importance of including 2nd Law analysis in the undergraduate syllabus. There are signs that the suspicion it engenders among some practising engineers has its origin in an inadequate introduction to the subject. In his own courses the author has found students to be stimulated by available energy analysis to face up to lack of understanding of fundamental principles.

DEFINITIONS AND PRESENTATION

Formal presentation of 2nd Law analysis must follow the 2nd Law and its corrolaries. However, it is usual to make reference to grading or usefulness of energy simultaneously with introduction to the 1st Law. In this way the phrase, "equivalence of work and heat", is kept in perspective and the grading of heat according to temperature level introduced. During discussion of irreversible processes it is also common to illustrate the connection between irreversibility in real processes and the loss of ability to recover work - or available energy - on bringing a system into equilibrium with its environment. At this level 2nd Law analysis may be introduced in the form of tutorial examples on calculation of the loss of available energy in a variety of flow and non-flow processes by direct integration over Carnot cycles between system temperature and surroundings, T_o, followed by reversible isothermal pressure equilibration with the surroundings.

These exercises lead naturally to expressions for available energy in closed and open systems the formal derivation of which is well known (5, 24). The availability of a closed system at rest in surroundings at pressure p_o and temperature T_o, A_{cs}, (5) is given by

$$A_{cs} = (U - U_o) + p_o(V - V_o) - T_o(S - S_o) \tag{1}$$

where unsubscripted properties internal energy, U, volume, V, and entropy, S, refer to the initial state of the system and subscript "o" refers to the value of these properties at (p_o, T_o). The corresponding availability, A_{sf}, (5) of mass m kg at a section in a steady flow system is

$$A_{sf} = (H - H_o) - T_o(S - S_o) + K.E. + P.E. \tag{2}$$

Here K.E. and P.E. are kinetic and potential energies at the flow section and are reduced to zero in the equilibrium (dead state) at (p_o, T_o). H is enthalpy.

In order to carry out an availability balance for a process it is necessary to consider the availability of work and heat flow. Clearly the availability of work, A_w, is just useful work, W_u, itself

$$A_w = W_u \quad \text{or} \quad A_w = W_u \tag{3}$$

The availability, δA_q of a flow of heat, δq, depends on the temperatures of the heat source, T, and the surroundings,

T_o, $\delta A_q = (1-T_o/T)\,\delta q$. Thus integrating over the whole range of heat supply

$$A_q = \int (1 - (T_o/T))\,dq \qquad (4)$$

Another useful concept is irreversibility, I, occurring in a process (5). Thus, in any real process in which the state of a closed system changes from 1 to 2 the balance of availability is

[Destruction of availability] = [reduction in availability of system] -[loss of availability in work] + [availability of heat supplied] (5)

with the corresponding rate equation for a steady flow system. Combining equations (1-5) it is easy to show that

$$I_{cs} = T_o\,\Delta S_u \text{ and } \dot{I}_{sf} = T_o\,\Delta\dot{S}_u \qquad (6)$$

where ΔS_u and $\Delta\dot{S}_u$ refer to the entropy changes of the universe affected by the process.

At this stage students will find equations (1), (2) and (6) more convenient in calculating available energy losses than the more direct Carnot cycle integration methods referred to above.

Availability changes and chemical reactions

In the very general case of systems containing multicomponent, multiphase, non-ideal mixtures the calculation of thermodynamic properties, and hence availability functions, is a major problem usually considered outside the realm of mechanical engineering courses. The discussion here is limited therefore to pure substances, perfect gas and air-vapour mixtures only. Apart from this of course the same treatment applies to more general mixtures. The author finds it convenient to introduce students to partial molar properties of mixtures and to derive the expression for the chemical potential of a constituent of a perfect gas mixture as a special case. This not only leads to an elegant derivation of the law of mass action but also is essential to any subsequent discussion of 2nd Law Analysis of industrial processes involving real fluid mixtures.

In combustion processes the change in availability from reactants to products is calculated using equations (1) and (2) using standard property changes, ΔU^o, ΔH^o and ΔS^o at 25 ^{o}C (25) and in the case of entropy, all reactants and products at a pressure of 1 atm. Corrections for non-standard conditions are made assuming perfect gas behaviour. Some authors favour the use of tables of standard chemical availability, ΔA^o, (26), of a compound defined, for steady flow, by

$$\Delta A^o = \Delta G^o \text{ (formation)} - \Sigma_{focp} \Delta G^o \text{ (formation)}$$

where the standard state of each pure constituent is 25°C and 1 atm. and the summation Σ_{focp} is taken over the fully oxidised combustion products. The latter are taken as reference state in the phases $H_2O(\ell)$, $CO_2(g)$, $SO_2(g)$, etc. Gibbs free energies, G, and the availability functions, A_{sf}, are of course identical apart from K.E. and P.E. terms only when the standard and surroundings temperatures are both 25°C. A different set of ΔA^o tables is required for each surroundings state.

Students may now be introduced to a whole range of practical problems involving combustion processes:- the relative effects of combustion and heat transfer irreversibilities in boiler furnaces, the effect on these of reducing air: fuel ratio and of preheat, more ambitious "professional exercises" such as the loss of availability in coal gasification and the comparison of source and amount of irreversibility between gas turbine combined cycle/coal gasification and conventional coal fired plant are also stimulating. Other practical worked solutions are available in the literature (7, 9-18, 21, 26, 27).

Availability relative to unrestricted equilibrium with surroundings

The approach above has been called the "flexible envelope" (7) or the "restricted equilibrium", (28) analysis. The dead state of the constituents of the system was assumed to be the surroundings pressure and temperature (p_o, T_o) with the system isolated from the surroundings otherwise. If this barrier were removed then complete equilibrium with the surroundings could be arranged reversibly. An additional net amount of work or availability could then be recovered. For example, in a gas-fired boiler there is a contribution to the availability of both the gas fuel and the combustion products which exists from their conceivable reversible mixing with the atmosphere. This mixing process terminates when the constituent gases are at their respective natural levels in the atmosphere. In a very thorough treatment (28) Haywood shows that, relative to the mixture in unrestricted equilibrium at the dead state in the atmosphere, the availability of a mixture per mole, $\bar{a}_{sf}$, at a section in a steady flow system is given by

$$\bar{a}_{sf} = \bar{h} - T_o\bar{s} + K.E. + P.E. - \Sigma x_i \mu_{io} \qquad (7)$$

x_i is the mole fraction of constituent i, whose chemical potential is μ_{io} at its mole fraction in the environment. The summation is over all constituents of the mixture. The corresponding availability per mole in a closed system is

$$\bar{a}_{cs} = \bar{u} + p_o\bar{v} - T_o\bar{s} - \Sigma x_i \mu_{io} \qquad (8)$$

In many applications the difference between restricted and unrestricted equilibrium treatments is negligible. As an exercise students may prove that this is so for combustion. Haywood (28) treats the problem of calculating the availability, relative to unrestricted equilibrium with the environment, of compounds not normally present in that environment.

More specific definitions of availability

In the presentation above the author has used the general term "availability" to include kinetic and potential energy terms in either closed or open systems. This is preferred by some authors (5, 8, 9, 24) but others (28) exclude them defining the function, B, (5) given by

$$B = (H - H_o) - T_o(S - S_o)$$

as the steady flow availability. It has become common to call B the "exergy" but more specifically (28) it is the steady flow exergy referred to restricted or "flexible envelope" equilibrium. Correspondingly the no-flow exergy is identical to A_{cs}, equation (1). Referred to unrestricted equilibrium with the surroundings there are two other exergies. It has become common to refer to the unrestricted equilibrium datum steady flow exergy, equation (7), less the K.E. and P.E. terms as "essergy". Some authors (33) have developed 2nd law analysis by treating essergy as a general form from which other less general availabilities follow.

2nd Law Performance Parameters

Many parameters to measure the performance of processes in terms of irreversibilities have been suggested. Keenan (6) defined six "coefficients of performance" logical in measuring the degree of reversibility achieved in closed system cyclic processes. Hedman (7) lists nine 2nd Law efficiencies which have been used in chemical processes. These include a number which consider only irreversibilities which it is practical to eliminate. The terms "2nd Law efficiency", η_{II}, (7, 18) or "effectiveness", ε, (5, 9, 24) are used generally to describe the degree of reversibility achieved although the latter might better be called "2nd Law effectiveness" to avoid confusion with the better known parameter used in heat exchanger design. Space does not permit reproduction here of all 2nd Law efficiencies which have been used in practice. Among steady flow processes it is convenient to distinguish the following separate categories in defining η_{II}

(i) Power producers and absorbers without combustion
(ii) Power producers with combustion
(iii) Indirect heat exchangers including boilers
(iv) Chemical processes.

As an example case (ii) is now discussed.

Power producers with combustion. Two 2nd law efficiencies are useful in this case.

$\eta'_{II\,pc}$ = (Useful power)/(Maximum (rev) power) and

$\eta^{2}_{II\,pc}$ = (Useful power + flux products availability)/ Maximum (rev) power)

The availability balance, Figure 1 is

$$\dot{A}_{fuel} + \dot{A}_{oxidant} = \dot{A}_{w} + \dot{A}_{out} + \dot{A}_{qr} + \dot{I}$$

Thus

$$\eta'_{II\,pc} = \dot{A}_{w}/(\dot{A}_{fuel} + \dot{A}_{oxidant})$$

and

$$\eta^{2}_{II\,pc} = (\dot{A}_{w} + \dot{A}_{out})/(\dot{A}_{fuel} + \dot{A}_{oxidant})$$

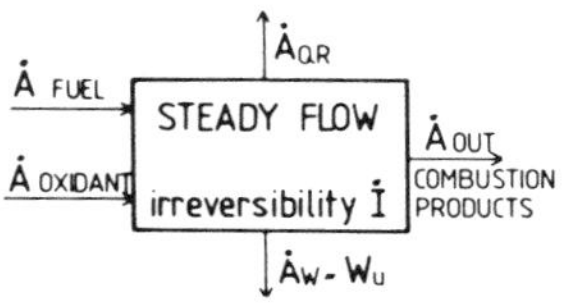

$\dot{A}_{QR}$
$\dot{A}$ FUEL
$\dot{A}$ OXIDANT
STEADY FLOW
irreversibility $\dot{I}$
$\dot{A}_{OUT}$
COMBUSTION PRODUCTS
$\dot{A}_{W} - W_{U}$

Figure 1

$\eta^2_{II\,pc}$ takes account of the available energy leaving in the combustion products while $\eta'_{II\,pc}$ does not. The distinction between these efficiencies is useful, for example, in describing the performance of electric utility gas turbines. $\eta'_{II\,pc}$ is more relevant for open cycle machines without exhaust heat recovery whereas in combined gas turbine/steam turbine cycles $\eta^2_{II\,pc}$ more correctly represents the gas turbine performance.

When an engine draws fuel and air and rejects combustion products to the surroundings, the denominator in the above efficiency expressions is just the negative of the Gibbs free energy of reaction, $-\Delta G$, at the surroundings temperature. Since $\Delta H \approx \Delta G$ for hydrocarbon fuels, $\eta'_{II\,pc}$ is very nearly equal to the thermal efficiency of the engine.

THERMOECONOMICS

Employers of graduates emphasise the importance of a grounding in costing methods. Appropriate and rewarding examples are to be found in costing energy streams in co-generation and chemical processing. (13, 17, 18, 19, 20, 26). Clearly costing must be based on the availability of the energy streams which is a true measure of their value. Costing on an energy basis regardless of its availability leads to obviously erroneous results. For example on this basis 1 kJ of energy is of equal value as electrical energy, high pressure or low pressure steam. Evans et al (19) developed availability accounting techniques applied to desalination. For each open system in the process, availability and "money balances" are applied to determine the unit cost of the main product. A balance of equipment costs and cost of irreversibilities may then be struck for the system. Because of the variability of economic conditions which may be applied, the solution of available energy and money balance equations is not unique. Students find the application to cogeneration of steam and electricity a helpful introduction to the subject.

Cogeneration of Steam and Electricity (13, (18), (26)

Gaggioli and Wepfer (18) studied the problem of attributing appropriate costs to cogenerated electricity and back-pressure steam. A simplified version of the problem is shown in Figure 2. The availability flows and losses/kg steam are included in the figure for the conditions of

Table 1. Costs quoted in the table are 1977 U.S. values converted to sterling. Having carried out an availability

Table 1: 10MW, 3.5 bar back pressure cogeneration

Boiler flowrate	21.3 kg/s
Turbine power	10 MW
Coal calorific value	28,200 kJ/kg
Coal cost	0.024 p/kg
Boiler cost	£5.6 m
Turbine cost	£0.39 m
Running hours/year	6000
Load factor	0.685
Amortisation factor	0.08

balance, Figure 2, a financial balance may be written for the boiler neglecting the cost of feedwater and assuming that the stack gases are valueless.

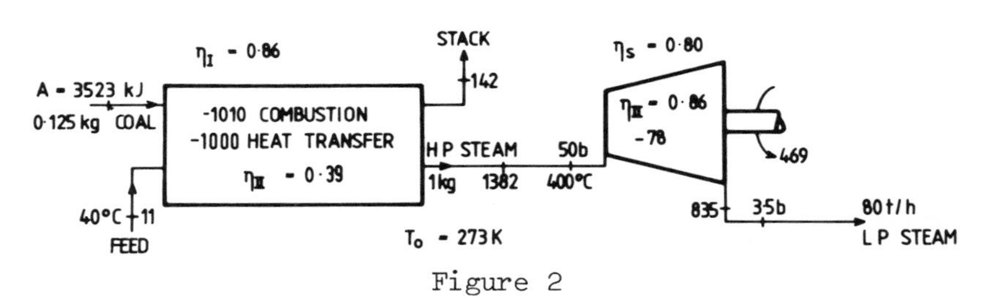

Figure 2

$$c_{hp} \times A_{hp} = c_f \times A_f + \text{boiler cost/kg steam}$$

where the unit costs, c, are referred to the respective availabilities, A of the streams/kg steam flowing. Thus the unit cost of producing HP steam, c_{hp}, is

$$c_{hp} = c_f/\eta_{II\,boiler} + (\text{boiler cost/kg steam})/A_{hp}$$

$$= 0.125 \times 0.024/(3523 \times 0.39) + 5.6 \times 10^6 \times 0.08/(21.3 \times 3600 \times 6000 \times 1382)$$

$$= £2.88 \times 10^{-6}/\text{kJ (HP steam availability)}$$

$$= 1.04 \text{ p/kWh HP steam availability}$$

The financial balance for the turbine is

$$c_{work} \times A_{work} + c_{\ell p} \times A_{\ell p} = c_{hp} \times A_{hp} + \text{turbine cost/kg steam}$$

In this equation there are two unknowns: the unit costs of turbine work, c_{work}, and LP steam, $c_{\ell p}$. Another financial constraint is necessary. In the "extraction" method of costing (18), the turbine work alone is charged for the capital cost of the turbine and the availability destruction in the turbine. Thus $c_{\ell p} = c_{hp}$ and

$$
\begin{aligned}
c_{work} &= c_{hp} \times (A_{hp}-A_{\ell p})/A_{work} + (\text{turbine cost/kg steam})/A_{work} \\
&= c_{hp}/\eta_{II turbine} + (\text{turbine cost/kg steam})/A_{work} \\
&= 2.88 \times 10^{-6}/0.86 + 0.39 \times 10^{6} \times 0.08/(469 \times 21.3 \\
&\quad \times 3600 \times 6000) \\
&= £3.49 \times 10^{-6}/\text{kJ(work)} \\
&= 1.26\text{p/kWh work}
\end{aligned}
$$

Of course if the LP steam were to be produced anyway a low pressure boiler, with lower η_{II}, would be necessary. In the "equality" method of costing this is recognised by charging the turbine capital equally to LP steam and work, $c_{\ell p} = c_{work}$. Then the cost of shaft work is given by

$$c_{work} = (c_{hp} \times A_{hp} + \text{turbine cost/kg steam})/(A_{work} + A_{\ell p})$$

Hence $c_{work} = c_{\ell p} = 1.10$ p/kWh of available energy. The cost of LP steam is substantially higher than by the extraction method of costing. Shaft work and LP steam may be costed in other ways too (18). In the "by-product work" method $c_{\ell p}$ is determined as if LP steam were produced by an LP boiler, with $\eta_{II\ell p} < \eta_{IIhp}$. This method results in much higher LP steam costs than the extraction or equality methods. Conversely, "by-product steam" costing leads to low steam cost and high work cost by charging work as if it were produced either from purchased electricity or by condensing turbine. Often the object of a study is to determine the additional costs of increasing power or process steam supply. Here "incremental costing" is used to charge the additional supply for the cost of producing it (16). These problems are examples of a range of exercises beneficial in linking thermodynamic and financial methods in design of thermal plant.

HEAT EXCHANGER PERFORMANCE OPTIMISATION

Optimising heat exchanger performances with respect to irreversibility and designing heat exchangers for a predetermined level of irreversibility are promising design tools (29, 30, 31). In the sense of 2nd Law analysis the object of heat exchanger design is to optimise "pumping" and "heat flow" irreversibilities to achieve minimum overall irreversibility. Heat exchanger surfaces with high heat flux: friction power expenditure - "high performance surfaces" (32) - are usually considered desirable. However, a surface with a lower "performance" may have lower irreversibility. These ideas and the associated design techniques comprise a valuable insight not only to the meaning of optimisation of heat exchangers but also to the role of the heat exchanger in power and process plants. For this reason the author feels that they should form part of the undergraduate heat transfer syllabus. As an example of presentation Bejan's design technique applied to gas-to-gas counterflow regeneration for prescribed irreversibility and minimum heat transfer surface area (31) is described below.

Gas-to-(same)-Gas Counterflow Regeneration Optimisation

For simplicity hot and cold gas capacity rates, C, mass flows, $\dot{m}$, inlet pressures, $p_1 = p_2$ and thermodynamic and transport properties are assumed equal. Hot and cold gas inlet temperatures are T_1 and T_2 respectively, Figure 3. The total rate of entropy increase in the exchanger, S, is

$$S = C[\ell n(T_3/T_1) + \ell n(T_4/T_2)] + (R/c_p)[\ell n(p_1/p_3) + \ell n(p_2/p_4)] \tag{9}$$

Defining the "entropy generation number", N_s, as (30)

$$N_s = S/C \tag{10}$$

and introducing the effectiveness, ε, into equation (9)

$$N_s = \ell n[1 + \varepsilon((T_2/T_1)-1)] + \ell n[1 - \varepsilon(1 - (T_1/T_2))] + (R/c_p)[\ell n(p_1/p_3) + \ell n(p_2/p_4)] \tag{11}$$

For the counterflow heat exchanger with unit capacity rate ratio, effectiveness and number of transfer units N_{tu} are related by (32)

$$\varepsilon = N_{tu}/(1 - N_{tu}) \text{ or } (1 - \varepsilon) = 1/(1 + N_{tu}) \tag{12}$$

In the limit of high ε and low $\Delta p/p$ equations (9, 10, 11, 12) lead approximately (31) to

$$N_s = [(T_2/T_1)^{\frac{1}{2}} - (T_1/T_2)^{\frac{1}{2}}]^2/N_{tu} + (R/c_p)[(\Delta p_1/p_1)+(\Delta p_2/p_2)] \quad (13)$$

The total irreversibility is a consequence of irreversible heat transfer, first term, and fluid friction, second term. It may be shown (31) that the entropy generation number/side, N_{si}, may be written

$$N_{si} = \tau/[(L_i/r_{hi})St_i] + (R/c_p)\ f_i(L_i/r_{hi})g_i^2 \quad (14)$$

where (L/r_h) is the ratio of flow passage length to hydraulic radius, f is its friction factor, g $(= G/2\rho p)^{\frac{1}{2}}$ where G is the mass velocity, St$(= h/\rho Vc_p)$, the Stanton number and $\tau = [(T_2/T_1)^{\frac{1}{2}} - (T_1/T_2)^{\frac{1}{2}}]^2$. It can be seen that N_{si} is a function of three independent flow parameters, Reynolds number, Re, flow length to hydraulic radius and mass velocity, g.

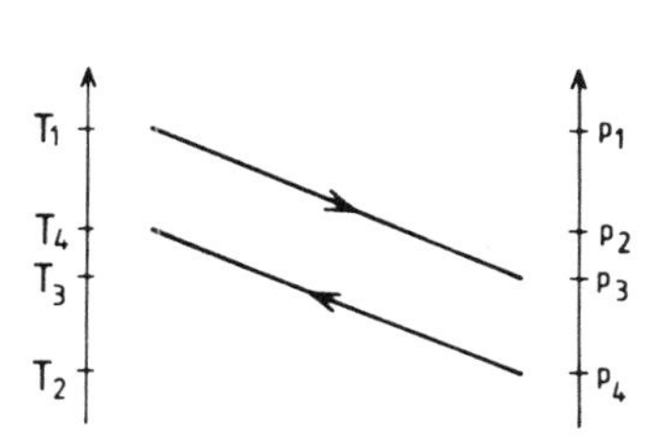

Figure 3

A value of (L/r_h) which minimises irreversibility can be deduced from equation (14), corresponding to an optimum balance of friction and heat flow irreversibilities. More interesting however, is design for fixed irreversibility, N_{si} and minimum heat transfer surface area. This can be shown to occur when $L_i/r_{hi} = 4(\tau/St)/3N_{si}$. The design method (31) makes use of standard experimental data on heat transfer in the form St $(Pr)^{2/3}$ and f as functions of Re. The more general design with unbalanced heat capacity rates is described by Bejan (31). In this case there is additional irreversibility due to the imbalance.

CONCLUSION

The lack of adequate introduction to and realistic practical examples of 2nd law analysis in many mechanical engineering thermodynamics courses has resulted in its sporadic use in industricl practice. The case for improving this situation is made in this paper. Some examples of the substantial body of knowledge and practical examples available in the literature have been described.

REFERENCES

1. Maxwell, J. C., "Theory of Heat", Longmans Green, London, 1871.

2. Gibbs, J. W., "Collected Works, Vol. 1", Yale University Press, 1948.

3. Darrieus, G., "The Rational Definition of Steam Turbine Efficiencies", Engineering, 130, pp.283-285, 1930.

4. Keenan, J. H., "A Steam Chart for Second Law Analysis", Mechanical Engineering, pp. 195-204, March, 1932.

5. Keenan, J. H., "Thermodynamics", Wiley, 1941.

6. Keenan, J. H., "Availability and Irreversibility in Thermodynamics", British Jl. Applied Physics, 2, pp. 183-192, 1951.

7. Hedman, B. H., "Application of the Second Law of Thermodynamics to Industrial Processes", Ph.D. thesis, Drexel University, June 1981.

8. Obert, E. F., "Concepts of Thermodynamics", McGraw Hill, 1963.

9. Obert, E. F., Gaggioli, R. A., "Thermodynamics", McGraw Hill, 1963.

10. Reistad, G. M., Gaggioli, R. A., Obert, E. F., "Available Energy and Economic Analyses of Total Energy Systems", Procs. 32nd Ann. Amer. Power Conf., 32, pp. 603-11, 1970.

11. Gaggioli, R. A., Yoon, J. J., Patulski, S. A., Latus, A. J., Obert, E. F., "Pinpointing the Real Inefficiencies in Power Plants and Energy Systems", Procs. Amer. Power Conf., 37, pp. 656-70, 1975.

12. Fehring, T. H., Gaggioli, R. A., "Economics of Feedwater Heater Replacement", Jl. Eng. for Power, 99, pp. 482-9, 1977.

13. Gaggioli, R. A., "Proper Evaluating and Pricing of Energy", Energy Use and Management, 2, pp. 31-43, 1977.

14. Gaggioli, R. A., Wepfer, W. J., Elkouh, A. F., "Available Energy Analysis for Heating, Ventilating and Air Conditioning. I - Inefficiencies in a Dual Duct System, II - Comparison of Recommended Improvements", ASME Symposium Vol. H00116 (1978 ASME winter meeting), pp. 1-30.

15. Wepfer, W. J., Gaggioli, R. A., Obert, E. F., "Proper Evaluation of Available Energy for Heating, Ventilating and Air Conditioning", Trans. ASHRAE, 85, pp. 214-30, 1979.

16. Gaggioli, R. A., Wepfer, W. J., Chen, H. H., "A Heat Recovery System for Process Steam Industries", Jl. Eng. for Power, 100, pp. 511-19, 1978.

17. Gaggioli, R. A., Fehring, R. H., "Economic Analysis of Boiler Feed Pump Drive Alternatives", Combustion, 49, pp. 35-39, 1978.

18. Gaggioli, R. A., Wepfer, W. J., "Available Energy Accounting a Cogenerative Case Study", A.I.Ch.E., 85th Annual Meeting, Philadelphia, 1978.

19. Evans, R. B., Crellin, G. L., Tribus, M., "Thermoeconomic Considerations of Sea Water Demineralisation", Chapter 2, Principles of Desalination, K. S. Spiegler (ed), Academic Press, 1966.

20. El-Sayed, Y. M. and Evans, R. B., "Thermoeconomics and the Design of Heat Systems", Jl. Eng. for Power, 92, pp. 27-35, 1970.

21. Smith, S. V., Sweeney, J. C., et al., "Thermodynamic Analysis of a Refinery Process", Vol. 1, Drexel University Report No. NSF-RA-N-75-105, June 1975.

22. I. Chem. E., "User Guide on Process Integration for the Efficient Use of Energy", I. Chem. E., 1982.

23. Maloney, D. P., Burton, J. R., "Using Second Law Analysis for Energy Studies in the Petrochemical Industry", Energy, 5, pp. 925-30, 1980.

24. Faires, V. M., Simmang, C. M., "Thermodynamics", Collier Macmillan, London, 1978.

25. Perry, R. H., Chilton, C. H., "Chemical Engineers' Handbook", McGraw-Hill, 1973.

26. Sussman, M. V., "Availability (Exergy) Analysis", Mulliken House, 1980.

27. Reistad, G. M., "Available Energy Accounting in the United States", Jl. Eng. for Power, 97, pp. 429-34, 1975.

28. Haywood, R. W., "Equilibrium Thermodynamics", John Wiley, 1980.

29. McClintoch, F. A., "The Design of Heat Exchangers for Minimum Irreversibility", ASME Paper, No. 51-A-108, 1951.

30. Bejan, A., "Second-law Analysis in Heat Transfer and Thermal Design", Advances in Heat Transfer, 15, pp. 1-58, Academic Press, 1982.

31. Bejan, A., "The Concept of Irreversibility in Heat Exchanger Design: Counterflow Heat Exchangers for Gas-to-gas Applications", Jl. Eng. for Power, 99, pp. 374-380, 1977.

32. Kays, W. M. and London, A. L., "Compact Heat Exchangers", McGraw-Hill, 1964.

33. Evans, R. B., "A Proof that Essergy is the Only Consistent Measure of Potential Work", Ph.D. thesis, Dartmouth College, 1969.

PUTTING THE SECOND LAW TO WORK

N. Hay

Department of Mechanical Engineering
University of Nottingham

The direct application of the Second Law to design has been very elusive. The potential that it could provide the key to some very elegant approach to design was always sensed but was not fully realised until comparatively recently. This elusive and powerful approach was identified as the optimisation of the design of thermal systems through minimising the rate of entropy generation, i.e. exergy loss. Methods of identifying and formulating the rate of entropy generation in terms of common design parameters such as heat exchanger tube diameter or solar collector operating temperature were developed leading to a rational economics basis for the optimisaition of a thermal system or piece of equipment. The approach is very powerful. Not only does it allow the optimum design to be identified but by comparing two non-optimum designs it is possible to choose between them the design nearer the optimum being the obvious better choice.

Some examples of the use of this 'Second Law Analysis' approach based on the rate of entropy generation for optimisation are demonstrated with particular reference to heat exchanger design. Other thermal systems are also mentioned. The approach follows that pioneered by A. Bejan (Ref. 1, 2), and is well suited for inclusion in the syllabuses for third year Thermodynamic courses.

We now outline how the concepts may be applied in a teaching context.

APPROACH

Optimisation of the design of thermal systems
through
Minimising the rate of entropy generation
to arrive at
The least irreversible design

Subsequent extension to economic analysis

METHOD

1. Derive an expression for the rate of entropy generation.

2. Introduce the pertinent design parameters into this expression.

3. Minimise the expression to yield relationships between the design parameters that will lead to minimum entropy generation, i.e. optimum design.

USEFUL EXTENSION

The optimum design may be used as a yardstick for assessing the relative merits of two non-optimum designs.

<u>Illustration</u>. Convective heat transfer volumetric entropy generation rate for a flowing viscous fluid with heat transfer (3-D).

$$\dot{S}'''_{gen} = \frac{k}{T^2}\left[\left(\frac{\partial T}{\partial x}\right)^2 + \left(\frac{\partial T}{\partial y}\right)^2\right]$$

$$+ \frac{\mu}{T}\left[2\left\{\left(\frac{\partial v_x}{\partial x}\right)^2 + \left(\frac{\partial v_y}{\partial y}\right)^2\right\} + \left\{\frac{\partial v_x}{\partial y} + \frac{\partial v_y}{\partial x}\right\}^2\right]. \quad (1)$$

1st term Entropy generation due to conduction.

2nd term Entropy generation due to fluid friction.

Thus

$$\dot{S}'''_{gen} = \dot{S}'''_{gen(heat\ transfer)} + \dot{S}'''_{gen(fluid\ friction)}.$$

Useful to define a parameter

$$\phi = \frac{\dot{S}'''_{gen(fluid\ friction)}}{\dot{S}'''_{gen(heat\ transfer)}} \tag{2}$$

If $\underline{v}(x, y)$ and $T(x, y)$ are known Equation (1) can yield the entropy generation distribution $\dot{S}'''_{gen}(x, y)$ throughout the flow, e.g. Poiseuille flow.

In practical engineering situations $v(x, y)$ and $T(x, y)$ cannot be determined, (turbulent flow, complex geometry). Bejan (1) got over this difficulty by reformulating the entropy generation in terms of the heat transfer and fluid friction at the <u>wall</u>,

i.e. in terms of St and f,

so that the extensive correlations available in the literature can be used in the approach.

For a flow passage a simple derivation yields

$$\dot{S}'_{gen} = \frac{\dot{q}'^2}{4T^2 c_p} \cdot \frac{D}{\dot{m}} \cdot \frac{1}{St} + \frac{2u^2}{T} \cdot \frac{\dot{m}}{D} f . \tag{3}$$

h.t. contribution f.f. contribution

Better h.t. → larger Stanton Number (St) → Smaller $\dot{S}'_{gen}$

Higher friction → larger f → larger $\dot{S}'_{gen}$

Since $Re \propto \frac{\dot{m}}{D}$, as Re increases $\dot{S}'_{gen\ ht}$ decreases while $\dot{S}'_{gen\ ff}$ increases. Hence there is a possibility for optimisation to minimise $\dot{S}'_{gen}$.

For a round pipe, putting $\partial S'_{gen}/\partial Re = 0$ after substituting the usual expressions for St and f, we get

$$Re_{opt} = 2.023\,(Pr)^{-0.071}\,(Bo)^{0.358}, \qquad (4)$$

where Bo is a non-dimensional duty parameter

$$Bo = \frac{\rho\,\dot{m}\dot{q}'}{\mu^{5/2}(kT)^{\frac{1}{2}}} \quad . \qquad (5)$$

Thus knowing the duty $\dot{m}$ and $\dot{q}'$ the tube diameter for minimum loss of energy can be found.

For any operating condition $\dot{S}'_{gen}$ can be related to the optimum condition

$$\frac{\dot{S}'_{gen}}{\dot{S}'_{gen\,min}} = f^{n}\left(\frac{Re}{Re_{opt}}\right) . \qquad (6)$$

Equation 6 is plotted in Fig. 1.

Increasing Re can result in an improvement or deterioration of the thermodynamic performance of the heat exchanger tube depending on whether Re > or < Re_{opt}.

Example. Data for a condenser are given. The tubes are to be 15 mm dia., 3 m long. Is it exergetically better to use a single pass or a double pass arrangement?

Data yield: Single pass 228 tubes 3 m long.
Double pass equivalent to 114 tubes 6 m. long

	Single Pass (SP)	Double Pass (DP)
Bo	6.21×10^{12}	12.4×10^{12}
Re_{opt}	5302	6793
Re (actual)	17610	35220
Re/Re_{opt}	3.32	5.18

Thus a single pass is better. However in order to get closer to the optimum Re should be reduced, hence a larger diameter pipe should be tried (see Fig. 1).

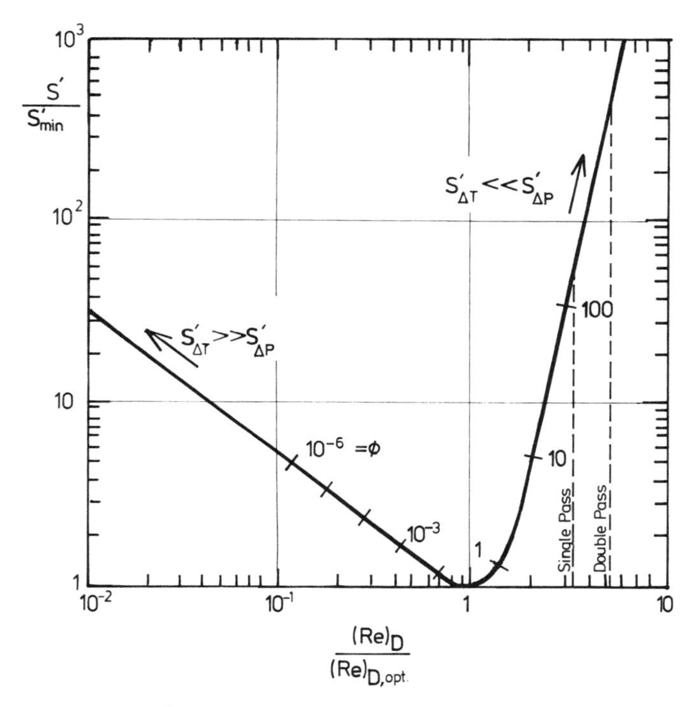

Figure 1. Relative entropy generation rate in a smooth tube.

EXTENSION TO ECONOMIC ANALYSIS

Irreversibility minimization alone is not a sufficient basis for preferring one design to another. One can introduce different costing for $S'_{\Delta T}$ and $S'_{\Delta p}$ so that the total cost associated with process irreversibility is

$$C_{S'} = C_{\Delta p} \; S'_{\Delta p} + C_{\Delta T} \; S'_{\Delta T}. \tag{7}$$

As equation (3) gives separately $S'_{\Delta T}$ and $S'_{\Delta p}$, costing can be done easily using equation (7).

Usually $S'_{\Delta p}$ reflects in electrical power consumption and is usually costed higher than $S'_{\Delta T}$ which consumes exergy which is not always as easily retrievable further on in the process.

ACKNOWLEDGEMENT

Figure 1 has been reproduced from Ref. 1 by kind permission of the publishers.

APPENDIX A
FURTHER APPLICATIONS

Optimisation of the design of:

- Solar collectors,
- Thermal insulation systems,
- Cryogenic supports,
- Heat transfer augmentation techniques,
- Heat exchangers,
- Thermal energy storage,
- Etc.

APPENDIX B
TEACHING OF THE APPROACH

Inclusion in final year thermo course is recommended. Minimum 6 Lectures to cover:

Exergy	(2)
Entropy generation	(1)
Applications to	
Heat exchanger tube	(1)
Solar collectors	(1)
Thermal insulation systems	(1)

Could be easily extended to 10 or 20 lectures if desired.

REFERENCES

1. Bejan, A., "Entropy Generation Through Heat and Fluid Flow", John Wiley, 1982.

2. Bejan, A., "Second-law Analysis in Heat Transfer and Thermal Design", Advances in Heat Transfer, Vol. 15, 1982, pp. 1-58.

PROCESS FEASIBILITY AND THERMODYNAMIC TEACHING

J. C. Mecklenburgh

Department of Chemical Engineering
University of Nottingham

SYNOPSIS

The teaching of thermodynamics to chemical engineers at Nottingham exploits the fact that process feasibility studies are largely based on thermodynamic considerations and not on kinetics. The amount of theory is kept to the minimum by linking the teaching and examples to the common process operations. The students are introduced to energy integration techniques while throughout emphasis is placed on the retrieval and use of physical data. A small first year project involving mainly thermodynamics introduces students to the idea that process design involves system design.

INTRODUCTION

It is generally accepted that thermodynamics is a difficult subject to learn and that there is more material available than time allows to be presented. The latter is especially true for chemical engineers who have to be familiar with both mechanical and chemical aspects of thermodynamics.

In order to decide what to teach and how to teach it the thermodynamics courses for chemical engineers at Nottingham are becoming increasingly based on the thermodynamics used in feasibility studies of a process plant.

FEASIBILITY

Those laws of thermodynamics which have the greatest influence on process engineering are the 1st and 2nd laws plus the associated law of the conservation of mass. The consequences of the second law place a finite limit on the amount of change achievable in the various process units.

For a process to be feasible it must, inter alia

a) Operate at or below the thermodynamic limitation.

b) Proceed at a reasonable rate so that the size of equipment needed for a given production rate, is economical.

An operation usually approaches its thermodynamic limit asymptotically with time. A very fast process such as boiling and condensation and some reactions, runs near this limit and the equipment size is determined by some considerations such as maintenance access. Most operations however proceed at a finite rate and are sized to meet a specified change (which must be below that thermoydnamically possible). Whatever the change chosen, the first law and the mass conservation law will apply.

DESIGN AND TEACHING STAGES

Most industrial projects have two stages for sanction. To obtain the first stage or design sanction the Board must be satisfied the project is feasible. They then approve expenditure for detailed design in the preparation for the second stage or construction sanction.

For the first sanction the design of the individual items is only as important as their contribution to the total capital cost. With a large number of items individual errors can statistically cancel out and the sizes and therefore the kinetics do not have to be known accurately. On the other hand the running costs such as those for chemicals and energy depend on mass and heat flows which are found largely by thermodynamic calculation. So design sanction needs accurate thermodynamics and rough kinetics.

The usual way of carrying out feasibility design is to first solve the thermodynamics and then 'adjust' the design by applying an empirical factor such as turbine or compressor efficiency, plate efficiency, temperature approach to reaction equilibrium, minimum temperature difference between heat exchange phases, discharge factors, etc. Theoretically many of these factors cannot be justified but practice has proved their worth.

For the detailed design of the items for construction purposes the kinetic models and the kinetic parameters such as heat and mass transfer coefficients and reaction rate constants have to be known accurately.

To understand the kinetics, particularly mass transfer and chemical reaction, the student must appreciate the meaning of thermodynamic equilibrium. Consequently the formal teaching of thermodynamics is done in the earlier years of a chemical engineering course and kinetics, especially the more complex aspects, in the later years. Thus it has been considered convenient in the Chemical Engineering Department at Nottingham to associate thermodynamics teaching with feasibility design and kinetics with detailed design.

THERMODYNAMIC FUNCTIONS IN PROCESS ENGINEERING

Usually gases are processed in continuous flow equipment and not in batch operations and so enthalpy rather than internal energy is applicable for energy balances. Liquids and solids are processed either continuously or batchwise but as there is no significant numerical difference between the two quantities, enthalpy can be used for all first law balances irrespective of phase.

The appropriate functions to define thermodynamic equilibrium are temperature difference for heat transfer, entropy for compression and Gibbs free energy for mass transfer and reaction. Helmholtz free energy is not applicable for the same reason as internal energy i.e. few batch gas operations. In practice it is found that the ratio of Gibbs free energy to absolute temperature is used more often (in Gibbs-Helmholtz type equations) than Gibbs free energy itself. Entropy is also useful as an algebraic quantity in the integration of the Gibbs-Helmholtz equation with variable enthalpy.

Only two Maxwell relationships are needed. The effect of pressure on entropy is found from the variation of Gibbs free energy and the effect of pressure on enthalpy is derived from the change of Gibbs free energy over temperature.

For mixtures, the partial molar quantities used are for volume, enthalpy and free energy (chemical potential). Chemical potential is used for deriving equilibrium relationships for mass transfer and reaction. Associated with these relationships are fugacity, pressure, activity and activity coefficient and equilibrium constants.

The numerical methods for energy integration[1] have proved more successful in both student understanding and industrial allocation than the functions called availability and energy.

APPROACH: THEORY v. APPLICATION

To be accomplished in engineering thermodynamics it is necessary to be able to

i) Use the theory to derive equations.
ii) Apply the equations.
iii) Find the relevant data.

Most students understand the ideas of energy balances and heat flowing down a temperature gradient. Most though find thermodynamics and particularly chemical thermodynamics difficult because the concepts of partial differentiation, entropy, free energy, chemical potential etc. are new to them and are difficult to visualise. The algebra can also become lengthy. However on talking to graduates in industry it seems that after applying thermodynamics to actual design problems the subject becomes a lot less daunting. Consequently there appears to be an advantage in emphasising applications over theory in teaching.

The primary purpose of teaching the theory is so that the students can apply the correct equations and methods from the ones they have met to a particular 'standard' problem. A second reason is so that they can derive new equations for an unusual situation. The latter requires a much greater understanding of thermodynamic theory. Most graduates unless they go into research are not going to meet an unusual situation. So it is best not to confuse them with a lot of theory to the detriment of their ability in applications but to keep to the minimum theory needed for understanding. The academic inclined will be able to extend their own knowledge of theory from the text books[2]. The others can usually find advice from within their own industrial organisations or failing that from the universities and colleges.

The theory needed includes first law balances, stoichiometry, functions of state, Carnot cycle approach to 2nd law,

Gibbs-Helmholtz, chemical equilibrium, Clausius-Clapeyron, non-ideal gases, corresponding states, Gibbs-Duhem, phase rule, cell emf's, variations of chemical potential, activity and solubilities. Many students do not find even this minimum list easy.

GASES

The two kinds of expansion usually met in a chemical plant are either 'isentropic' as in compressors and turbines or 'isenthalpic' as through reducing valves. Constant temperature, pressure and volume expansions are not usually found on a plant though isothermal operation represents the 'ideal' compressor.

Internal combustion engines are not important in the process plant except for waste gas turbines. When heat from a reaction is to be converted into work it is done indirectly via steam raising. Direct use within an engine would make the reactor too complicated.

For many process engineering calculations the assumption of perfect gas behaviour is sufficiently accurate. This is especially true if a 'turbine efficiency' or 'approach to reaction equilibrium' is to be subsequently applied. The possible cases when non-ideal behaviour is needed are for some energy balances at high pressure. Students are taught though to use available thermodynamic tables and charts in preference to the perfect gas laws as this usually leads to easier calculation. When non-ideal behaviour is indicated, and thermodynamic tables are unavailable then the student is referred to the universal compressibility, fugacity, enthalpy, etc. charts given in such references as Perry[3] and Reid and Sherwood[4]. The students are encouraged to use these charts to check the non-ideality of a high pressure system.

For computer calculations involving non-ideal gases the Redlich-Kwong equation of state is recommended to the student with a note about more complex relations discussed by Reid and Sherwood. Non-ideal gas mixtures are treated using the simple Kay's Rule for average critical properties with the student referred to Reid and Sherwood for more complex rules. For manual calculations component fugacities are found from either the Lewis-Randall rule or Joffe's equation which makes use of the compressibility charts etc. For computer work the student is again referred to Reid and Sherwood. The algebra of non-ideal gas thermodynamics is tedious and so is difficult to teach, especially as it has only limited value in improving design accuracy.

VAPOUR LIQUID EQUILIBRIUM

The ideal behaviour is Raoult's law but many systems deviate from this. The Van Laar equation is a convenient and simple correlation of activity coefficients suitable for teaching. As before, more complex equations are given in Reid and Sherwood. The Van Laar equation can successfully deal with azeotropes both homogeneous and heterogeneous and so can give estimates of the likelihood of immiscibility and the mutual solubilities. For multicomponent equilibrium the idea of excess free energy is introduced in conjunction with the two term Van Laar (or Margules) equation which is simpler to apply than the conventional three term Van Laar.

The student is taught to test experimental data using the Gibbs-Duhem relationship. However the important point to put across to the students is that they must spot azeotropes as their existence or otherwise can affect very greatly the separation system chosen for the plant. Subject to this, the vapour-liquid data does not have to be too accurate in order to design a distillation plant that will work satisfactorily.

Solubility data for the common gases in water is readily available (e.g. Perry). The student needs to know how to use this data for his particular operating pressure and temperature. Henry's coefficients are the usual vehicle for this for the 'permanent' gases. For gases such as ammonia and hydrogen chloride, enthalpy composition diagrams are used. (Perry).

SOLID-LIQUID EQUILIBRIUM

As with gases, aqueous solubilities are easily found (Perry) and it is a matter of changing them to appropriate operating conditions in order for example to calculate boiling and freezing point changes and the effects on cell emf's. The vehicles are activities. However there is the problem of notation to confuse the students because there are several different ways of expressing concentration. This means there are several definitions of activity and of standard chemical potential. The introduction of various superfixes to distinguish the last (stars, squares, daggers, double stars etc.) amuses but does not edify the students.

For the mineral alkalis and acids with large heats of solution, enthalpy-composition diagrams are available (Perry).

SOURCES OF DATA

If the emphasis is on application of the equations then problems are more realistic if they are set in terms of real chemicals. This means having actual data. Sources ready to hand are Perry[3], Reid and Sherwood[4], the Rubber Handbook[5], and Rogers and Mayhew[6] and the students are required to familiarise themselves with these publications.

There are always deficiencies and it is necessary to be able to estimate data. The standard text for this is Reid and Sherwood and there are also methods in Perry. Usually estimation methods are empirical and their education value is limited to critical properties and the law of corresponding states. Consequently the students are given some simple convenient methods and told to consult Reid and Sherwood for more complex but accurate methods. They are given group contribution schemes for critical properties, 'heats' and free energies of formation, liquid specific heats and gas specific heat coefficients. They are provided with equations for latent heat (Riedel, Watson), gas specific heat capacities (cubic equations), vapour pressure (Antoine) and activity coefficients (Van Laar) and a table for liquid density (Lydersen et al.). As already said they have for non-ideal gas behaviour either generalised charts or the Redlich-Kwong equation. The students are taught to consider how the accuracy of their data and models affects the accuracy of their design. This is done by introducing them to simple statistical and sensitivity analysis.

CURRICULUM - 1ST YEAR

The teaching of thermodynamics at Nottingham not only includes courses under that title. For example a first year course on Chemical Process Principles contains an important section on mass and energy balances. This shows the students that chemical engineering is very much concerned with systems in which the two conservation laws play leading parts. It is convenient in this course to introduce the enthalpy composition diagrams for the mineral alkalis and acids and use of tie lines and the Lever Rule.

In the first year course entitled Thermodynamics, the first term is spent mostly on the first law with the students being taught to look up or estimate and use densities, heats of formation, latent and specific heat capacities. In the second term, the second law is introduced and the teaching is concerned with expansion in general and turbines, compressors and refrigerators in particular. Towards the end of this course chemical equilibrium with ideal gases and vapour-liquid equilibrium with Raoult's law are introduced.

An important feature of the second term is a mini-project based on a chemical process. It is primarily an exercise on mass and energy balances but the students are given simple sizing and costing algorithms and data so that they can finish the project with a simple layout and economic assessment. This gives chemical engineering realism to the exercise and shows the students the importance of mass and energy balances in feasibility studies. Each student has his own set of conditions so that they can help each other but cannot copy.

There is a large thermodynamic element in the exercise not only in the energy balances. For many reactions and separations the amount of change is given. But for turbines and compressors and for those operations where Raoult's law can be applied or enthalpy-composition diagrams can be used (e.g. quenchers, scrubbers, and two phase reactors) the students are asked to calculate the amount of change.

CURRICULUM - 2ND YEAR

One of the main courses is Unit Operations which covers the design of distillation columns, solvent extractors and evaporators. The methods (e.g. McCabe-Thiele) are based on the idea of equilibrium stages and are therefore thermodynamic in nature. About half the entitled thermodynamics course services the above in teaching the students how to find and then arrange equilibrium data between phases. This means considering non-ideal mixtures and introducing chemical potential. This is a partial molar quantity and the idea of this can be introduced via partial molar volumes and stocktaking of process plant inventories.

A further quarter of the thermodynamics course is devoted to chemical equilibrium particularly at high pressure and to electrolysis. The final quarter is concerned with process synthesis in general and energy integration in particular.

After struggling with the difficult and unfamiliar terms of thermodynamics, students find the idea of maintaining positive temperature differences straightforward. The topics covered include determining minimum energy requirement, removing heat flows from across the pinch, use of heat pumps and matching of services. The heat exchanger matching part of the process is only illustrated by simple circuits (3-4 streams only) as flowsheeting large exchanger systems is an acquired skill.

To give a realistic example on energy integration, over half the examples set throughout the year are concerned with the same plant which consists of an endothermic and an exothermic reactor, both under pressure, and a distillation column. This plant provides examples on data collection, reaction equilibrium, condensation with inerts, distillation, compression, steam raising, turbines and refrigeration. The answers from these sheets provide the data for the examples on energy integration. [The other examples not concerned with the plant cover errors in data, mixing effects, heterogeneous azeotropes, immiscible solutions, salt solutions and electrolysis].

Energy integration emphasises the importance of thermodynamics in feasibility studies and educationally as well as commercially is a welcomed development.

CURRICULUM - 3RD AND 4TH YEARS

Thermodynamics does not formally appear on the timetable in the final years, though is used in the courses on reactors and multicomponent distillation. In a fourth year course on Process Design, energy integration is again covered as it is felt that a second go when the students are more mature will be beneficial.

The design project at Nottingham is set and partially supervised by industry. It covers two terms of the fourth year and is tackled by groups of 3 or 4 students. After the first term they are required to produce a first report covering the selection of the process route, mass and energy balances and preliminary cost estimates. This can involve in some projects considerable thermodynamics. The groups then divide for the individuals to design the various items and the final report with the detailed design, operation and safety instructions and more accurate costings are handed in at the end of the second term. This final year project thus follows closely the sequence of design and then construction sanction.

CONCLUSION

Thermodynamics teaching can benefit from the recognition that process engineering projects usually have first design and then construction sanction stages and that thermodynamics is associated with the former and kinetics with the latter. Courses need only contain the theory and thermodynamic functions associated with common plant operations. Emphasis can be placed on the successful application of thermodynamic results to plant design and the ability to find, estimate and use data can be developed. In this way the subject may become less formidable to most students. Furthermore mini-projects can be set in the earlier years of the course which can introduce the idea of designing chemical plant as a system without the students first having to be taught extensive kinetics.

REFERENCES

1. I.Chem.E. Working Party, "User Guide on Process Integration for the Efficient Use of Energy", I.Chem.E., 1982.

2. Denbigh, K. G., "Principles of Chemical Equilibrium", Cambridge, 3rd ed., 1971.

3. Perry, R. H., and Chilton, C. H., (eds.), "Chemical Engineers Handbook", McGraw-Hill, 5th ed., 1973, Section 3.

4. Reid, R. C., Prausnitz, J. M. and Sherwood, T. K., "The Properties of Gases and Liquids", McGraw-Hill, 3rd ed., 1977.

5. CRC, "Handbook of Chemistry and Physics", CRC Press, annual.

6. Rogers, G. F. C., and Mayhew, Y. R., "Thermodynamic and Transport Properties of Fluids", Oxford, 3rd ed., 1980.

TEACHING THE EXERGY METHOD TO ENGINEERS

T. J. Kotas

Department of Mechanical Engineering
Queen Mary College
University of London

SYNOPSIS

The shortcomings of the traditional methods of thermodynamic analysis are pointed out. A brief outline of an alternative, new method of thermodynamic analysis known as the Exergy Method is presented. Recommendations are put forward for integrating the material of the Exergy Method into a three year course of Engineering Thermodynamics. Advantages of the Exergy Method are listed.

INTRODUCTION

One of the main areas of application of Engineering Thermodynamics is the study of effectiveness of thermodynamic processes. The traditional techniques which are used for this purpose are of two types:

(i) application of the energy balance to the system under consideration, usually, with a view to determining 'unaccounted for' heat transfer between the system and the environment, and

(ii) calculation of a criterion of performance relevant to the system under consideration.

The energy balance treats all forms of energy as equivalent i.e. without differentiating between the different grades of energy crossing the boundary of the system. Thus, e.g. heat transfer to the environment ('heat loss') from a pipe carrying high temperature steam will be treated in the same way as low grade thermal energy rejected in the condenser of a steam plant.

The results of application of the energy balance to a cryogenic system can be very baffling, particularly to a student, since in this area of application the loss of thermal energy ('production of cold') is a desirable effect whilst the gain of thermal energy as a result of, say, heat transfer from the environment is to be avoided.

Further, the energy balance provides, in general, no information about internal losses. Should we apply the energy balance to an adiabatic system such as a throttling valve, a heat exchanger or a combustion chamber, we could be led to believe that these processes are free from losses of any kind.

The calculation of a criterion of performance appropriate to the process under consideration can be of help in the assessment of its degree of thermodynamic perfection. However, this is not always the case. For example, there are no traditional criteria by which to assess the performance or the thermodynamic losses of processes occurring in a throttling valve, steam ejector, an open type (mixing) feed heater or an adiabatic combustion chamber. In the case of other processes the criteria of performance may be available but the information which they provide regarding the performance of the system may be found incomplete or inadequate. For example, the heat exchanger effectiveness provides no indication of the effect of pressure losses on the performance of the heat exchanger. The overall thermal efficiency of a CHP plant, in which the heating, $\dot{Q}$, and power, $\dot{W}_X$, outputs are simply added to yield the numerator i.e.

$$\eta_{TH} = \frac{\dot{Q} + \dot{W}_X}{(\text{Rate of energy input})} \qquad (1)$$

gives in fact no indication of the performance of the plant but mainly an assessment of the boiler flue losses of the plant.

As shown above, the traditional methods of process analysis are so ineffective because they are based, in the main, on the First Law of Thermodynamics. The Second Law is brought in into the formulation of only some of the criteria of performance, e.g. isentropic efficiencies.

However, as is well known, it is the Second Law which governs the limits of convertibility between different forms of energy and what follows from it, also the relative grades or quality of the different forms of energy. Hence, as will be seen, it is the failure to take into account the changes in the quality of energy occurring during a particular process, which makes the traditional methods of thermodynamic analysis so unsatisfactory.

THE EXERGY METHOD AND THE CONCEPT OF EXERGY

The exergy method provides an alternative to the traditional methods of thermodynamic analysis. The basic concept of this method, as its name implies is exergy which may be loosely defined as a universal measure of the work potential of the different forms of energy in relation to a given environment. Alternatively, exergy is a measure, expressed in terms of work, of the quality, or the capacity to cause change, of a given form of energy.

As most of the important industrial systems are of the steady flow type, we shall confine this discussion to the control region method of analysis. Accordingly, the energy transfers for which we have to determine the corresponding exergy transfers are

(a) Work

(b) Heat transfer

(c) Energy associated with streams of matter entering or leaving the control region.

As follows from the above definition of exergy, the exergy flux rate $\dot{E}^W$ associated with a given work transfer rate (power) $\dot{W}_X$ is

$$\dot{E}^W \equiv \dot{W}_X \tag{2}$$

The exergy associated with a given heat transfer rate called thermal exergy flux $\dot{E}^Q$ is determined by using the concept of a reversible heat engine as shown in Fig. 1.

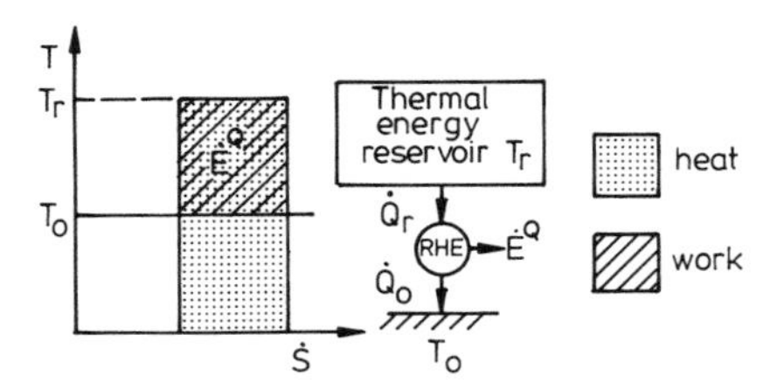

Fig. 1 Thermal exergy flux

Hence we get

$$\dot{E}^Q = \dot{Q}_r \; \tau \tag{3}$$

where

$$\tau = \frac{T_r - T_o}{T_r} \tag{4}$$

As will be seen from (4) τ is a form of Carnot efficiency in which T_o is the temperature of the environment. τ is called exergetic temperature.

To determine the exergy associated with a stream of matter E^M we must resort to the concept of idealised devices such as reversible expanders and compressors, semi-permeable membranes etc. in order to reduce reversibly the stream from its original state to a state of equilibrium with the environment, as shown in Fig. 2.

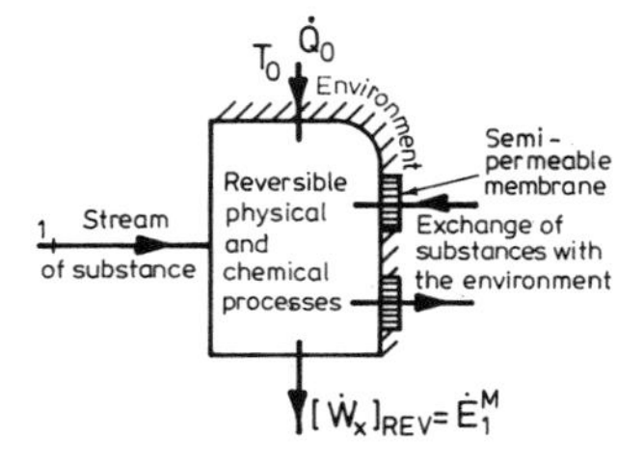

Fig. 2. Determination of the exergy flux of a stream of matter.

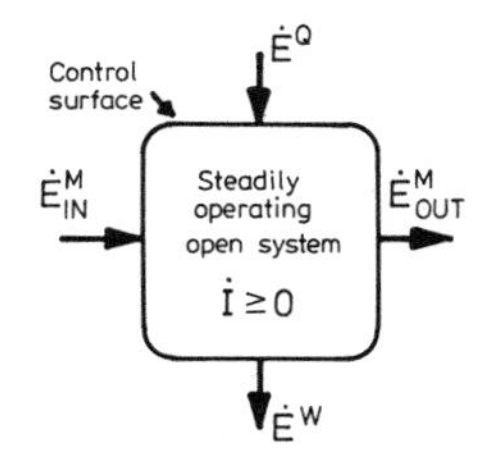

Fig. 3. Energy balance for an open system.

EXERGY BALANCE AND IRREVERSIBILITY RATE

Fig. 3 shows the exergy streams $\dot{E}^W$, $\dot{E}^Q$, $\dot{E}^M_{in}$ and $\dot{E}^M_{out}$ crossing the control surface. If the process taking place within the control region is reversible, the total exergy inflow will be equal to the total exergy outflow, i.e. there will be no loss of work potential. However, if the process is a real one, i.e. irreversible, some of the inflow exergy will be lost and hence the outflow of exergy will be smaller than the inflow by this amount. Thus, with reference to Fig. 3 we can write

$$\underbrace{(\dot{E}^M_{in} + \dot{E}^Q)}_{\text{Exergy inflow}} - \underbrace{(\dot{E}^W + \dot{E}^M_{out})}_{\text{Exergy outflow}} = \dot{I} \qquad (5)$$

where $\dot{I}$ is the irreversibility rate, or the rate of loss of exergy, of the process. Expression (5) which is called the exergy balance can also be obtained by combining the steady flow energy equation with the expression for the entropy production rate. The exergy balance may be said to express the law of degradation of energy in real processes.

The irreversibility rate is a measure of the thermodynamic imperfection of a process. It can be calculated for any plant component, a group of components or the whole plant depending on how the control surface is drawn. Since $\dot{I}$ obeys the additive law we can write (see Fig. 4)

$$\dot{I}_T = \sum_n \dot{I}_i \qquad (6)$$

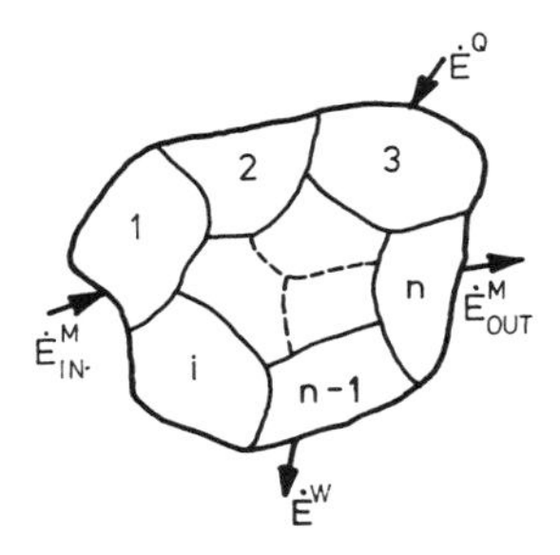

Fig. 4 A multi-component open system

As follows from expression (6) all components contribute, to a different degree, to the total plant irreversibility. Those components which make the largest contribution can be easily identified.

As an alternative to the calculation of $\dot{I}$ from the exergy balance, we can use the Gouy-Stodola relation

$$\dot{I} = T_o \dot{\Pi} \tag{7}$$

where

T_o = environmental temperature

$\dot{\Pi}$ = entropy production rate in the control region.

When considering an improvement in the performance of a plant component, it is important to take into account the constraints imposed by physical and economic factors on the reduction in $\dot{I}$ which is possible. If the minimum value of $\dot{I}$ imposed by such constraints is denoted as $\dot{I}_{UNAVOIDABLE}$ [1, 2] we can write

$$\dot{I}_{AVOIDABLE} = \dot{I} - \dot{I}_{UNAVOIDABLE} \tag{8}$$

RATIONAL EFFICIENCY AND THE COEFFICIENT OF STRUCTURAL BONDS

Rational efficiency ψ is a criterion of performance based on the concept of exergy. It can be calculated for any plant or plant component for which a useful output can be expressed in terms of exergy. Rational efficiency is defined for a control region by the following expression [3]

$$\psi = \frac{\text{Exergy output rate}}{\text{Exergy input rate}} \tag{9}$$

With the aid of the exergy balance the following alternative expression for rational efficiency can be obtained

$$\psi = 1 - \frac{\dot{I}}{\text{Exergy input rate}} \tag{10}$$

The total range of values of the rational efficiency lies within the following limits

$$0 < \psi < 1 \tag{11}$$

Hence, as will be seen, the closer the value of ψ is to unity, the closer is the plant to reversible operation.

The coefficient of structural bonds, CSB, is an exergy related concept which is defined as follows

$$\sigma_{k,i} = \left(\frac{\partial \dot{I}_T}{\partial \dot{I}_k}\right)_{x_i = var} \tag{12}$$

where

$\dot{I}_T$ = irreversibility rate of the system under consideration

$\dot{I}_k$ = irreversibility of the k-th component of the system

x_i = parameter of the system which produces the changes.

The CSB is found useful in the study of the bonds which exist between plant components and the plant as a whole. It also finds application in the thermoeconomic optimization of plant components. The CSB may be said to illustrate the principle of non-equivalence of exergy and exergy losses within a plant.

TOOLS AND AIDS FOR EXERGY ANALYSIS

The Grassmann diagram is a pictorial representation of exergy flows and losses. It can be very useful in representing, often complex, forms of interaction between plant components. It can also be applied to single components to show the relative values of exergy fluxes and different forms of irreversibility. As an example, the Grassmann diagram for a heat pump is shown next to the traditional Sankey diagram in Fig. 5. As can be seen, far more information about the plant performance is conveyed by the former than the latter type of diagram.

Exergy-enthalpy (ε-h) diagram can be very convenient for representing thermodynamic processes as well as thermodynamic cycles. When these are plotted on ε-h charts of particular thermodynamic fluids, changes in exergy and enthalpy can be read off directly. Fig. 6 shows [4] a simple vapour-compression refrigeration cycle plotted on an exergy-enthalpy chart for ammonia.

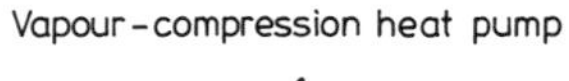

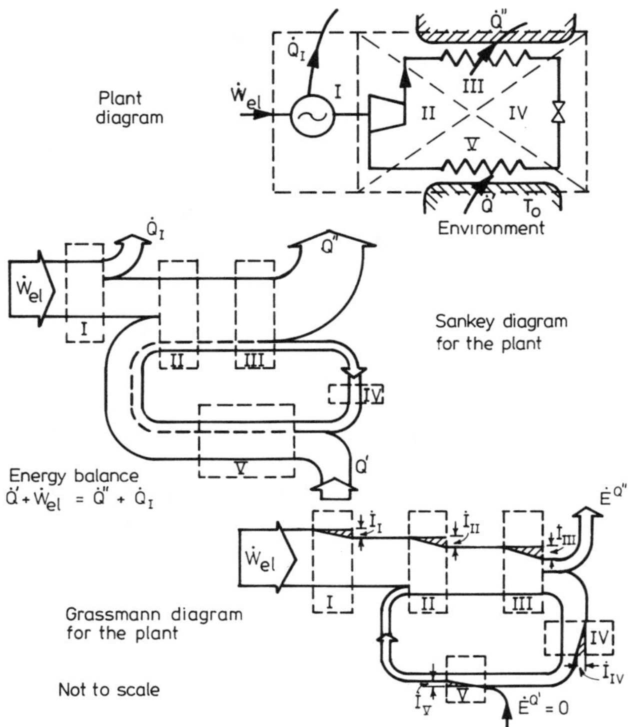

Fig. 5 A vapour-compression heat pump

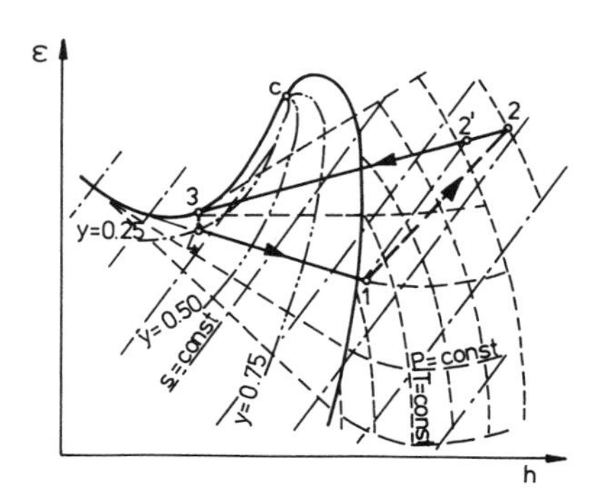

Fig. 6 Exergy-enthalpy chart for ammonia

Exergetic temperature-enthalpy (τ-H) diagram can be used for evaluating the irreversibility incurred in heat exchange processes. An example of application to the heat exchange process in the Linde air-liquefaction process with and without auxiliary refrigeration [4] is illustrated in Fig. 7. As the area enclosed between process lines, shown shaded, represents the process irreversiblity, the effect of changes in the heat exchange process can be conveniently studied.

Tables of standard molar chemical exergy, $\tilde{\varepsilon}^{o}$, can be used to facilitate computations. The molar exergy of a substance can be decomposed into the physical and chemical components denoted by $\tilde{\varepsilon}_{ph}$ and $\tilde{\varepsilon}^{o}$ respectively. Thus we can write

$$\tilde{\varepsilon} = \tilde{\varepsilon}_{ph} + \tilde{\varepsilon}^{o} \qquad (13)$$

Values of $\tilde{\varepsilon}^{o}$ for a number of chemical substances have been computed and are available in tabulated form [4, 5]. The use of such tables facilitates considerably calculations in chemical processes.

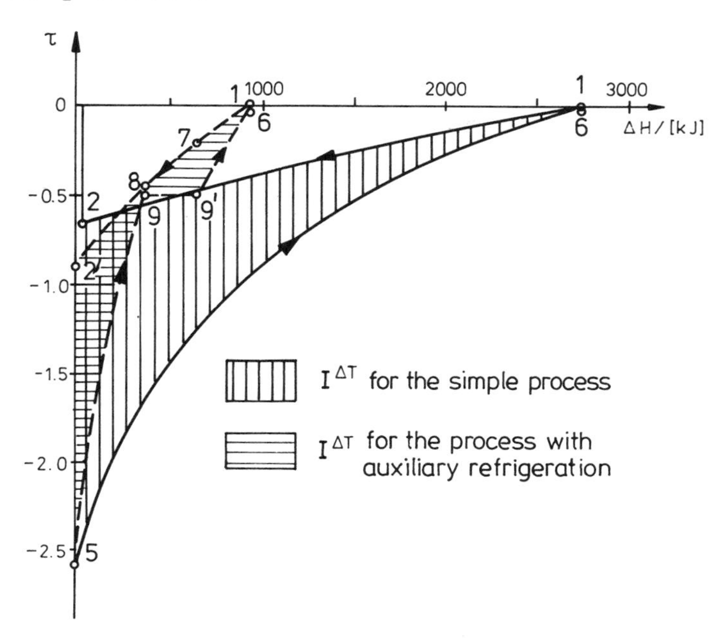

Fig. 7 Evaluation of heat exchange process irreversibility from a τ-H diagram.

Tables of mean, molar, isobaric exergy capacity, $\tilde{c}_p^{\varepsilon}$, for ideal gases can facilitate enormously the calculation of the physical component of exergy. $\tilde{c}_p^{\varepsilon}$ is defined as follows

$$\tilde{c}_p^{\varepsilon} = \frac{(\tilde{\varepsilon} - \tilde{\varepsilon}^o)_p}{T - T^o} \tag{14}$$

and is tabulated against temperature for a variety of common ideal gases [4].

THERMOECONOMIC APPLICATIONS OF EXERGY

One can view the irreversibility rate of a plant as the exergy consumption rate necessary to maintain a particular process rate of a plant of a given design and given capital cost. Since, in general, the more efficient a plant is the more expensive it is, there is an inverse relationship between the irreversibility rate and the process rate on one hand and the capital cost of the plant on the other hand. This relationship forms a basis of exergy related techniques for the optimisation of the costs of plant components and for other thermoeconomic applications [4, 6, 7].

RECOMMENDED PLAN FOR TEACHING THE EXERGY METHOD

The Exergy Method, in the form in which it has been outlined above, can be taught as a compact, self-contained course to the more advanced, say, final year undergraduate students, postgraduate students, practising engineers or indeed anyone who is already familiar with the fundamentals of Applied Thermodynamics and the traditional techniques of application of this discipline to thermal or chemical systems or both.

With regard to the question of the method of teaching undergraduate students, it is the present author's view that the concepts and techniques of the Exergy Method should be integrated into a course of Engineering Thermodynamics, to complement and reinforce the traditional work. The incorporation of the subject matter into the usual three year course of Engineering Thermodynamics would give the students adequate period of time over which to assimilate properly the concepts and the techniques of the Exergy Method.

The first year of such a course is usually devoted to the fundamentals of Applied Thermodynamics. This is therefore the time when the ground work should be prepared for the teaching of the Exergy Method proper in the second and the third year. The introduction of the Second Law through the Entropy Postulate is most helpful since it leads naturally to the formulation of the concept of entropy production. The calculation of the entropy production for a system (or entropy production rate in steady flow processes) in some simple numerical examples will give the students valuable experience of a quantitative form of assessment of the degree of thermodynamic perfection of a process. It will further provide them with an opportunity to handle such basic concepts as *isolated system, environment, thermal energy reservoir, control surface*, etc. The concept of entropy production finds also application, as a subsidiary quantity, in the derivation of some fundamental exergy relations.

In the second year course the simpler concepts of the Exergy Method can be introduced and the methods of analysis applied to some of the less complex physical processes such as compression and expansion processes, heat transfer, mixing and simple separation processes. Multi-component plants such as the more elementary types of power plants and refrigeration plants can then be analysed. In preparation for the introduction in the third year of the concept of chemical exergy, the maximum work of a chemical reaction and the concept of chemical potential should be dealt with.

The third year course should start with the extension of the exergy analysis to include chemically reacting systems. The required depth of treatment of this aspect of analysis might be greater for Chemical Engineering students than for those studying for a career in different areas of Mechanical Engineering. The third year course should also include in-depth treatment of the application of the Exergy Method to some of the following areas according to particular specialization:

1. Expanders and compressors
2. Heat exchangers
3. Refrigeration and cryogenic processes
4. Distillation processes
5. Chemical processes
6. Combustion
7. Power generation
8. Thermoeconomics

Quite clearly, because of the different ways in which Engineering Thermodynamics is taught at different universities and other higher institutes of engineering education, the scheme for teaching the Exergy Method outlined above is intended only for general guidance.

CONCLUSIONS

It is hoped that the brief outline of the Exergy Method given above has given some indications with regard to its capabilities, advantages and the breadth of problems to which it can be applied.

The following are some of the more important advantages which the Exergy Method offers:

(1) It promotes a better understanding of the nature of thermodynamic losses and of the associated causative factors.

(2) It provides means of quantitative assessment of the thermodynamic losses of different types (produced by different causes) and their contribution by different parts of the plant.

(3) It provides means of investigation, through the CSB's, of the mutual interactions, or structural bonds, existing between different plant components and the plant as a whole.

(4) It provides means of assessing limits to the possible improvement in the performance of the plant and its components.

(5) It offers techniques of iterative thermoeconomic optimization of selected plant components.

(6) It teaches the student the practical implications of the Second Law in a more effective way than the traditional approach through the study of its various statements and corollaries.

REFERENCES

1. Gardner, J. B. and Smith, K. C., "Power Consumption and Thermodynamic Reversibility in Low Temperature Refrigeration and Separation Processes", Advances in Cryogenic Engineering, Plenum, New York, 1960.

2. Flower, J. R. and Linnhoff, B., "Thermodynamic Analysis in the Design of Process Networks", CACE '79 Conference, 214th event of the EFCE, Montreux, Switzerland, 8-11 April, 1979.

3. Kotas, T. J., "Exergy Criteria of Performance for Thermal Plant", Int. J. Heat and Fluid Flow, Vol. 2, No. 4, 147-163.

4. Kotas, T. J., "Exergy Method of Thermal Plant Analysis", to be published by Butterworths, September, 1985.

5. Szargut, J. and Petela, R., "Egzergia" (in Polish) Wydawnictwa Naukowo-Techniczne, Warsaw, 1965.

6. El-Sayed, Y. M. and Evans, R. B., "Thermoeconomics and the Design of Heat Systems", Trans. of the ASME, Journal of Engineering for Power, January 1970, 27-35.

7. Beyer, J., "Structure of Thermal Systems and Economic Optimization of System Parameters", (in German) "Energieanwendung", 1974, Vol. 23, No. 9, 274-279.

THERMODYNAMICS APPLIED TO ENERGY ENGINEERING

G. E. Andrews

Department of Fuel and Energy
University of Leeds

INTRODUCTION

The Department of Fuel and Energy at Leeds University is unique in British Universities and has been in existence since 1906. Throughout this period it has taught courses in applied thermodynamics. The courses are specifically aimed at the thermodynamics of energy generation and utilisation. As 95% of energy is derived from the combustion of fossil fuels this process forms a key feature of the thermodynamic teaching. The applications-oriented teaching results in related thermodynamics, fluid mechanics and heat transfer courses which are different from conventional mechanical engineering courses.

COMBUSTION THERMODYNAMICS AND FLUID MECHANICS

Fundamentals

Combustion stoichiometry calculations are applied to the real fuels derived from oil and coal as well as the simpler pure gaseous fuels. The vital importance of the ratio of the fuel and air flow rates in the control of the combustion process is emphasised by studying the influence of this ratio on the exhaust gas composition, including combustion generated pollutants. The limits of fuel and air mixture ratios over which a flame can propagate are studied in terms of the basic flammability limits as well as the practical problem of flame stability.

One of the most important thermodynamic calculations in combustion is that of the adiabatic flame temperature. These calculations are first introduced without considering dissociation and then extended to consider the influence of dissociated species in high temperature gases. During the teaching of flame temperature calculations, the thermodynamic properties of high temperature gases are dealt with. The influence of real operating conditions, high inlet air pressures and temperatures, are included. The applications range from simple burner flames, through engines, to the very high temperature fuel/oxygen flames in rockets.

Laminar Flames

Laminar flames are used extensively in the gas industry for most domestic and small boiler heating applications. Basic aerodynamic properties of premixed laminar flames such as burning velocity, flame speed in explosions and flame stability are considered in detail from a theoretical and experimental viewpoint. Laminar flames represent a complex interaction of fluid flow, chemical kinetics and thermodynamic transport processes. The differential equations necessary to provide a solution to this problem are discussed. Laminar diffusion flames are also studied in a similar manner and the additional problems of flame stability, flame length and soot formation are introduced.

Turbulent Flames

Most large scale applications of combustion involve turbulent flames for gas, liquid and pulverised coal flames. The detailed analysis of these flames is based on an initial course which discusses the wide range of available burner and furnace designs. Following a detailed derivation of the Navier-Stokes equation with turbulence, the additional complexities of flame propagation in turbulent flow are discussed. Theories of turbulent burning velocity and stirred reactors are reviewed and used together with global reaction data to predict flame extinction. Finally, jet theory is introduced and applied to the calculation of turbulent diffusion flame length.

HEAT TRANSFER

Efficient heat transfer and heat recovery is essential for the efficient use of energy. Heat transfer from high temperature gases are analysed for conduction, convection and radiation processes. Specialised aspects of the course relate to heat exchanger design, temperature measurement of high temperature gases and solids, heat transfer during boiling and condensation and evaporation, distillation and drying processes. Heat transfer within furnace enclosures is analysed in detail including the problems of radiative interchange and the emittance and absorption of radiation by high temperature combustion product gases. The practically important problem of radiation from pulverised coal flames is one of the most complex radiative heat exchange problems and methods of calculation are studied.

INTERNAL COMBUSTION ENGINES

In addition to the conventional treatment of the thermodynamics of the spark ignition, compression ignition and gas turbine engines, their combustion processes are analysed. This draws on the combustion teaching already outlined. Mixture formation, variable thermodynamic properties of the mixture, ignition advance and turbulent flame propagation in spark ignition engines are investigated. For compression ignition engines the physical and chemical processes involved in the ignition delay are analysed together with the high pressure turbulent diffusion flame phenomena and the important problem of particulate emissions. These latter problems are also relevant to gas turbine combustion where the additional problems of flame stability and future fuels are considered. For all three engines heat transfer problems are discussed with a particular emphasis on the gas turbine, where water cooling is not possible and air film cooling processes have to be used.

CONVENTIONAL THERMODYNAMICS

The detailed coverage of combustion processes and related aerodynamics and heat transfer does not occur by omission of more conventional thermodynamics. The thermodynamic laws and cycles together with the thermodynamic properties of steam and power generating steam cycles and cooling towers are all studied in detail. Other courses include refrigeration processes, heat pumps, process mass and heat balance and the thermodynamics of process design. The specialised lecture courses are included by placing a lower emphasis on production, strength of materials and dynamics than in a Mechanical Engineering course.

DISCUSSION IN SESSION 5

"APPLICATIONS OF THERMODYNAMICS TO DESIGN ASSESSMENT"

Edited by K. W. Ramsden

Cranfield Institute of Technology

On the paper by B. Linhoff "Thermodynamics Applied to Energy Engineering"

This paper had previously been published in the Proceedings of the Royal Society in 1983 and was an Esso Energy Award paper in 1981. The paper emphasises the importance of "Linking" individual pieces of equipment very carefully into an overall process network.

Professor P. T. Lansberg of Southampton University commented that since the paper describes the Carnot cycle quasi-statically, the work it produces is given out infinitely slowly. He further emphasised the point made implicitly by Professor Linhoff that the power input of such an engine is zero.

He added that for given temperatures of heat reservoirs and with assumptions about the heat flux from reservoirs to working substance, the power output of such an engine can be maximised.

In the simplest case, according to Lansberg, a lower than Carnot efficiency occurs, typically

$$\eta_{power} = 1 - \sqrt{\frac{Tcold}{Thot}} < \eta_{energy} = 1 - \frac{Tcold}{Thot}$$

This, he maintains, is a good example to show the difference between power efficiency and energy efficiency.

On the two papers:

By B. M. Burnside of Heriot Watt University
"Relevance of Second Law Analysis in a Mechanical Engineering Syllabus"

and D. T. J. Kotas of Queen Mary College, London
"Teaching the Exergy Method to Engineering Students"

These two papers were presented jointly and the Chairman, Professor G. D. S. MacLellan of Leicester University deferred discussion until both papers had been presented.

Substantial discussion took place with contributions from Professor R. S. Silver of Glasgow University, from Dr. S. P. S. Andrew of I.C.I., Professor Linhoff of UMIST and Professor W. A. Woods of Queen Mary College.

In particular, Mr. D. H. Bacon of Plymouth Polytechnic asked the authors that since there appears to be two approaches to thermo-economic design, namely,

(1) Gagglioli (United States)
(2) Beger (Germany)

do they both give the same results when applied to overall plant design or does one have advantages over the other.

Dr. Burnside in reply said:-
"The author must own up to not having studied the German and Russian literature on 2nd Law Analysis as thoroughly as the literature in English. However, he gets the impression that, apart from details of technique, the approaches are much the same based on original ideas of Maxwell, Gibbs, Keenan and Bosnjakovic. The heat exchanger design method of Bejan, which was preceded by that of McClintoch, is an additional technique rather than an alternative (see section 4 of the author's paper).

Concerning presentation to undergraduates, the author has found the programme outlined in his paper very satisfactory. In particular he finds it simpler to use "available energy" as a general term - there are very good historical reasons for doing so (1). The term "exergy", if it must be used, is used to describe the available energy at a point in a flow referred to restricted, or "flexible envelope", equilibrium with the surroundings. The term "essergy" then is the corresponding available energy referred to unrestricted equilibrium with the environment. From the point of view of English language speakers this approach is historical - recognising the original contributions of Maxwell and Gibbs and giving credit to the German language speaking contribution to 2nd Law Analysis. Presumably, in translation phrases like "available energy of a closed system relative to restricted equilibrium with the environment" would be rendered in German by replacing the words "available energy" by "exergie".

Another approach which has become popular is to use the term "exergy" as a general term in English to replace "available energy". Then we talk, for example, of "exergy of closed systems referred to restricted equilibrium with the environment", in English. The author can have no objection to this in the German language, but it does seem anomalous in English. The corresponding situation in reverse would be to use the words "available energy" in German!

Dr. Kotas replied by saying:-
"There are two main approaches to thermoeconomic optimization.

(i) The 'autonomous method' developed in the U.S.A. by Tribus, Evans, El-Sayed and propagated by others, depends on isolating a sub-system (or zone) from the rest of the plant and treating it as an autonomous system. The success of this depends on the correct determination of unit costs of exergy streams crossing the zone boundary.

(ii) The 'structural method' which has been originated by Beyer makes use of the concept of the coefficient of structural bonds (see Kotas, Session 5) to link irreversibility changes in the sub-system under consideration to the corresponding changes in the irreversibility of the plant as a whole. This enables the thermoeconomic optimization of a sub-system to be carried out on its own, in terms of the local parameters.

The two methods, provided they are correctly applied, should give the same results. The first method is more suitable for application to systems which can be divided into relatively few zones, and in special areas of application such as exergy costing in co-generation plants. The advantage of the second method is that it can be applied to plants with many components. This is due to the fact that the coefficients of structural bonds can be calculated on a computer using the graph theoretic approach.

The 'Bejan approach' as presented by Dr. Hay is not intended, as far as I can see, for thermoeconomic optimization of the operating conditions of a component.

The 'Linhoff approach' deals with the synthesis and optimization of heat-exchanger networks and is particularly applicable to chemical plants with large numbers of heat exchangers when heat transfer over a finite temperature difference is the dominant form or irreversibility."

On the paper by Dr. N. Hay of Nottingham University, "Putting the Second Law to Work"

Professor M. J. French of Lancaster University congratulated the author on his closing remarks concerning the importance of not being too ambitious in what is taught and added emphasis to the importance of this.

He continued by saying that "In a simple case like a heat exchanger, where cost is being optimised and exergy from different sources and hence of different costs is involved, exergy is of no special help and it is better to work directly in the separate costs, as I have done in by book+, reserving exergy for more difficult cases."

Dr. T. J. Kotas of Queen Mary College said that "The use of the entropy generation method for the study of heat transfer processes is valuable as means of mapping out local variation in dissipation. When it comes to the design of heat exchangers of optimum geometry it is far better, in my view, to use the concept of exergy. Exergy has a more readily interpreted physical and economic significance and the two components of irreversibility (exergy loss) $I^{\Delta P}$ and $I^{\Delta T}$ corresponding to the entropy generation quantities $s^{1}_{\Delta P}$ and $s^{1}_{\Delta T}$ can be easily costed by means of thermoeconomic techniques. I can tell you from our own work in this area that the use of separate values for the unit costs of $I^{\Delta P}$ and $I^{\Delta T}$ affects only marginally the optimum geometrical parameters of the heat exchanger i.e. in this case the thermodynamic optimum and the economic optimum are very close."

+ FRENCH, M. J., Engineering Design. Heinemann, London 1971.

On the paper by Dr. J. C. Meklenburg of Nottingham University, "Process Feasibility and Thermodynamic Teaching"

Mr. R. W. Haywood of Cambridge University drew attention to the author's statement that "the advent of numerical methods had meant that there is no need to use the functions called availability and exergy" and asked whether the author meant that there was no need for those concepts in addition to no need for those functions.

In reply, Dr. Mecklenburg agreed that the concept of availability and exergy are useful but in practice, numerical methods he considered had proved more useful.

Professor M. J. French of Lancaster University congratulated the author on an interesting paper. He also drew attention to the author's closing remarks on the importance of keeping the load on the student's understanding to a reasonable level. It is always important, he said, to bear this in mind when teaching thermodynamics.

SESSION 6

SYLLABUS DEVELOPMENT

Chairman - Dr. J. D. Lewins

Secretary - Dr. D. C. Anderson

THERMODYNAMICS AND IRREVERSIBILITY

R. S. Silver

University of Glasgow

IRREVERSIBILITY IN INTRODUCTORY THERMODYNAMICS

The present paper deals only with the non-relativistic range, where we take constancy of mass and constancy of energy as separate fundamental principles. The science of thermodynamics now explains all the phenomena of change reasonably well in terms of rearrangement of mass and energy subject to these overall constancy constraints. But man as an economic agent also desires to make deliberate change. Thus thermodynamics as the science of how changes can occur is the fundamental science of economics. It is essential that it should be well understood by those using it, and so the quality of its teaching to engineers is vital.

The ways in which thermodynamics is usually taught contain elements arising from the difficulties which occurred historically in reaching the concept of the conservation of energy, which is not yet 150 years old. That is only about six generations, and roughly also about only six generations of definitive text-books. It *was* difficult to sort things out, and thermodynamics teaching evolved a fairly rigid structure of defined concepts and attitudes. These were intended to simplify the understanding, but some of them merely encapsulate the historical difficulties. The conceptual structure contains superfluous and redundant elements which are not necessary when you develop the intrinsic logic of the subject from the starting point of conservation of mass and of energy.

One relic from the historical situation is that the conventional approach to thermodynamics still gives the separate phenomenologies of work and of heat, as modes of energy transmission, equal weight in trying to develop the subject. This is all the more sad since about 25 years ago it was decided, once and for all, to give the work mode priority in the definition of energy units - the joule, the work action of one newton through one metre. That priority should have been matched in the teaching of the subject. The logical position is quite clear; given the hypothesis of conservation of energy you can start with one process of energy transmission and deduce the necessary existence of another. It is equally clear that owing to its adoption for the basic energy unit, the work mode should be the starting point. Moreover it is also evident that, while the existence of an other mode of transmission can be deduced, there has to be an argument which can uniquely identify that other process as heat.

The development of these ideas occupies the first two chapters of my text-book (1). I shall not attempt to summarise it here but draw attention to the fact that the identification argument depends critically on the inclusion of frictional dissipation. If the friction term is omitted from the relevant equations you might arbitrarily choose to regard the deduced process as identifiable with the heat process, but only the inclusion of the friction term determines the identity beyond dispute. That fact is implied in the interpretation of Joule's famous experiment.

But in conventional teaching friction is excluded from the basic equations of the subject and so its logical development is undermined. For practical applications, frictional dissipation is brought in as if it were something extraneous, capable of being dealt with by "fudge factor" devices. This works adequately for quite a wide range of mechanical engineering applications simply because in many of these - notably in the central theme of power generation - the frictional dissipations are a relatively small proportion in the energy interplay. But it becomes quite useless in many applications to electrical, electronic, and chemical engineering, where frictional effects are not relatively small.

The basic equations which I develop for the energy transmitted by heat or by work differ formally very little from those in conventional texts. They are as follows, when kinetic and potential energies are included.

1. Work output from reference mass unit
   (a) Net or total

$$\Delta W_t = p\,dV - d\phi - d(v^2/2) - \Delta W_f \qquad (1)$$

   (b) Dynamic

$$\Delta W_d = - V\,dp - d\phi - d(v^2/2) - \Delta W_f \qquad (2)$$

2. Heat input to reference mass unit

$$\Delta Q = d\,U + p\,d\,V - \Delta W_f \qquad (3)$$

The thermal equation (3) applies always, whether the work output is given by (1) or (2). Later in the development, once the existence of entropy as a property is established, the thermal equation is given in the alternative form

$$\Delta Q = T\,dS - \Delta W_f \qquad (4)$$

The important formal difference in these equations from those of conventional thermodynamics is in each case the presence of the friction term ΔW_f. Note also that I use Δ instead of d for the increments of energy transmission. It surprises me that authors and teachers who are correctly punctilious about distinguishing between processes and properties, nevertheless use the same symbol d for the increments of both. You may not like my Δ, and mathematical purists will object to having Δ and d in the same equation, but some distinctive symbol is justified.

There is however an importance difference implicit in these equations quite apart from the form. Real processes are not only irreversible, friction being always present, but they are also non-equilibrium. Thus the property values p, T, U, S, V, H etc. which appear in my equations are non-equilibrium values. This is a major difference from conventional presentations. Some texts read as if properties and their numerical values had no meaning except at equilibrium. This concept of a property which exists meaningfully at one precise situation and ceases, like the Cheshire cat, to exist for a displacement to either side, is something which I find totally unacceptable for both philosophic and practical reasons. Every engineer knows that he can measure quantities of the nature of the thermodynamic properties in real processes where equilibrium conditions are not satisfied. These measurements must have significance.

The question of these non-equilibrium, or quasi-equilibrium, values is discussed to some extent in my textbook ([1]), but is given much more precise development in a paper published in 1980 ([3]). For the present discussion it is sufficient to bear in mind that all properties appearing in my equations are to be understood as non-equilibrium values.

ONSAGER-BASED IRREVERSIBLE THERMODYNAMICS OF THE STEADY STATE

Because of the accepted limitations of conventional thermodynamics to reversible equilibrium, an entirely separate discipline for irreversible thermodynamics has grown up over the last fifty years. It is based on the work of Onsager, and notable contributions to its development have been made by distinguished workers including Prigogine, De Groot, and Meixner. For convenience I shall henceforth refer to that general corpus of work by the initials OIT.

The basis of OIT lies in statistical theory of small departures from equlibrium. The relevant equations can be solved for such small departures and enable the treatment of steady-state flow situations. By steady-state flow is meant the general situation when fluxes of mass and of energy between a source and a sink are maintained so that at every cross-section of the flow the conditions are constant in time but vary in space. Thus in the OIT treatment it is assumed that <u>local</u> equilibrium exists <u>at</u> each cross-section, although there are gradients of properties <u>through</u> the cross-sections. The solutions of the statistics for such small perturbations from equilibrium lead to the definition of certain coefficients, usually designated λ-coefficients. The following summary shows briefly how these concepts of OIT are applied to a range of steady-state flow phenomena.

Fluxes J_s of entropy and J_m of mass are considered. The vector flux J_s is said to have a conjugate force vector grad T while the vector flux J_m has a conjugate force vector grad ψ. The rate of entropy generation per unit volume of the flow regime is div J_s, and is asserted to be given by the equation

$$T \operatorname{div} J_s = - J_s . \operatorname{grad} T - J_m . \operatorname{grad} \psi \tag{5}$$

However it is also asserted that the action of the driving forces is coupled so that entropy flux J_s and mass flux J_m are each responsive to both grad T and grad ψ by the following relations

$$\left.\begin{aligned} J_s &= \lambda_{11} \text{ grad } T + \lambda_{12} \text{ grad } \psi \\ J_m &= \lambda_{21} \text{ grad } T + \lambda_{22} \text{ grad } \psi \end{aligned}\right\} \tag{6}$$

Two important points characterise the λ-coefficients. In the first place they are equilibrium properties, constituting a second set additional to the more familiar equilibrium properties. This result arises of course from the assumption of local equilibrium in neighbouring cross-sections. It means that the λ are constants, in the sense that they are independent of the gradients and of the fluxes, although they will depend on pressure, temperature, and constitution just as do the usual equilibrium properties. Secondly, important interrelations are found to exist between the cross-coupling coefficients. These inter-relations, known as the Onsager reciprocal relations, also depend for their derivation on the local equilibrium assumption. The reciprocal relation in the above example is

$$\lambda_{21} = \lambda_{12} \tag{7}$$

The application of OIT to any specific steady flow problem depends entirely on the constancy of the λ-coefficients and on the validity of the reciprocal relations, and the question must therefore be asked "How reliable can this method be in conditions far from equilibrium, since the whole thing depends on a statistical analysis valid only for small perturbations?" One favourite example of OIT application is to the theory of the thermocouple. It is acceptable that the near-equilibrium basis of OIT should serve for the thermocouple used for temperature measurement in a potentiometric circuit and hence with a low load. But when you want to consider thermoelectric refrigeration, where the loads become substantial, confidence in a "near-equilibrium" theory may be unfounded. In the many important fields of engineering development which involve such matters as plasma flow, MHD, multiphase flow, membrane separation processes, continuous flow chemical reactions, and the like, where the phenomena are highly irreversible and grossly far from equilibrium, the utility of OIT must be doubtful.

Its practical utility is also severely limited by the insistence on the use of equations involving coupling. The experimental measurement of coefficients appearing in the equations becomes exceedingly awkward, and usually impossible in practice, and if measurements made do not turn out to be constants, the theoretical basis has vanished and you can do nothing about it.

A method which can deal with cases far from equilibrium and can express the phenomena in terms of uncoupled fluxes and forces would be of much greater practical use. Such a method might have developed naturally if conventional thermodynamics had not so rigidly held to reversible equilibrium. It follows that my approach, in which I include frictional resistance in the basic treatment and assume non-equilibrium property values, can perhaps lead to a more practical system of irreversible thermodynamics. The next section gives a brief account of my treatment.

GENERAL THEORY OF IRREVERSIBLE THERMODYNAMICS OF THE STEADY STATE

The theory is given in a short paper in 1979 (2) and a full account in 1980 (3). The latter deals with the general case of several fluxes and counterfluxes of mass of different species at different individual pressures and temperatures, reacting as they pass through a cross-section. The full theory cannot be given here, but the overall phenomena consist of a constant mass flux J_m, a heat flux J_q, and a related entropy flux J_s. In this Section I treat these fluxes on the basis described in Section 1 and derive the OIT equation given in Section 2. I also show how my preferred treatment can remove the restriction to near equilibrium conditions.

In a steady state flow considered in vector terms, div J_m is zero. Thus the work equation corresponding to equation (2) of Section 1 is

$$\dot{W}_d = - J_m.[V \text{ grad } p + \text{grad}(\phi + v^2/2)] - \dot{W}_f \quad (8)$$

where $\dot{W}_d$ and $\dot{W}_f$ are respectively the rates of work output and frictional dissipation per unit volume.

Correspondingly the heat equation (4) becomes

$$- \text{div } J_q = J_m.T \text{ grad} S - \dot{W}_f \quad (9)$$

Equations (8) and (9) give the energy equation for steady flow viz.

$$- \operatorname{div} J_q = \dot{W}_d + J_m . \operatorname{grad}(H + \phi + v^2/2) \qquad (10)$$

The total entropy flux is

$$J_s = J_q/T + J_m S \qquad (11)$$

from which we obtain

$$T \operatorname{div} J_s = -(J_q/T).\operatorname{grad} T + \operatorname{div} J_q + J_m . T \operatorname{grad} S \qquad (12)$$

Substitution of equation (9) in (12) gives

$$T \operatorname{div} J_s = -(J_q/T).\operatorname{grad} T = \dot{W}_f \qquad (13)$$

Then using equation (8) we find

$$T \operatorname{div} J_s = -(J_q/T).\operatorname{grad} T - J_m .[V \operatorname{grad} p + \operatorname{grad}(\phi + v^2/2)] - \dot{W}_d \qquad (14)$$

In most applications $\dot{W}_d$ is zero, but it can in any case be represented in terms of a potential ξ so that equation (14) becomes

$$T \operatorname{div} J_s = -(J_q/T).\operatorname{grad} T - J_m .[V \operatorname{grad} p + \operatorname{grad}(\phi + v^2/2 - \xi)] \qquad (15)$$

Introducing the Gibbs potential G, equation (15) can be expressed alternatively as

$$T \operatorname{div} J_s = -(J_q/T).\operatorname{grad} T - J_m . S \operatorname{grad} T + \operatorname{grad}(G + \phi + v^2/2 - \xi)] \qquad (16)$$

By taking the two terms in grad T together equation (16) becomes

$$T \operatorname{div} J_s = -J_s .\operatorname{grad} T - J_m .\operatorname{grad}(G + \phi + v^2/2 - \xi) \qquad (17)$$

Equation (17) is the OIT relation in Section 2 equation (5) where

$$\psi = G + \phi + v^2/2 - \xi \qquad (18)$$

This shows that my approach can lead to the same result as OIT. However it is not necessary to form equation (17) and

I regard (15) or (16) as the terminal result. I express either in the form

$$T \operatorname{div} J_s = -(J_q/T).\operatorname{grad} T - J_m.\operatorname{grad} \psi' \qquad (19)$$

where ψ' is a function such that

$$\operatorname{grad} \psi' = V \operatorname{grad} p + \operatorname{grad} (\phi + v^2/2 - \xi)$$

$$\text{or } \operatorname{grad} \psi' = S \operatorname{grad} T + \operatorname{grad}(G + \phi + v^2/2 - \xi) \qquad (20)$$

In that development no constraint to equilibrium is anywhere implied. The OIT move from equation (16) to (17) does not itself introduce any constraint. It is made only because the assumed near-equilibrium theory of OIT requires coupled flux/force conjugate relations as in equation (6) and hence forces the transition to J_s. That requires the λ-coefficients to be constant and so unnecessarily restricts the application of irreversible thermodynamics.

In contrast by stopping at the equivalent equations (15) and (16), and using either in the general form ((19), I can use uncoupled flux/force conjugate relations simply as

$$J_q/T = - \alpha \operatorname{grad} T \qquad (21)$$

$$J_m = - \beta \operatorname{grad} \psi' \qquad (22)$$

Now there is no requirement for the coefficients α and β to be constant. They may be functions of the gradients and therefore of the fluxes, and if, physically, coupling does exist, it will show up in experimental values of α and β.

The utility will be obvious to any mechanical engineering thermodynamicist. When thermal and potential energy effects are negligible equation (22) becomes simply

$$J_m = - \beta (V \operatorname{grad} p + \operatorname{grad} v^2/2) \qquad (23)$$

This is familiar, with β the resistance coefficient being a function of Reynolds Number. It is equally obvious when field potential energies are included. Further when thermal effects are present the heat transfer is correspondingly included with α being a function of Reynolds Number, Nusselt Number, and Prandtl Number.

These equations are still applicable when chemical potentials are present and hence they can be used on the whole range of irreversible thermodynamics for which OIT has been set up. Because the parameters are coefficients in uncoupled equations they are much more readily obtained from experimental measurement. Because they are not asserted to be constants they are much more adaptable and functional relations can be determined. Finally they are not restricted to near-equilibrium conditions.

My first papers on this procedure were rejected, on the grounds that the use of uncoupled equations is entirely forbidden by the fundamental statistical theory of near-equilibrium conditions. This objection is removed by a theorem in vector algebra. If a set of vectors satisfies the coupled form of equations (6) and meets the constraint (5), an envelope of such sets can be generated by varying one linking parameter. If the cross-coefficients are equal for any set in the envelope they are equal for every set, though the equal values are different for different sets. Finally there will always be one set for which the equal cross-coupling coefficients value is zero - i.e. there will be an uncoupled set. Hence the objection is not valid, for if the cross-coupling coefficients are equal an uncoupled set is available, and if they are not equal the whole OIT theory collapses.

The second obstacle was of course the whole problem of the significance of property values far from equilibrium already mentioned in Section 1. This was dealt with fully in my 1980 paper ([3]). I consider the completely general case of several fluxes and counterfluxes of mass of different chemical species at different individual pressures and temperatures, reacting and changing constitution as they pass through a cross-section. The exergy of unit mass of such a condition is expressible as a function of all the individual component properties relative to a reference temperature and pressure. In the usual definition of exergy that reference base is the environment, but it can be taken as an arbitrary base. Using the fact that even in highly non-equilibrium conditions experimental sensors obtain temperature and pressure values T,p at a cross-section, I define the local disequilibrium η as the exergy of unit mass at a cross-section with reference to T,p at the same cross-section. The expressions for $\bar{\eta}$ are obtained, and I then suggest that in the steady state a uniform value of $\bar{\eta}$ is established axially. This constraint is sufficient to establish the overall equations. Naturally the value of grad ψ' in the full equation corresponding to equation (20) includes all the features of the multi-component, multi-stream, situation.

The treatment being general, the OIT case of local equilibrium is included as a special case when $\overline{\eta} = 0$. The non-equilibrium properties which appear in my equations can therefore be related to the OIT parameters. The important correspondences are as follows. I denote the general steady-state non-equilibrium value of entropy by $\overline{S}$. That is the value for a finite disequilibrium $\overline{\eta}$. The OIT local equilibrium corresponds to quasi-equilibrium where $\overline{\eta} = 0$, and I designate the entropy value for that condition as $\overline{S}_o$. This corresponds directly to the "entropy transport parameter" S^* familiar in OIT. Many important effects in thermodiffusion, thermomigration, thermomagnetics, and thermoelectrics, are explained by OIT in terms of the difference $(S^* - S_o)$. These same effects are all accounted for in my treatment in terms of the corresponding difference $(\overline{S}_o - S_o)$, but more importantly, the discussion can be extended well beyond the near-equilibrium limitation, in terms of the general difference $(\overline{S} - S_o)$.

CONCLUSIONS

Thus what began with the mere intention of introducing irreversibility into thermodynamics teaching has pointed the way to a comprehensive unified treatment of irreversible non-equilibrium thermodynamics. It is worth commenting that it explains incidentally why "ordinary" mechanical engineering thermodynamics has "worked" satisfactorily, i.e. why the facts of heat of transport, the phenomenon of pressure gradient generated by temperature gradient, and the like, have not led to difficulties in "ordinary" applications. The answer is interesting. All of these effects depend on the difference $\overline{S} - S_o$, and in a single component single phase fluid the order of magnitude of that difference can only be of the order of $\sigma^2/2T$, where $\sigma^2/2$ is the variance of the velocity distribution. Even with high turbulence in supersonic gas flows $(\overline{S} - S_o)/S_o$ is at most of order 2%. Thus in single component flow the effects are unlikely to be noticeable. That is why the "fudge factor" procedure has been serviceable.

However for multi-component flows, such as can occur with combustion gases and plasmas, and in the presence of electromagnetic potentials, the values of $\overline{S} - S_o$ can be much greater. The method which I have proposed should prove useful in analysing such cases. It should also help for multi-component flows through porous media and in membrane separation processes, and in thermoelectric refrigeration devices, where, although the flow is slow, high gradients of pressure, potential, and temperature are required to handle industrially economic loads. Hence such processes are also highly irreversible and far from equilibrium.

From the point of view of this Symposium, the most useful aspect of this work is perhaps that it should enable the development of teaching thermodynamics as a unified whole. The economic and industrial utility of the subject depends increasingly on its ability to handle real processes, and that cannot be done without a coherent account of irreversibility.

REFERENCES

1. Silver, R. S., "An Introduction to Thermodynamics" (with some new derivations based on real irreversible pro-- cesses), Cambridge University Press, 1971.

2. Silver, R. S., "Coupling and Uncoupling in Irreversible Thermodynamics", J. Phys. A: Math. Gen., 1979, 12, L141-6.

3. Silver, R. S., "Collinearity and Disequilibrium in Irreversible Thermodynamics of the Steady State", J. Phys. A: Math. Gen., 1980, 13, 3253-3274.

LIST OF SYMBOLS
SPECIFIC PROPERTIES (PER UNIT MASS)

G Gibbs Potential
H Enthalpy
S Entropy
$\overline{S}$ Mean entropy of multi-component unit mass with finite disequilibrium
$\overline{S}_o$ Value of $\overline{S}$ when disequilibrium is zero
S_o^o Entropy at equilibrium
S^* Onsager-based entropy transport parameter
U Internal energy
V Volume
ϕ Field potential
ψ Combined potential $G + \phi + v^2/2 - \xi$

ψ' Corresponds in proposed treatment to ψ
ξ Work potential if device present
η Disequilibrium
$\sigma^2/_2$ Kinetic energy of velocity variance

Intensive Properties

P Pressure
T Temperature
v Velocity

Energy Transmissions

ΔQ Energy transmitted to unit mass by heat process
ΔW_t or
ΔW_d Energy transmitted from unit mass by work process
ΔW_f Energy dissipated against friction per unitmass
$\dot{W}_d$ Rate of work output per unit volume
$\dot{W}_f$ Rate of frictional dissipation per unit volume

Vector Fluxes

J_m Mass
J_q Heat
J_s Entropy

Coefficients

λ, λ_{11}, λ_{12}, λ_{21}, λ_{22}
Coefficients in Onsager-based irreversible thermodynamics
α Generalised thermal conductivity
β Generalised mass conductivity

CAN THERMODYNAMICS BE MADE MORE SIMPLE?

M. D. Dampier

Department of Mathematics
University of Leicester

A method of handling the elementary mathematics of classical thermodynamics using some simple ideas drawn from vector calculus is developed. It is contended that, with no appreciable increase in mathematical difficulty, this method gives the student a greatly improved total view of the thermodynamics of any system he is studying, and by clarifying the mathematical structure frees his attention for the more important physical questions.

INTRODUCTION

The beginner studying thermodynamics has two difficulties to overcome. In the realm of interpretation he has to make something of the concept of a quasistatic process although, not being in possession of a genuine dynamics of heat, he is in no position to answer clearly the question, "What terms are being neglected when a real process is replaced by its quasistatic approximation, and when is this justified?" And then in the realm of the theory proper he is faced with such an abundance of different variables and partial derivatives that the essential simplicity of the theory is lost, and he is in no position to answer the question, "What equations between the thermodynamic quantities of a system are to be expected? and how do we know when we have exhausted the possibilities?"

In an attempt to deal with the second of these difficulties the following method of handling the thermodynamic equstions has been developed. Whether or not it actually achieves a real simplification must remain to some extent a matter of taste.

COORDINATES IN THE PHASE SPACE

In the classical thermodynamic theory we study various functions defined on a certain space. This space - the phase space or state space - represents the totality of equilibrium states of the system, and the functions are usually known as functions of state. The phase space has no intrinsic geometric structure, and none of the functions of state can be preferred absolutely as a coordinate in the space. For the common simple case of a homogeneous fluid the phase space can be described by the coordinates (p, T), or (p, V), or (V, T), where p is the pressure, T the temperature, and V the volume of the fluid; or we can use (S, V), where S is the entropy, and so on. No pair of functions is of more fundamental significance than any other pair. The essential point in this case is that the phase space is two-dimensional and so any two independent functions defined on it can be used as coordinates (independent variables). Difficulties can arise with particular proposed choices, as for example in the case of water when (p, V) cannot be used as global coordinates since two distinct states can be found having the same values of p and V.

The continual changes from one to another set of independent variables arising from the fact that none of the physically significant functions have any preferred role as coordinates is the source of confusion to the beginner. The fundamental idea of the method given here avoids this by choosing once and for all arbitrary coordinates (x_1, x_2) in terms of which all functions of state can be, at least theoretically, expressed. Naturally (x_1, x_2) can be specialized to (p, T) or some other pair if required, but that turns out to be unnecessary. We do not need to assume that a single pair of coordinates are adequate for the whole space but will not follow that up here. The extension to higher dimensional phase spaces is similar.

To the mathematician the phase space is a differentiable manifold, and the thermodynamic theory turns out to involve the study of certain differential forms defined on the manifold. Such a statement, however, will not help the beginner. Fortunately in the cases of elementary interest we can express all that is required in a mathematical form which will already be familiar to the student from his knowledge of two-and three-dimensional space. We utilize certain ideas drawn from vector calculus in order to abbreviate the writing of partial differential expressions and to provide a practical calculus for handling the thermodynamic equations.

MATHEMATICAL IDEAS

We assume for definiteness that we are dealing with the two-dimensional case. Once the space has been coordinatised we can introduce the gradient vector of any function in the usual way

$$\nabla p \equiv \left(\frac{\partial p}{\partial x_1}, \frac{\partial p}{\partial x_2}\right). \tag{1}$$

Let Γ be any smooth curve in phase space given parametrically by $x_1(\tau)$, $x_2(\tau)$, then along Γ any function of state becomes a function of τ. Denoting by a dash the derivative with respect to τ of any function of τ, we can define the tangent vector, $\mathbf{t}$, to Γ by

$$\mathbf{t} \equiv (x_1', x_2'). \tag{2}$$

Then by the chain rule for partial differentiation we have

$$p' \equiv \nabla p \cdot \mathbf{t} \tag{3}$$

where we use the scalar product dot notation for what is strictly the bracket product between a vector and a linear form although we need not worry about this since we can regard the dot product as a mere shorthand for the expression $\frac{\partial p}{\partial x_1} x_1' + \frac{\partial p}{\partial x_2} x_2'$.

Now suppose, p, V, T are three functions of state. Let Q be any point in the phase space and let Γ be any curve passing through Q and such that, $T' = 0$ at Q, then

$$\left(\frac{\partial V}{\partial p}\right)_T = \frac{V'}{p'}. \tag{4}$$

Thus

$$\left(\frac{\partial V}{\partial p}\right)_T = \frac{\nabla V \cdot \mathbf{t}}{\nabla p \cdot \mathbf{t}}, \tag{5}$$

where $\mathbf{t}$ is such that $\nabla T \cdot \mathbf{t} = 0$.

Since the space is two-dimensional, any three vectors defined at a point must be linearly dependent, so that ∇p, ∇V, and ∇T must satisfy an equation

$$\alpha\nabla p + \beta\nabla V + \gamma\nabla T = 0, \tag{6}$$

where α, β, γ are scalars which are not all zero. In fact we shall for simplicity usually assume without further ado that in the cases of interest α, β, γ are all non-zero. We shall see that equations of this form are fundamental in thermodynamics. Obviously only the ratios of α, β, γ are significant; these ratios being in fact various partial derivatives. Application of (5) to (6) readily gives for example

$$-\frac{\alpha}{\beta} = \left(\frac{\partial V}{\partial p}\right)_T, \quad -\frac{\gamma}{\beta} \left(\frac{\partial V}{\partial T}\right)_p, \tag{7}$$

which could alternatively be read off directly from the chain rule equation

$$\nabla V = \left(\frac{\partial V}{\partial p}\right)_T \nabla p + \left(\frac{\partial V}{\partial T}\right)_p \nabla T, \tag{8}$$

which must be merely a rewritten form of (6).

So long as we keep to just the three functions p, V, T, we can without ambiguity denote $\left(\frac{\partial V}{\partial p}\right)_T$ by V_p etc. The relations between the various ratios of α, β, γ immediately give the two standard results

$$V_p P_V = 1 \quad \text{etc.} \tag{9}$$

$$P_T V_p T_V = -1 \quad \text{etc.} \tag{10}$$

Finally we introduce in the usual way the vector product operation, although in this two dimensional case it simply gives a scalar result,

$$\mathbf{a} \wedge \mathbf{b} = a_1 b_2 - a_2 b_1, \tag{11}$$

and the curl

$$\nabla \wedge \mathbf{a} = \frac{\partial a_2}{\partial x_1} - \frac{\partial a_1}{\partial x_2}. \tag{12}$$

We will need the two elementary results

$$\nabla \wedge (\phi\mathbf{a}) = \nabla\phi \wedge \mathbf{a} + \phi\nabla \wedge \mathbf{a}, \tag{13}$$

$$\nabla \wedge \nabla\phi = 0. \tag{14}$$

In terms of the $\wedge$ operation we can put down a further form for the partial derivatives e.g.

$$\left(\frac{\partial V}{\partial p}\right)_T = \frac{\nabla V \wedge \nabla T}{\nabla p \wedge \nabla T} \tag{15}$$

This last form is the most convenient for generalization to the n-dimensional case.

TWO-DIMENSIONAL THERMODYNAMICS

We begin with the case of the homogeneous fluid. Its thermodynamics depends upon the fact that if Γ is any smooth curve in the phase space that does not cross a discontinuity such as a change of phase then

$$U' = TS' - pV' , \tag{16}$$

where U is the internal energy, the other variables are as before, and the dash denotes differentiation with respect to the parameter of the curve. The case of discontinuities requires a finite form of the equation which it would not be appropriate to tackle at this stage. Equation (16) gives

$$\nabla U \cdot \mathbf{t} = T\nabla S \cdot \mathbf{t} - p\nabla V \cdot \mathbf{t}, \tag{17}$$

and since this is true for all $\mathbf{t}$,

$$\nabla U = T\nabla S - p\nabla V \tag{18}$$

Equation (18) is the fundamental equation although it does not figure so prominently in the theory as the corollary obtained by taking the curl of both sides and using (13) and (14) to get

$$\nabla T \wedge \nabla S - \nabla p \wedge \nabla V = 0 \tag{19}$$

This equation comprises the Maxwell relations, the various standard forms can be obtained by specializing the variables. One of the advantages of the present method lies in this elegant and convenient form of the Maxwell relation and in the ease with which it may be used in deductions as shown below.

The aim of most of the thermodynamic calculations for the simple fluid is to connect quantities with the equation of state. In differential form this equation is

$$\nabla V - V_T \nabla T - V_p \nabla p = 0 \tag{20}$$

with alternative forms available in terms of other partial derivatives. In some cases it is convenient to introduce the expansion coefficient, β, and the isothermal compressibility, k, by

$$\beta V = V_T , \quad kV = - V_p . \tag{21}$$

The equation (20) becomes

$$\frac{1}{V} \nabla V - \beta \nabla T + \kappa \nabla p = 0. \tag{22}$$

From the point of view adopted here we would regard k, β as being defined by (22), i.e. they are the coefficients that appear in the p, V, T equation when ∇V is given coefficient $\frac{1}{V}$ and the signs are as shown. Since in general we do not possess the equation of state in closed form, (22) is a more realistic way of writing (20), showing that we have to deal with two constitutive coefficients k, β whose functional dependence on state coordinates is unknown.

Further constitutive coefficients are of importance, the most significant being the heat capacities. These are defined as the coefficient of ∇T in an equation

$$T\nabla S - C\nabla T + \ldots = 0, \tag{23}$$

where various choices are possible for the third variables. So for the usual C_V and C_p,

$$T\nabla S - C_V \nabla T + \lambda \nabla V = 0 \tag{24}$$

$$T\nabla S - C_p \nabla T + \mu \nabla p = 0 \tag{25}$$

The coefficients λ, μ are readily found by using the Maxwell relation; for example from (25)

$$T\nabla S \wedge \nabla T + \mu \nabla p \wedge \nabla T = 0 , \tag{26}$$

since $\nabla T \wedge \nabla T = 0$, whilst from (20)

$$\nabla S \wedge \nabla p - V_T \nabla T \wedge \nabla p = 0. \tag{27}$$

Hence, using the Maxwell relation and the antisymmetry of $\wedge$, we obtain

$$TV_T = \mu \; . \tag{28}$$

A similar method shows

$$-Tp_T = \lambda. \tag{29}$$

Thus

$$T\nabla S - C_V \nabla T - Tp_T \nabla V = 0 \tag{30}$$

$$T\nabla S - C_p \nabla T + TV_T \nabla p = 0 \tag{31}$$

It will be instructive to give some examples of the use of these equations. First, subtraction of (31) from (30) gives

$$(C_p - C_V)\nabla T - Tp_T \nabla V - TV_T \nabla p = 0, \tag{32}$$

which on comparison with (30) shows that

$$C_p - C_V = TV_T p_T, \tag{33}$$

Secondly, if we eliminate ∇T from (30) and (31) we obtain

$$(C_p - C_V)\nabla S - C_p p_T \nabla V - C_V V_T \nabla p = 0, \tag{34}$$

which immediately gives the adiabatic compressibility, since

$$\left(\frac{\partial p}{\partial V}\right)_S = - \frac{C_p}{C_V} \frac{pT}{V_T} = \frac{C_p}{C_V} p_V . \tag{35}$$

And thirdly, as an example of a second-order result, we apply to (31) the same technique we used in deriving the Maxwell relation, obtaining

$$\nabla \left(\frac{C_p}{T}\right) \wedge \nabla T = \nabla_T \wedge \nabla p, \tag{36}$$

which easily leads to

$$\left(\frac{\partial C_p}{\partial p}\right)_T = - T \left(\frac{\partial^2 V}{\partial T^2}\right)_p \tag{37}$$

This more or less exhausts the sort of thing that can be done in thermodynamics for this simple type of system. Of course other possible equations can be written down and their coefficients may have some experimental significance. For example.

$$\nu\nabla U - \nabla T + J\nabla V = 0, \tag{38}$$

in which J is the Joule coefficient. Since from the basic equation (18) and the C_V equation (30) we immediately deduce that

$$\nabla U - C_V\nabla T + (p - Tp_T)\nabla V = 0 \tag{39}$$

we have directly that

$$\nu = \frac{1}{C_V}, \quad C_V J = p - Tp_T. \tag{40}$$

It is not hard to see that besides the fundamental equation (18) only the equation of state (20) and one of the heat capacity equations, say (31) need be written down. The Maxwell relation reduces the four constitutive coefficients to three, and then the coefficients in any other equation can always be expressed in terms of the three basic ones. Higher order equations can be obtained by using the curl operator.

THREE-DIMENSIONAL THERMODYNAMICS

The method outlined so far will do no more than can be done by direct manipulation of partial derivatives, but it gives a more systematic approach to these manipulations. In particular it goes a long way towards answering the question posed in the introduction, "What equations between the thermodynamic quantities of a system are to be expected, and how do we know when we have exhausted the possibilities?" To support this contention we consider a three-dimensional case, namely that of a fluid with significant surface tension γ. The fundamental equation becomes

$$\nabla U = T\nabla S - p\nabla V + \gamma\nabla A \tag{41}$$

where A is the surface area of the fluid. The Maxwell relation takes the vectorial form

$$\nabla T \wedge \nabla S - \nabla p \wedge \nabla V + \nabla\gamma \wedge \nabla A = 0 \tag{42}$$

Partial derivatives are given by formulae like

$$\left(\frac{\partial p}{\partial A}\right)_{S,V} = \frac{(\nabla p \wedge \nabla V) \cdot \nabla S}{(\nabla A \wedge \nabla V) \cdot \nabla S} \tag{43}$$

Now to proceed in a systematic manner we divide the six functions that appear on the right hand side of (41) into two groups of three and write down as our basic equations those giving the gradients of one set in terms of the gradients of the other. Systematically it is best to select S, V, A as one set and T, p, γ as the other, following the division already given in (41). But since the constitutive coefficients need to be as accessible as possible to experiment we need to be guided more carefully. In fact we write

$$\nabla S = \frac{C}{T}\nabla T + \ell\nabla p + m\nabla A \tag{44}$$

$$\nabla V = \beta V\nabla T - kV\nabla p + n\nabla A \tag{45}$$

$$\nabla\gamma = \gamma_T\nabla T + \gamma_p\nabla p + \gamma_A\nabla A \tag{46}$$

where we can identify the coefficients C, β, k as respectively, the heat capacity at constant pressure and area, the volume coefficient of expansion at constant pressure and area, and the isothermal constant area compressibility. The coefficients ℓ, m, n can be given interpretations in terms of partial derivatives, but more important they can be eliminated by Maxwell's relation. We have from (44), (45), (46) that

$$\nabla T \wedge \nabla S = \ell\nabla T \wedge \nabla p + m\nabla T \wedge \nabla A \tag{47}$$

$$\nabla V \wedge \nabla p = \beta V\nabla T \wedge \nabla p + n\nabla A \wedge \nabla p \tag{48}$$

$$\nabla\gamma \wedge \nabla A = \gamma_T\nabla T \wedge \nabla A + \gamma_p\nabla p \wedge \nabla A. \tag{49}$$

Hence, by Maxwell's relation (42),

$$(\ell + \beta V)\nabla T \wedge \nabla p + (m + \gamma_T)\nabla T \wedge \nabla A + (n - \gamma_p)\nabla A \wedge \nabla p. \tag{50}$$

Thus, since p, T, A are independent, we deduce

$$\ell = -\beta V, \quad m = -\gamma_T, \quad n = \gamma_p \tag{51}$$

So for example

$$\left(\frac{\partial V}{\partial A}\right)_{p,T} = n = +\left(\frac{\partial\gamma}{\partial p}\right)_{A,T} \tag{52}$$

In other words if there is surface absorption the surface tension must depend upon pressure, a well-known result.

We can say that unless we bring in further variables or proceed to higher order relations we have exhausted all the independent equations between the constitutive coefficients.

Enough has now been said to give the essence of the method and to outline its advantages. The extension to phase changes and higher-order transitions will be given elsewhere.

THERMODYNAMICS: A NEW TEACHING APPROACH

D. R. Croft, P. W. Foss and M. J. Denman

Department of Mechanical and Production Engineering
Sheffield City Polytechnic

This paper outlines the content and presentation of a course suitable for students in both technology and business studies programmes. The course reflects the considerable guidance from industrialists in manufacturing and process industries; it has two main thrusts: energy conversion and energy conservation.

The aims of the course are:

1. to produce a student who can "manage" the use of energy; that is, he can design processes, devices and systems which make the most effective use of energy.

2. to provide a course, the content of which must satisfy at least one of the criteria.

   (i) it must have direct relevance to realistic engineering situations;

   (ii) if it is theoretical foundation work, then it must under-pin work described in (i) above.

INTRODUCTION

It was the experience of one of the authors that he was introduced to Thermodynamics by the Senior Professor of the Department. There was no need to take notes since the Professor delivered his lectures from a lectern which supported a book (written by him); all that was required was the insertion of a tick and the date, on a copy of the book acquired by the student. One of the more illuminating comments was that entropy "could be considered as the chance of finding an object in nature at that temperature"; another was a demonstration of the second law using a chain. First holding the chain at arms length with the chain hanging down to the floor, the Professor then raised the chain in two ways; one, by simply raising his arm, secondly by waggling his hand such that the top link of the chain moved in a horizontal circular path. Eighty students remained mystified.

One week, a research student deputised for the Professor. His approach was to quickly revise the non-flow and steady-flow energy equations, explain that entropy was a derived property and point out that a great many cycles were basically "2 isentropics wedged between 2 lines of constant pressure". The final pearl - that the performance of any plant could be thought of as "what you get, compared to what you have to pay for". Another lecture on fluid properties and the use of tables, and eighty students felt prepared for the next chapter of the Professor's book.

Since that time no author seems to have been willing to follow the sort of approach adopted by the research student although references (9) and (10) are steps in that direction. The subject of Thermodynamics must now be seen in a wider context since people in general are concerned with the environmental effects of energy conversion and the increasing costs of energy supply. Industry now is much more concerned with energy costs and the need for energy conservation to reduce such costs.

Given the limited time allocated to this paper, the authors have chosen to illustrate the content and presentation of the course by reference to the areas of energy conversion and energy systems modelling.

ENERGY CONVERSION

The introduction to energy conversion is through the use of a novel graphic device, the energy conversion triangle. This device shows the types of conversion processes available and also illustrates the dilemma between high cycle efficiency and power density requirements.

Thring (2) in 1962 used an energy conversion diagram to illustrate a new route for electrical production, magneto-hydrodynamics. His diagram was incomplete and only showed one route for stored energy to mechanical energy. The authors have expanded this diagram and re-named it the Energy Conversion Triangle. Commercially produced conversion diagrams are available but they are usually too complex for use in introductory lectures.

The Energy Conversion Triangle is best taught using an interactive teaching method. Part of the diagram is drawn and labelled with the three principal energy forms (Fig. 1). The students are then asked to name types of stored energy and devices to directly convert one form of energy to another.

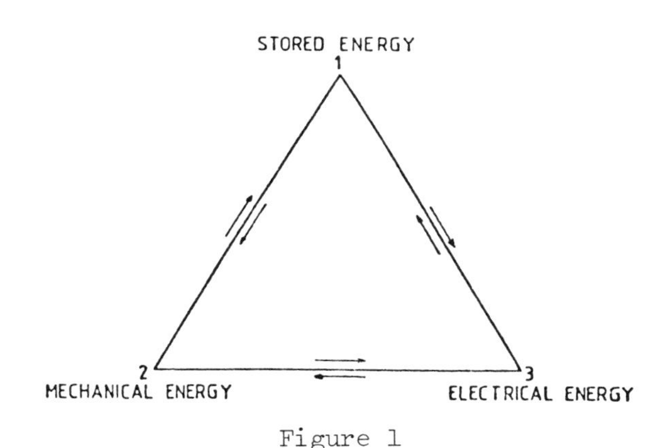

Figure 1

Typical replies from the students detail the following stored energies and devices for direct 'routes' along the diagram;

Stored Energy:	Chemical (fossil fuels)	Solar
	Nuclear	Wave
	Potential (high level reservoirs, pressure)	Wind
	Kinetic (rotating flywheel)	

Route 1-2	Water turbine	Route 2-1	Pump
	Salter duck		Wave generator
	Windmill, aerogenerator		Propeller

Route 2-3	Dynamo Alternator	Route 3-2	Electric motor Linear motor
Route 1-3	Fuel cell Battery	Route 3-1	Electrolysis Electro-deposition

If students mention routes involving heat these are discounted as the call is for direct routes.

The next stage is to complete the diagram with devices which employ the 'heat routes' (Fig. 2).

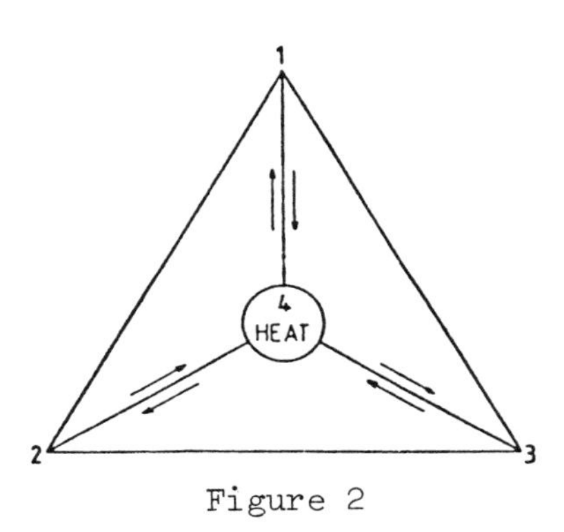

Figure 2

Again, typical replies from students detail the following devices for heat routes along the diagram;

Route 1-4	Solar panel (non-voltaic) Combustion Nuclear fission/fusion	Route 4-1	Dissociation
Route 4-2	Steam/gas turbine Piston engine	Route 2-4	Friction
Route 4-3	Magneto-hydrodynamics Thermoelectricity Thermionics	Route 3-4	Electrical resistance heating Induction heating

Conversion Efficiencies

Students are then asked the key question: What are the efficiencies of conversion from one form to another?

Discussion will reveal that many routes have theoretical efficiencies of 100% and achievable efficiencies greater than 90%, i.e. electric motors, pumps, water turbines, batteries, etc. The route that many students estimate incorrectly is the one used by conventional thermal power stations which have generating efficiencies of less than 35%. At this stage notes can be given to the students with an explanation for the low efficiencies for routes 1-2-4, 1-4-3.

Points to be Made to Emphasise the Significance of the Heat Route

Mechanical energy can be completely and continuously transformed into heat by friction and similarly electrical energy can be transformed to heat with 100% efficiency. Heat on the other hand cannot be converted to mechanical or electrical energy with 100% efficiency. The second law of thermodynamics states that some energy must inevitably be rejected. The highest achievable efficiency for conversion of heat to mechanical or electrical energy via routes 1-4-2 and 1-4-3 is given by the Carnot efficiency.

$$\eta = 1 - \frac{T_1}{T_2}$$

where T_1 = absolute temperature of heat sink

T_2 = absolute temperature of heat source

When fuel was relatively inexpensive and plentiful an inefficient route could be chosen as the fuel cost would only be a fraction of the total cost. As fuel prices have risen the efficiencies of power stations have improved to offset the additional fuel cost element. Table 1 gives the average overall efficiencies of thermal power stations in the U.K.

Table 1 Overall thermal efficiencies of U.K. power stations

YEAR	OVERALL THERMAL EFFICIENCY
1932	17.08
1937	20.05
1950	21.45
1960	26.53
1970	28.30
1980	31.68

At this stage lectures are given in the basic thermodynamic cycles, the Carnot and Rankine cycle, etc. This leads to the modifications made to the Rankine cycle to improve efficiency. It is then shown that the generating efficiencies of modern thermal power stations are unlikely to improve appreciably in the near future while the cost of fossil fuels will increase rapidly and the availability will decrease.

The next question to be answered is "why do we continue to use such an inefficient route?" In answering this question the concept of power density can be introduced.

Power Density

The power density is defined as the rate at which energy crosses a square metre of heat exchanger surface. When students are asked to estimate the power density of a fossil fuel boiler most if not all, have no idea. With a little practice and help the students can arrive at a reasonable guestimate. The larger power stations in the U.K. are rated at 2,000 MW electrical output, hence the students can deduce a nominal fuel input of 6,000 MW, or 1,500 MW for each of the boilers. A boiler would comprise the furnace, superheater, reheater and economiser. The major difficulty for the students it to estimate the amount of steam tube inside the boiler combination. Having been shown cross-sectional views of the boiler many students will accept a value of the order of 100,000 m^2. In practice, power densities of 50,000 watt/m^2 can be achieved with fossil fuels.

Students can then make comparisons with other devices. In the U.K. the daytime average for solar radiation is approximately 200 watt/m^2. Thus the students can see that solar power stations would have to cover immense areas to have the same power output as conventional ones and that the savings in fuel would not justify the initial capital cost at current fuel prices. This illustrates that plant with high power densities are therefore compact resulting in low capital and operational costs.

Nuclear power stations have power densities similar to fossil fuel power stations, the fuel handling charge is greater but the overall running costs are less due to the low cost of the fuel.

Other devices which can be discussed are windpower, which only has a power density of 400 watt/m^2 and wave power which can have power densities of several thousands of watt/m^2 over the first several metres of depth of water. Fuel cells now have generating efficiencies approaching 70 % and can have power densities of 600 watt/m^2. Unlike wind and wave generators, they can operate at maximum power anytime and do not rely on natural forces which can vary considerably.

ENERGY MODELLING

The type of energy conversion device used does not depend solely on the most economical method. Other issues must be considered, security of supplies, overseas trading balance, environmental effects, safety, etc. Many students have preconceived ideas on these issues, the two most popular being (i) the waste of valuable resources (fossil fuels) to obtain heat when solar energy is free and inexhaustible, and (ii) that nuclear power is dangerous and should not be used. Both these arguments are valid to a greater or lesser extent but the students must be aware of the consequences of restricting fossil fuels for premium use and the banning of nuclear power. The answer to question (i) is largely one of economics and to (ii) is achieved by the use of energy modelling.

A model of the U.K. energy usage and supply (based on references 5 and 6) has been computerised and is so designed that the student can alter various input parameters and see what consequences result in the provision of energy supplies in the future. If say nuclear power were progressively run down over a twenty year period, and was replaced by coal burning power stations, the student would then be aware of the logistical consequences of his planning. Could enough new mines be opened, how much coal would have to be transported, could enough power stations be commissioned in time? etc. The energy model will not give definitive answers to our future energy needs as it relies on many assumptions to our present and future energy patterns, but it does illustrate the need for a mixed fuel policy. Its main virtue is that it allows the student to see the consequences of his ideas, inputted as forecasting patterns.

Methodology of the Model

In the model, energy demands are allocated to four sectors of the economy; the industrial sector, commercial and services sector, transport sector and domestic sector. These sectors require energy supply for heating purposes, energy supply for power and energy supply for transport. The sectors and supplies of fuel are shown in Fig. 3, secondary choices of fuel being used to top up when a first choice fuel reaches its maximum supply. Electrical energy for the four sectors can be generated from fossil, nuclear and renewable resources, while heat can be supplied from gas, oil, coal, solar, electricity, combined heat and power installations and in the future by synthetic oil and synthetic natural gas.

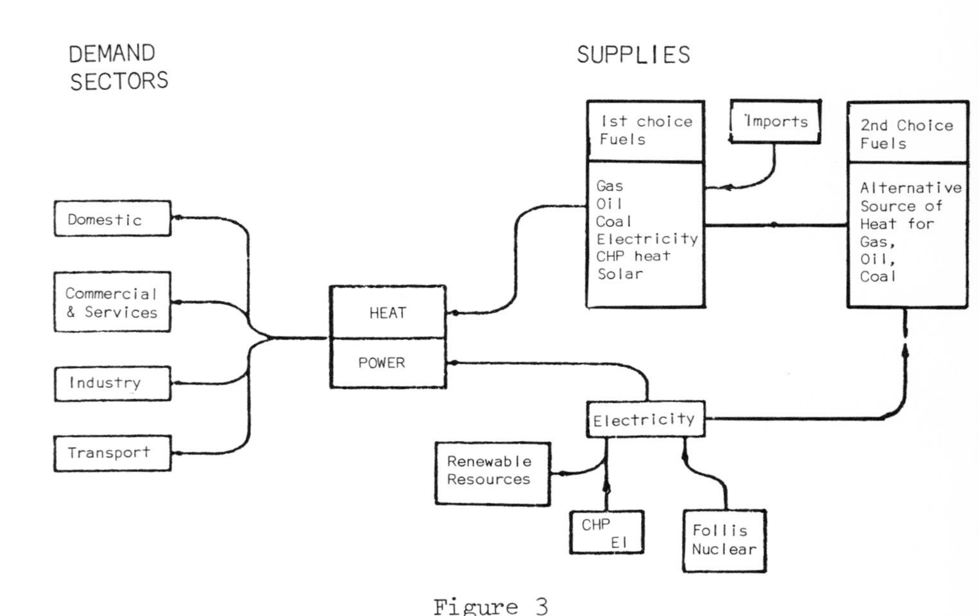

Figure 3

The link between energy demands and available resources is made by employing a set of 'market shares'. The market shares are user variables and will depend on the forecast of the person using the model.

The computer model evaluates the demand from the four sectors and allocates supplies from the five main fuels (gas, oil, coal, nuclear, renewables). The basic calculation of the model is repeated at time intervals, stretching to the year 2000. The 'forecasting' as such involves changes to market shares, conversion efficiencies, etc. and user accessible variables which prescribe the four demand sectors. Many variables are allocated internally in order for the programme to operate quickly. Only the advanced students are encouraged to change all the variables.

Any given fuel supply is limited by some factors or combinations of factors, so that the model is constrained to simulate the real situation. E.g. coal, oil and gas supplies are limited by the lesser of (i) available reserves or (ii) available plant (coal mines, oil and gas wells), plus maximum imports (which are prescribed by the user).

The allocation of supplies to demand will depend on logistical constraints, i.e. electricity will be allocated to those demand sectors that can use nothing else; while substitutes (synthetic oil and gas) will have to be used if the energy source is not sufficient to meet demand. A sequence of allocations of resources to demands is necessary and this may entail up to twenty allocations.

The output from the model will give at yearly intervals, up to the year 2000, the following output:

(i) total demand for energy (joule/year) in each of the following four sectors:- domestic, commercial and services, industry, transport.

(ii) the supply of energy (joule/year) from each of the following five sources:- oil, gas, coal, nuclear, renewables.

(iii) imports of fuels (if any) for the fuels:- coal, oil, gas.

The model at Sheffield City Polytechnic is based on the U.K. economy but by suitable alterations to the limiting factors and input data, a model can be constructed to suit other developed countries or third world countries. To facilitate the initial use of the computer model by inexperienced students, two "views" are stored. One view has variables and trends set to a 'low energy' future, while the other has a more flagrant use of energy which results in substantial imports of fuel to meet demand towards the year 2000.

Suitable student exercises can be framed around maximising industrial or commercial activity without increasing energy demand: using all or no nuclear power to generate electrical power: the impact of CHP on fuel supplies and reserves.

These exercises illustrate that there are no definitive answers to the problem of energy supply, and that a balanced view should be taken by students to these problems. They also remind the student that energy demands and supplies for one sector are intimately related to those in other sectors and most importantly that technological decisions have social implications. The use and misuse of energy affects everybody all the time.

REFERENCES

1. Croft, D. R., and Denman, M. J., "Teaching Energy". Conf. on Energy Education, University of Rhode Island, 1981.

2. Thring, M. W., "Magnetohydrodynamic Power Production", I.E.E. Journal, 1962, Vol. 8.

3. Chapman, P., "Fuel's Paradise", Penguin, 1979.

4. Hoyle, F., "Energy or Extinction", Heinemann, 1977.

5. Energy Paper No.39, "Energy Technologies for the U.K.", H.M.S.O.

6. Energy Modelling, Special Energy Policy Publication, I.P.C. Science and Technology Press, Ltd. 1974.

7. O'Callaghan, P. W., "Design and Management for Energy Conservation", Pergamon, 1981.

8. Payne, G. A., "The Energy Manager's Handbook", Westbury House (IPC), 2nd Ed. 1980.

9. Eastop, T. D. and McConkey, A., "Applied Thermodynamics for Engineering Technologists", Longman, 3rd Ed. 1978.

10. Open University: Courses T231 (Units 10 - 16 incl.) and T233. Both concerned with Thermofluids.

THERMODYNAMIC NOMENCLATURE

J. D. Lewins

University of Cambridge

A survey questionnaire, reproduced at Annex A, was issued to all authors contributing to the Workshop to elicit some sample response to the question of reviewing thermodynamic nomenclature. I am grateful to those who have relied, many going far beyond a simple YES/NO and some of their suggestions are incorporated here. I hope more will come out of our discussion.

Why should we review nomenclature? As several remarked, there is a good deal to be said for leaving well alone. But this is to imply that our present nomenclature is indeed well. There will be plenty of voices in discussion to argue the principle we know so well in Cambridge, the 'principle of unripe time', that I should put only the arguments <u>for</u> change.

Our discipline grew out of nineteenth century physics and technology with even earlier roots in caloric theory. It is not only that we continue to employ terms, such as heat capacity, which reflect caloric theory, but that the discipline remains unusual in being capable of a base in some generalised macroscopic origins and leading by largely qualitative, logical arguments to its major constructs in the concept of entropy. The style of argument used is reminiscent of Greek geometers. Is it a good thing that we retain a nomenclature that isolates us from other disciplines?

I think not. The argument that there is an entrenched usage that should not be disturbed is an argument I suggest on the behalf of the teacher not the taught. What must a student of thermodynamics think when he comes from a lecture on linear systems into a thermodynamic lecture to be told that 'in a system, no matter is transferred'. The intelligent student is going to question the intelligence of a lecturer who so cuts himself off from the fundamental concepts of systems engineering. And our difficulty is because thermodynamics developed these terms first and, being overtaken by newer disciplines, is in danger of being swept away by them.

A second but perhaps insufficient reason for change is simply that a logical structure of names must be appealing and make learning easier. If indeed so naive a statement is insufficient, I believe the force of the argument is compounded if we recognise the imminent arrival of intelligent computer systems. If again thermodynamics is not going to be left out of the benefits of artificial intelligence, it behoves us to employ a co-ordinated language to relate problems of thermodynamics to a wider canvas. Imagine the occasion when a speaking artificial intelligence is to learn the fundamentals of thermodynamics. Computer: "What is heat?" Our thermodynamicist: "Heat is a mode of energy transfer" Computer: "Goodness, then what is heat transfer that you told me about last week - is that energy transfer squared?"

If we compare the problem to two recent measures of rationalisation, I believe it is neither so large nor so impossible as some make out. The adoption of SI units was slow and painful - but not painful to students and practicing engineers who have in fact reaped the benefit of the stress their mentors went through. The revision of chemical names such as ethanol for ethyl alcohol is another example where chemists gritted their teeth and acknowledged the value of providing a logical nomenclature for their discipline. Not easy but not impossible.

These are the reasons in general I suggest for a thorough review of thermodynamic nomenclature being needed. Now let me turn to the details of the questionnaire.

1. Sign Convention

YES/NO 11 to 4 YES/NO 13 to 5

A recognition of the undesirability of a conflict between sign conventions and a strong willingness to use the 'Chemists Convention' of positive work in. Useful comments were made about the need to be able to argue in terms both of work out and work in and the increased difficulty of understanding such things as 'lost work' theorems which would now become increased work supply theorems. Perhaps we should allow discussion of both modes of work (and both modes of heat, in and out) while writing the formal equation as $\Delta E = Q + W$. I note one further conflict, that in extending the argument to continuous control surfaces, the direction convention has the normal vector pointing outwards.

2. Adiabatic/Adiathermal.
YES/NO 12 to 5 YES/NO 11 to 2
YES/NO 1 to 1 YES/NO 2 to 6.

A strong majority wish to retain the use of adiabatic despite its murky provenance and adiathermal was not supported. At the same time, many (but not all) expressed overall a wish to use straightforward english words in preference to new words of Green and Latin origin, even at the expense of a phrase in English where one greek word suffices. (Others like their isochorics and their isobarics). Some referred to the problem as one of not introducing jargon. I feel that a discipline is entitled to invent words to mean something which originates in the discipline if no ordinary words mean the same thing. It is jargon when we use words which have no natural root (e.g. adiabatic) and we are only institutionalising jargon when we accept usage of inappropriate words.

3. Heat Capacity.
YES/NO 9 to 12 YES/NO 8 to 13
C_v YES/NO 7 to 3 C_p YES/NO 7 to 3
Latent Heat YES/NO 4 to 6

We all accept the meaning of C_v etc. in fundamental terms but the suggestion was that the nomenclature involving 'heat' was inappropriate. Views were divided. Rather than 'coefficient', there was a suggestion to use

internal energy change with temperature at constant volume etc.

or

terminology based on rate

Overall, a majority in favour of something more rational than 'sensible' and 'latent heats'. Professor Dampier's paper (in this session) promotes the suggestion that C_p, C_v etc. should be regarded as entropy coefficients at constant x analogous to expansion coefficients and compressibility coefficients.

$$C_x \equiv T\left(\frac{\partial S}{\partial T}\right)_x$$

4a. Phase Change Symbols

YES/NO 16 to 1 YES/NO 5 to 9

Several pointed out that the subscript f was probably from the German flussigkeit for gas; many thought the difficulty of confusing 1 with l would remain. Haywood has already used 'enthalpy of evaporation' for latent heat. We should perhaps note the chemists usage of Δh_x rather than h_{fg} with a subscript e for evaporation, s for sublimation, f for fusion etc. More complicated cases (two solid phases say) would need a pair of subscripts identifying the phases concerned but the notation would imply constant pressure-temperature.

4b. Fluid (see above)

Most take fluid to encompass liquid and gas and again most take it to encompass one-phase fluids and two-phase fluids, i.e. liquid-gas mixtures. A further question arises over the meaning of 'vapour'. To many, vapour is something fluid, not an ideal gas and implying something near the saturation curve. This raises difficulties of precision. Certainly at low enough pressures and temperatures one can enter the ideal gas region immediately from the saturation curve; does this make an ideal gas a vapour? It is because of this that one cannot say that vapour is the one-phase fluid that is below its critical point but not a liquid.

For some years in Cambridge we have used the term vapour precisely to mean only the two-phase gas liquid mixture, whether mixed or separated by a meniscus in a way that is not described as such by the thermodynamics of simple systems. At Annex B I attach a copy of our teaching note.

5. Reversible

YES/NO 6 to 1

Only a minority want to change 'reversible' and these on the whole preferred effacable or non-dissipative. As one war weary veteran remarked, "it wouldn't matter what you changed it to. They wouldn't start to understand it."

6. Energy and Vigour

YES/NO 2 to 12

No enthusiasm for a duplication of units. Some sought to meet the problem by insisting that one spoke of whether the power station was producing work or waste heat; to either of these you attach the units of energy or power. I don't think this quite meets the difficulty. "How much energy is the power station emitting?" Answer: "It depends what you mean by energy..."

On reflection, I would not pursue distinct named units (nor 'vigour'). The logical answer is to use exergy. The power station produces 4000 MW energy and 1000 MW exergy. Does this mean we need a word for rate of exergy transfer to complement power as rate of energy transfer?

8. Universal Gas Constant. This was the examiner's disaster and I apologise particularly to Yon Mayhew for ruining a perfectly valid question with wrong units and poor proof-reading. We certainly have different conventions for the universal gas constant: R_M is the Royal Society usage. One ingenious reply used $N_o k$, the Avogrado number and the Boltzmann constant. Of course the problem stems from our difficulty in accepting a unit of mass and a unit of matter defined partly in mass terms. I think I will leave Dr. Mayhew to sort this muddled question out at our meeting.*

9. Other Points. Passing reference was made to the rational need to replace the term 'kilogram'; I ruled this outside the scope of a thermodynamics meeting! Professor Gurney drew my attention to the irrationality of using lower case p for pressure and upper case T for temperature. Lower case t is necessary for time so I suppose rationality would dictate P and T. Similarly, I have difficulty of how to distinguish at the blackboard between V or v for speed and V or v for volumes. The German usage is for c (celerity) for speed and a bold vector **v** for velocity. Then we would use c_s perhaps for the speed of sound. Professor Landsberg commented on a long standing problem of the partial differential notation which is so acute in thermodynamics. His paper at the Workshop enlarges on this point but Appendix C is a recent internal note on this difficulty as we experience it in Cambridge.

attach:	Appendix A	Questionnaire
	Appendix B	Pure Substance Diagram
	Appendix C	Partial Differentials

---

* see the following paper.

ANNEXURE A - NOMENCLATURE SURVEY

from: J D Lewins
Engr Dept
U Cambridge

THERMODYNAMIC WORKSHOP
Cambridge Sep 1984

Thermodynamic Nomenclature Changes

In a paper to the meeting, I have suggested the need to reconsider some of the currently used terminology and sign convention of engineering thermodynamics. These problems arise from the history of thermodynamics with its overtones of the calorific theory of heat.

It seems to me that our meeting is an opportunity to take general soundings on perceived problems with a view perhaps to establishing a consensus amongst teachers of thermodynamics. This note, therefore, is a survey to determine what those attending the Workshop would

(a) recognise as a problem
(b) recommend for improvement.

Our final session will allow time for presentation of the summarised results of this census and a discussion to identify common ground.

Please answer the following questions and return the paper to me by 1 August so that I may summarise the response of the meeting. Please feel free to write in alternative suggestions where a simple YES/NO seems inadequate and to suggest your own improvements.

Sign convention

The major problem of sign convention is the sign for work. Mechanical engineers in the UK are accustomed to treat work produced by a system as positive. Chemists and engineers on the Continent adopt the sign convention that work in, like heat in to a system, adds energy and is positive. The Open University it seems has adopted the chemists/continental (CC) convention.

The effect of the two conventions is essentially a matter of writing

$$\Delta E = Q - W \quad \text{or} \quad \Delta E = Q + W$$

Do you agree that the discrepancy is undesirable? YES/NO

Would you be willing to accept the CC convention? YES/NO

Adiabatic

Most writers take adiabatic (from the greek: no through path) to mean 'without transfer of heat'. The word is credit by Maxwell to Rankine (although I have not been able to find the reference in Rankine's papers). Early writers and physicists (e.g. Pippard) use adiabatic to mean both 'without heat' and 'reversible'.

Thermodynamic Nomenclature

Since 'diathermal' is accepted as meaning 'with the passage of heat' we might use 'adiathermal' for 'no heat passage'.

Do you support retaining the term adiabatic? YES/NO

If Yes, do you mean 'no heat' YES
or 'no heat and reversible' (isentropic) YES

If No, do you support 'adiathermal' YES/NO

Heat Capacity

We have adopted the terminology 'heat capacity' and 'specific heat capacity' in the UK (not yet in the USA where specific heat is commonly used). Even so, the terminology is not felicitous since the change might be brought about by work in an irreversible process.

Coupled with this is the need for a term to describe what is still commonly called 'latent heat'. i.e. the enthalpy change between phases at constant pressure/temperature (i.e. the old concept of 'sensible' and 'latent' heat).

Do you accept Heat Capacity? YES/NO
Do you accept Latent Heat? YES/NO

If No, would you accept terminology based on 'coefficients'?

e.g.

c_v is the '(temperature) coefficient of internal energy (at constant volume)' with temperature, at constant volume, specified where necessary. YES/NO

c_p is the '(temperature) coefficient of enthalpy (at constant pressure)' with temperature, at constant pressure, specified where necessary. YES/NO

h_{ph} is the 'phase coefficient of enthalpy (at constant pressure)' with constant pressure specified where necessary. YES/NO

Phase Change Symbols - liquid to gas

Many texts write changes from liquid to gas as, for example, h_{fg}, where f presumably stands for fluid. But both liquids and gases are fluids and the symbol is thereore illogical. It may have arisen in avoiding the subscript l (liquid) which on a typewriter might be confused with 1 (unity). Modern typewriters and printing should prevent any such confusion while at the blackboard we can use a script ℓ.

Do you use the term Fluid to encompass liquid and gas? YES/NO
Do you prefer phase changes as? h_{fg}/h_{lg}

Reversible

Reversible in thermodynamics means of course more than it does in plain english. A 'reversible' process restores the system and its environment. Several writers have sought a better word or phrase, including Bridgman with 'recoverable' and the use of 'effacable' in a similar context in Cambridge.

Do you seek to change 'reversible'? YES/NO

If Yes, do you favour

Replaceable	
Recoverable	Effaceable
Restorable	Non-dissipative

Power/Energy (thermal) and Energy (electrical)

A custom has grown up of refering to MW(e) and MW(t) to distinguish between electrical power and thermal power output of power stations. Thermodynamicists recognise this as the distinction between work output and heat output. The terminology is not felicitous and does not follow SI practice. Should we suggest a better usage?

I note the analogous situation in health physics in which the effect of radiation needs to be expressed both as a physical deposition of energy and as a biological effect. The SI therefore accepts two subsidiary units, the gray (physical) and sievert (biological) which are both measured in units of joule/kg. The connection between the two is then a Relative Biological Efficiency, i.e. a non-dimensional efficiency factor analogous to thermal efficiency for converting heat to work.

So one way forward would be to introduce subsidiary named units for energy and power that would distinguish between heat and work. We might take the opportunity of honouring further thermodynamicists. One suggestion is to retain joule and watt to refer to energy and its rate of change (i.e. in a system) and to introduce new terms for energy transfer, distinguishing between heat and work

Transfer as Work units: gibbs (energy - J) newcomen (power - J/s)
Transfer as Heat units: rumford (energy - J) nernst (power - J/s)

Do you accept a need to distinguish heat and work? YES/NO

Energy and 'Vigour'

The idea of energy has grown from Newtonian mechanics and its generalisation to conservative systems where N, the Newtonian energy, satisfies

$$\Delta N = -W$$

if indeed we are willing to continue the 'work out is positive' sign convention. In Newtonian mechanics, energy is a single valued concept.

But in the development of thermodynamics, where we elaborate the first law to

$$\Delta E = Q - W$$

we are accustomed to say that heat and work are manifestations of the transfer of energy. Yet this generalised energy, E, is different from N in that it is a two-part concept: energy has a quality (the temperatuire at which it passes to a system) as well as quantity. We currently lack a word to express this second aspect and to denote the general transferrence and the particular transferrence quantified in joules.

In the analogous transfer of matter, we already have established adequate terminology since we speak of matter being transferred quantified as mass (kg) but with the understanding that quality is also significant: 1 kg of lead is distinct from 1 kg of water. Thus 'mass' is not misunderstood since we have the distinct word 'matter'. Unfortunately we have at present only the one word 'energy' to denote a concept and its quantification.

I suggest the term 'vigour' could be used to denote that aspect of the transfer of energy that leads to work as opposed to heat or energy in general. It would be used for transfer of energy or a rate of transfer. This use would be analogous to the distinction made between energy and exergy content of a system but in the context of the transfer between systems.

Do you favour a term to clarify the two-part concept of energy transfer? YES/NO

If Yes, what suggestions have you?

Universal Gas Constant and Relative Molecular Mass (Suggested by Y Mayhew)

The gas constant, R, and the molecular (universal) gas constant, R_M are conventionally connected as $R_M = MR$ where M is the relative molecular mass. If this latter term is to be believed, it should be non-dimensional, being the ratio of the mass of the molecule to the mass of one atom of C-12. Yet R is conventionally measured in J/kg and R_M in J/kmol. This implies that we are measuring a relative molecular mass in mixed units, kmol/kg.

Would you expect M to be dimensionless? YES/NO

If YES, can you suggest acceptable units for R and R_M?

Additions

Please indicate any further points of nomenclature or symbols you wish to raise.

Annexure B

The p-v-t surface
for a pure substance which contracts on freezing

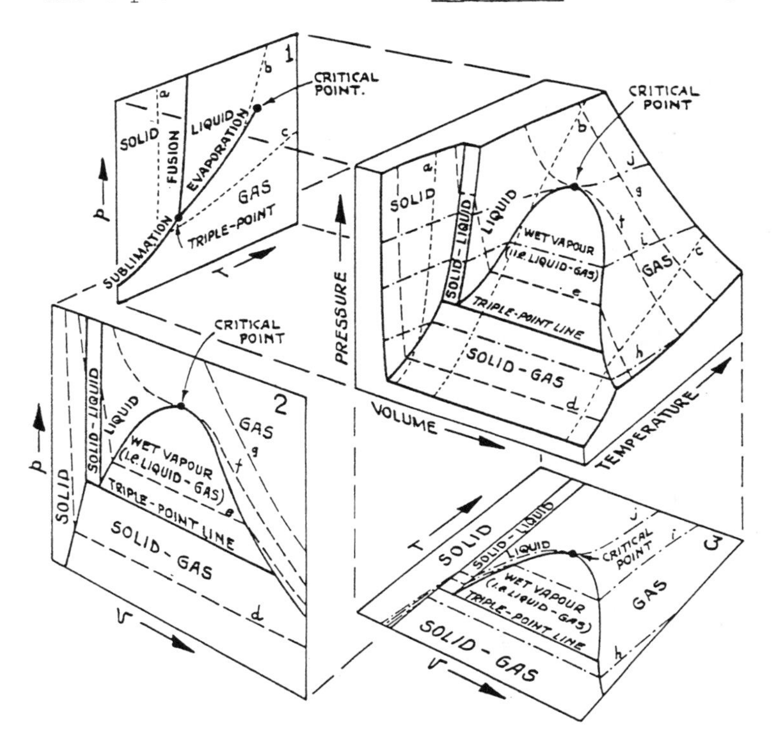

Isochores (constant volume lines) a,b,c ----------------

Isotherms (constant temperature) d,e,f,g — — — — — —

Isobars (constant pressure) h,i,j — . — . — . —

The p-v-T surface
for a pure substance which expands on freezing

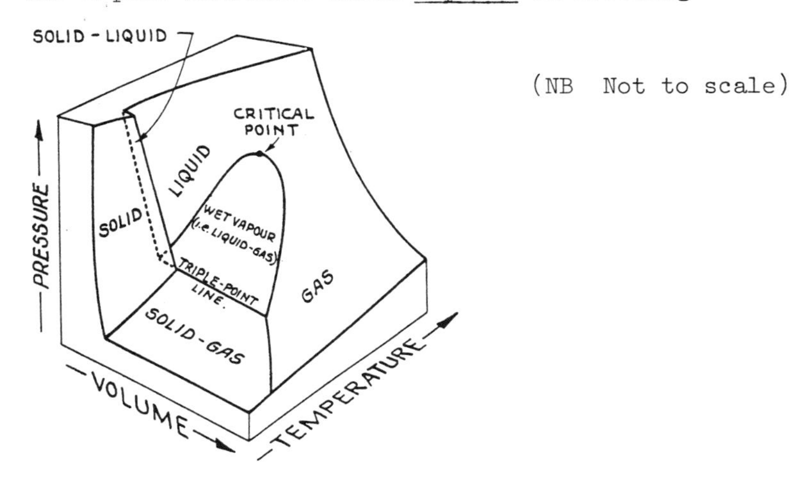

(NB Not to scale)

Notes on Terminology

1. In the labelling of the various zones of the p-v-T surface, the three phases in which a (pure) substance can occur are called solid, liquid and gas. There is no division between gas and liquid above the critical point. The two phases merge imperceptibly into one another.

2. The term vapour is reserved for use in that zone in which liquid and gas are present as a mixture, i.e. in the zone above the triple-point line, bounded by the saturated-vapour lines.

3. In some texts, the term gas is used to refer to states of the substance corresponding to temperatures greater than the critical, i.e. to the zone above the isothermal line through the critical point. The zone between this line and the saturated-vapour line is then referred to as vapour or superheated vapour in this description of the system. As there is no scientific justification for this

division at the critical temperature, it has not been adopted here. It follows therefore that, in the system presented,

(i) A wet vapour is a mixture of liquid and gas. The limiting cases are (a) when the last trace of gas condenses and the state become saturated-liquid, (b) when the last trace of liquid evaporates and the state becomes saturated - or dry-vapour or, as is frequently used, dry saturated vapour.

(ii) Unit mass of a wet vapour of dryness fraction x consists of a mass x of dry saturated vapour and a mass (1-x) of saturated-liquid. The moisture content is (1-x) and is usually expressed as a percentage.

(iii) When a vapour becomes dry, it becomes a gas and thereafter, when superheated, i.e. at a temperature higher than the saturation temperature corresponding to its pressure, is referred to as a gas.

4. The substance 'H_2O' deserves special mention because of the time-honoured names which have been adopted for its different phases, namely ice, water and steam. H_2O in the vapour state is referred to as wet steam and, in the limiting cases, as saturated-water and saturated - or dry - or dry saturated steam. H_2O in the gaseous phase is referred to as superheated steam. H_2O of course expands on freezing.

5. Surfaces for mixtures (i.e. not pure substances) are more complex than shown here.

Annexure C

PARTIAL DIFFERENTIALS IN MATHS AND THERMO

Teachers in the Heat Group expressed concern at our examination review meeting about the student facility or lack of it with partial differential expressions. We ask mathematics lecturers to consider how they might help us. Many students showed very poor understanding, such as writing unbalanced equations of the form

$$\delta U = \frac{\partial U}{\partial S}\Big)_p + \frac{\partial U}{\partial p}\Big)_S$$

Perhaps there is no remedial work possible here! Some other points where a better liaison between mathematics and thermodynamics might help are referred to below.

Change of Dependent Variable

Thermodynamics takes two independent variables for its description but frequently changes the selected pair. In conventional P.D. the change would be shown by writing a new function. Thus instead of $z = z(x,y)$ if we chose a variable $v = v(x,y)$ one would transform z to perhaps $w = w(x, v)$. Then the partial differentials

$$\frac{\partial z}{\partial x}\Big), \frac{\partial w}{\partial x}\Big),$$

are clearly different.

In thermodynamics, where z is perhaps internal energy and the transformed variable w is still internal energy, it is natural to retain the same symbol U and write the function $U=U(x,v)$ say with the essential need to use a notation to show which second variable is being kept constant, i.e.

$$\frac{\partial u}{\partial x}\Big)_y, \frac{\partial u}{\partial x}\Big)_v$$

Can this point be made in maths teaching?

The Maxwell relations of thermodynamics are based on the four functions $U=U(S,V)$, $H=H(S,p)$, $F=F(T,V)$ and $G=G(T,p)$. Can the discussion of 'curls' include the derivation of these relations as a condition for a potential function to exist, the vanishing of its curl? The differential relations required are on the Thermodynamics Data Card.

The area transformation properties of the determinant in Abelian or linear transformations passes over to the Jacobian. Can this be illustrated again in thermodynamic terms on the basis of the equivalence of heat and work, i.e.

$$\frac{\partial(T,S)}{\partial(p,V)} = 1$$

Manipulation in plane diagrams would be good practice. For example, given the Clausius-Clapeyron equation for change of pressure with temperature in a two-phase region, find the slope of the saturated gas line in T-s and h-s diagrams, expressing the result in terms of the specific heat capacity at constant pressure and the Joule-Thomson coefficient.

There may well be other areas where mathematics teaching could take note of the needs of thermodynamics teaching.

Solution. Take $s = s(p,T)$ so $\delta s = \left.\frac{\partial s}{\partial T}\right)_p \delta T + \left.\frac{\partial s}{\partial p}\right)_T \delta p$

Then

$$\left.\frac{\partial s}{\partial T}\right)_{x=1} = \left.\frac{\partial s}{\partial T}\right)_p + \left.\frac{\partial s}{\partial p}\right)_T \left.\frac{dp}{dT}\right)_{sat} \qquad \text{known from Clausius-Clapeyron}$$

and from $h = u+pv$, $du = Tds - pdv$ we have

$$dh = Tds + vdp \quad \text{(i)} \left.\frac{\partial s}{\partial T}\right)_p = \frac{c_p}{T} \quad \text{and (ii)} \left.\frac{\partial s}{\partial p}\right)_T = (\mu_T - v)T$$

$$\rightarrow \left.\frac{\partial T}{\partial s}\right)_{x=1} \qquad \text{now known}$$

Then

$$\left.\frac{\partial h}{\partial s}\right)_{x=1} = T + v\left.\frac{\partial p}{\partial s}\right)_{x=1} = T + v\left.\frac{\partial T}{\partial s}\right)_{x=1} \left.\frac{dp}{dT}\right)_{sat}$$

$$\text{nb} \qquad c_p \equiv \left.\frac{\partial h}{\partial T}\right)_p, \quad \mu_T \equiv \left.\frac{\partial h}{\partial p}\right)_T$$

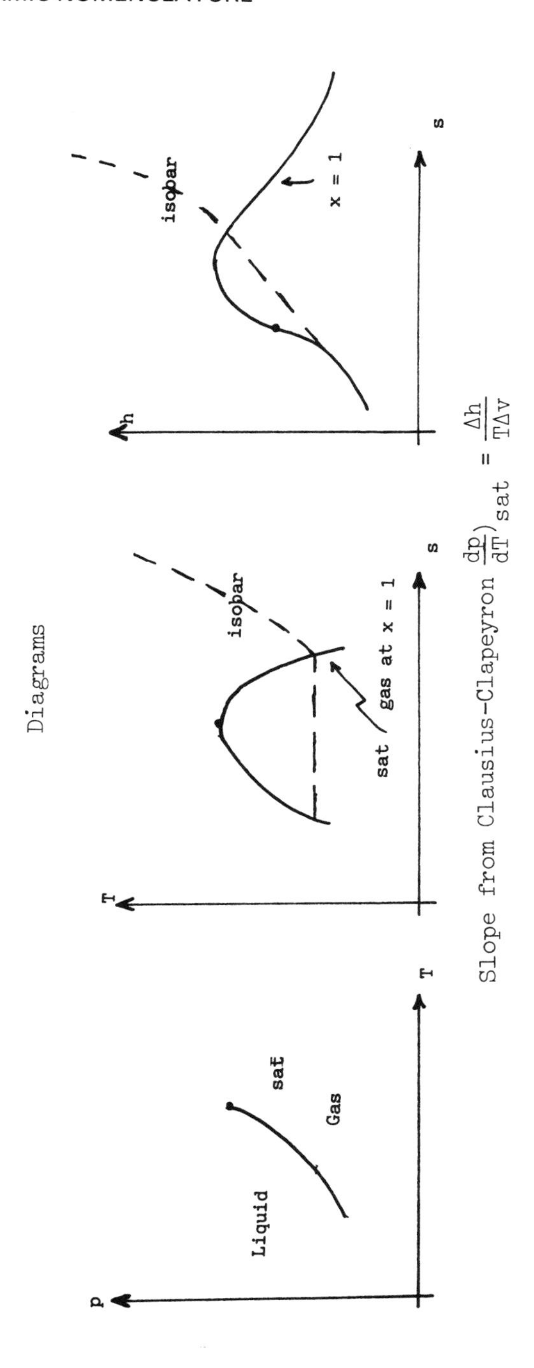

Diagrams
p
Liquid
sat
Gas
T
T
isobar
sat
gas at x = 1
s
h
isobar
x = 1
s
Slope from Clausius-Clapeyron $\left.\frac{dp}{dT}\right)_{sat} = \frac{\Delta h}{T\Delta v}$

THERMODYNAMIC SIGN CONVENTION AND NOMENCLATURE

Y. R. Mayhew

Department of Mechanical Engineering
University of Bristol

J. W. Rose

Department of Mechanical Engineering
Queen Mary College
University of London

The following notes constitute a partial response to the paper of the same title by Jeffery Lewins. They do not necessarily answer the questions posed, but are intended to emphasise certain principles which provide guidance on how some problems of Lewins should be dealt with or be disposed of.

PHYSICAL QUANTITIES, NUMBERS AND UNITS

Measurement essentially consists of operations in which the ratio of one physical quantity to another of the same kind is determined. It follows that a physical quantity, or the symbol used to denote it, must be understood to represent a *product* of a numerical value and a unit

$$\text{physical quantity} = \text{numerical value} \times \text{unit}$$

The following principles should be noted.

(a) Operations on equations involving physical quantities, numerical values, and units, should follow the ordinary rules of algebra. It is, for example, legitimate to write

$$\frac{\text{unit}}{\text{physical quantity}} = \frac{1}{\text{numerical value}}$$

(b) Neither any physical quantity, nor the symbol used to denote it, should imply a particular choice of unit.

(c) Particulars concerning a physical quantity must be quoted with the quantity, or the symbol used to denote it, not with the unit.

In particular the following References should be noted in this connection: Ref.(1) pp.6-7, Ref.(2) pp.602-8, Ref.(3) pp.1-5, Ref.(4) pp.7-9 and footnote † on p.12, Ref.(5) p.515, and Ref.(6) pp.1 and 58. Ref.(7) provides international support for principle (a).

The authors are at one with McGlashan, Ref.(6), who applies principle (c) when he allows the use of the joule, J (definition 1 J = 1 N m), for both scalars (e.g. energy) and vectors (e.g. moment). Similarly we might say 'rate of increase of area, A = 10 mm^2/s' and 'kinematic viscosity, ν = 10 mm^2/s'. Refs. (3) and (4) quote examples of good and bad practice with regard of pressure, e.g. it is correct to say that

the gauge pressure = 5 x 10^6 N/m^2

and incorrect to say that

the pressure = 5 x 10^6 N/m^2 gauge

According to principle (c), the onus of explaining the nature of the physical quantity is on the part of the statement to the left of the = sign.

An equally objectionable practice goes back to pre-SI, c-g-s times. In those days the derived unit for kinematic viscosity, cm^2/s, was denoted by stokes, but the stokes was barred, by custom, from other physical quantities having the same derived unit, such as thermal diffusivity. There are many groups of physical quantities for which the derived unit within each group is identical. Principle (c) demands that, where a special name is coined, it can be used for all quantities having the same dimensions.

The authors therefore totally reject the proposals of Lewins' sections "Power/Energy (thermal) and Energy (electrical)" and "Energy and 'Vigour'". In this context the authors like to mention one important breach of principle (c) resulting from the use of the temperature scale symbol ^oC. The ^oC is not a pure unit, but a unit plus datum indicator (i.e. indicating that the temperature is measured from a datum which lies 0.01 K below the triple point of water). When we write the temperature difference between two Celsius temperatures T_{C2}= 45 ^oC and T_{C1} = 15 ^oC as

$$T_{C_2} - T_{C_1} = 45\ ^{\circ}C - 15\ ^{\circ}C = 30\ K$$

we cannot attach the symbol $^{\circ}C$ to the 30. Is it not time that we applied principle (c) to this situation by writing $T_{C2} = 45\ K$ and $T_{C1} = 15\ K$, leaving it to the symbol T_C to explain that the quantity is Celsius temperature? The above equation would then be written

$$T_{C2} - T_{C1} = 45\ K - 15\ K = (45 - 15)\ K = 30\ K$$

SIGN CONVENTION, AND ADIABATIC, ISENTROPIC AND REVERSIBLE PROCESSES

On balance, the authors would welcome general adoption of the sign convention for work and heat used by physicists, chemists and engineers on the Continent, i.e. both quantities are regarded as positive when the associated energy transfer is into the system. This change would require some directive 'from above', such as a recommendation from the Engineering Council and adoption in their examinations. It would be desirable to carry US engineering teaching with us, at least in the long run.

Adiabatic should continue to mean $Q = 0$ at the boundary of a system, and isentropic should mean that $\Delta S = 0$ for a system. In practice isentropic $\equiv$ adiabatic and reversible, although it is possible to conceive a process undergone by a system which is dissipative but with just sufficient negative heat transfer such that $\Delta S = 0$. It is best to leave the finer points of this to the teacher rather than lay down rigid rules.

Reversible, unqualified, can strictly only be applied to an isolated system (system plus surroundings, or 'universe'). We can overcome the difficulty by using terms such as 'internally reversible' or 'non-dissipative'. Here again we may leave the choice to the teacher.

HEAT CAPACITY

The objection to specific heat is that it uses 'heat' as part of a name for a property. The move to *specific heat capacity*, which is quite a mouthful, improves the situation slightly but not sufficiently. The option is perhaps to stick to *specific heat* and make sure that students understand the objection, or to invent a new term. Any new term is likely to be a mouthful also, e.g. enthalpy-temperature coefficient for $(\partial h/\partial T)_p$ and integral energy-temperature coefficient for $(\partial u/\partial T)_v$.

PHASE-CHANGE SYMBOL

There seems to be no strong reason for departing from u_{fg} and h_{fg}: after all, it is merely a question of convention, and the suffix fg is used very widely. Quite apart from possible confusion between the letter 'l' and the number '1', lg can also mean logarithm.

Jeffery Lewins mistranslates the German word 'Flüssigkeit': it means liquid, not gas. The Germans have no word for fluid (liquid and gas) but have adopted the English word in recent years. The Germans and Russians use a very undesirable convention: i' for h_f, i" for h_g, and a totally different letter, usually r, for h_{fg}.

Latent heat ought to be banned because it is ambiguous (e.g. u_{fg} or h_{fg}?). Terms such as enthalpy or internal energy of fusion, sublimation and evaporation should be adopted instead. Alternative, more precise, terms would become too long and elaborate.

AMOUNT OF SUBSTANCE AND RELATIVE MOLECUALR AND ATOMIC MASS

It is generous of Jeffery Lewins to admit that he has made a bit of a mess of Yon Mayhew's original contribution, which was concerned not with the molar (universal) gas constant, but with the relation between amount of substance and relative molecular, or atomic, mass. (Incidentally, here the German language possesses a more succinct term, viz. 'Stoffmenge' meaning literally matter-quantity.)

What must be understood is that two concepts, easily confused, exist side by side: relative molecular (atomic) mass M_r (A_r) which is a dimensionless quantity, and molar mass M which has dimensions. Because of the way the unit of amount of substance is defined, the two quantities happen to be numerically equal when M is measured in kg/kgmol (or g/mol), but not when M is measured in the derived base unit, kg/mol.

Few books explain the relation between the concepts involved, Ref.(2) pp. 151-2, Ref. (6) pp.10-4, and Ref. (8) pp.54-5, being some known to the authors. To put the record straight, Mayhew's original note, based mainly on Ref. (2), is attached as an Appendix.

REFERENCES

1. "Quantities, Units and Symbols", a report by the Symbols Committee of the Royal Society, 1975.

2. Rogers, G. F. C., and Mayhew, Y. R., "Engineering Thermodynamics, Work and Heat Transfer", Longman, 1984.

3. Rose, J. W., and Cooper, J. R., "Technical data on fuel", British National Committee, World Energy Conference, 1977.

4. "Guide to the Preparation of Papers", Institution of Mechanical Engineers, 1984.

5. Terrien, M. J., and de Boer, J., Discussion of "The Rational Treatment of Temperature and Temperature Scales", by R. W. Haywood, I.Mech.E. Proceedings, 1967-68, Vol.182, Part I, No. 21.

6. McGlashan, M. L., "Physico-chemical Quantities and Units", Royal Institute of Chemistry, 1971.

7. ISO 31/0 - 1974(E), "General Introduction to ISO 31", International Standards Organisation, 1974.

8. "The International System of Units", HMSO, 1982.

APPENDIX
AMOUNT OF SUBSTANCE AND RELATIVE MOLECULAR AND ATOMIC MASS

The physical quantity relative molecular or atomic mass ('molecular or atomic weight'), M_r or A_r, and the physical quantity molar mass, M, are still treated in a cavalier, imprecise manner, and one still finds units such as kgmol (or kg mol, kg-mol, etc.) floating around in otherwise respectable books. Unlike for other base units, there has never been an internationally-accepted definition of the unit of amount of substance until SI adopted the mole as one of the base units in 1971.

A confusion arises from the improper use of M_r (or A_r). For example, one often finds the following expression written down:

$$\text{specific gas constant } R = \frac{\text{molar(or universal) gas constant } R_M}{\text{relative molecular mass } M_{r,X}}$$

For example, for oxygen O_2

$$R_{O_2} = \frac{8.3144 \dfrac{kJ}{kmol\ K}}{31.999} = 0.2598 \frac{kJ}{kmol\ K}$$

which makes nonsense of the units. The denominator should, of course, quote the molar mass M_{O_2} = 31.999 kg/kmol, which puts everything right.

Let us look at the relation between M and M_r. The unit of amount of substance, the mole - unit symbol mol, is defined as follows:

> The mole is the amount of substance of a system which contains as many elementary entities as there are atoms in 0.012 kilogram of carbon 12.
>
> When the mole is used, the elementary entities must be specified and may be atoms, molecules, ions, electrons, other particles, or specified groups of such particles.

The number of entities in 1 kmol of carbon 12 each of mass $m(^{12}C)$, or by the above definition in 1 kmol of any other substance X having entites of mass m(X), is the Avogadro (or Loschmidt) number

$$N_A = \frac{0.012\ kg}{m^{12}(C)}\ mol^{-1} = \frac{12\ kg}{m(^{12}C)}\ kmol^{-1}$$

The mass per kmol of any substance X, called the <u>molar mass</u> M of the substance, is therefore given by

$$M = N_A m(X) = \left| \frac{12\ kg}{m(^{12}C)}\ kmol^{-1} \right| m(X) = \left| \frac{m(X)}{\frac{1}{12}\, m(^{12}C)} \right| \frac{kg}{kmol}$$

The ratio $m(X)/\frac{1}{12}\, m(^{12}C)$ will be recognised as the definition of <u>relative atomic mass</u>, A_r, or <u>relative molecular mass</u>, M_r, depending upon whether X is an atomic or a molecular species. We can write down the relation

$$M = A_r\ kg/kmol \qquad or \qquad M = M_r\ kg/kmol$$

Thus M is numerically equal to A_r or M_r when measured in kg/kmol (or g/mol), and is a dimensional quantity, while A_r or M_r are dimensionless quantities.

ONE WORLD - ONE THERMODYNAMICS

P. T. Landsberg

Department of Mathematics
University of Southampton

The view is advocated that the basic ideas of thermodynamics are the same for all students, engineers, chemists and physicists, while the applications used should differ. What are these basic ideas? It is suggested that they are the four (phonetic) 'F's - phase space, Pfaffians, fluids and fluxes, including some statistical interpretation and a clear understanding of inexact differentials. Sample arguments are given.

INTRODUCTION

Taught to mechanical and chemical engineers, chemists, physicists, oceanographers, and sometimes to biologists and mathematicians, thermodynamics occupies a place equal to that of classical mechanics as regards its ubiquitous nature. Academics in different departments teach their own brand of thermodynamics. In this respect the subject differs from mechanics which is often left to the physicists or applied mathematicians to teach in "service" courses. A reason for this is that while the examples used in mechanics are usually rather circumscribed, thermodynamic examples are drawn from different fields which are best appreciated by students in that particular field and may not be thought relevant or even understandable by students in other fields.

But is there a core of thermodynamic theory which everyone should know? My view is that this is so, and what should be known are the four (phonetic) F's: Phase space, Pfaffians, fluids and fluxes. I have here used Pfaffians as an abbreviation for partial differentials, integrating factors and the elementary part of the mathematics involved.

PHASE SPACE

This device is useful in the discussion of working cycles, in statistical mechanics (quantum or classical), in quantum mechanics, and elsewhere, so that one might as well use it in thermodynamics. Once fully explained it will be useful elsewhere, so that the time invested is well spent. In the first few lectures one needs to deal only with equilibrium states and how to take a system from one to the other. Now the phase space is the resting place of these equilibrium states, and the quasi-static change is a sequence of equilibrium states and so is represented by a curve in phase space. It clearly may be traced in either direction. If one writes $d'Q = C_v \, dT + \ell_v \, dv$, for example, one is simply talking about an increment of a curve in the phase space. We need not call it a "reversible process" of the old days. All students should be taught this.

One has to talk about adiabatic linkages of pairs of points. These are the adiabatic "processes" of the old days and bring in the totality of points (i.e. states) any two of which can be linked adiabatically. The quasistatic adiabatic changes normally fill a whole volume of phase space. Sets of points may be brought into the discussion merely as a conceptional tool. There is no need to execute heavy mathematics with them [1,2].

The purity and clarity of these concepts seem to me to make them suitable for all classes of students of thermodynamics, certainly including engineers. For chemists and chemical engineers one might even extend the use of appropriately defined sets of points in phase space [3,4].

This device avoids the need to introduce the reversible process. How should it be defined in any case? One may offer four inequivalent definitions.

(i) A process which is reversible must take place infinitely slowly. [The discharge of a capacitor through a high resistance suggests that an infinitely slow process need not be reversible.]

(ii) No changes of any kind must remain in the surroundings of a system when a reversible process is followed by the same process taken in the opposite sense. [What about an ideal pendulum? It conforms to the definition (ii), but not to (i).]

(iii) A reversible linkage is a sequence of equilibrium states. [This is the concept advocated here, but without the use of the term "reversible", which would be banished except for historical purposes.]

(iv) A reversible process is one which can be exactly reversed by an infinitesimal change in the external conditions.

Some authors adopt (iii), e.g. Guggenheim did so from his earlier writings to his later ones ([5,6]), but he stuck to the term "process" which suggests movement in one direction, whereas one is really dealing with a linkage or a sequence of states, no direction being implied. The difference can be brought out perhaps by imaging that instead of talking about a line one talks about a point which moves on the line. The moving point implies a direction of movement and is not needed to discuss a line. Other authors favour definition (ii), e.g. Denbigh ([7]) and Planck ([8]), or (iv) which was used by Pippard ([9]). There is no unanimity in this matter, nor is there complete clarity ([10]). That is why the concept of a quasistatic linkage, or if necessary, a quasistatic process, is preferable.

PFAFFIANS

There is no hope of understanding thermodynamics unless one has a sound grasp of a few key concepts from the theory of functions of several variables, notably the properties of partial differentiation, exact and inexact differentials and integrating factors. The importance of the path-dependence or path-independence of line integrals is a good intuitive way of approaching these matters. This can be done by talking about work in mechanics and has the advantage that a student is taught to think about physical concepts already familiar to him while he absorbs some new mathematical ideas. Thereafter one can introduce the thermodynamic ideas of internal energy U and entropy S, using the newly acquired mathematical concepts. Thus U and S are functions of position in the phase space, and the integrals $\int dU$ and $\int dS$ are therefore path-independent. There are other quantities of crucial importance which are path-dependent: the integrals of heat absorbed and of work performed, $\int d'Q$ and $\int d'W$. How does one know they are path-dependent? One makes use of the beautiful idea that one example is enough to establish the possible path-dependence. So one simply integrates the work done by a gas between two states A and B for two different very simple paths 1 and 2 (Fig. 1).

$$W_1 = \int_A^B p_1 dv = p_1(v_2 - v_1), \qquad W_2 = \int_A^B p_2 dv = p_2(v_2 - v_1)$$

As they are unequal dW is an inexact differential, written therefore as d'W, and one cannot associate a value of W with a point in phase space. Since dU = d'Q - d'W, it then follows at once that d'Q is also inexact. Thus Q is not a function of state either.

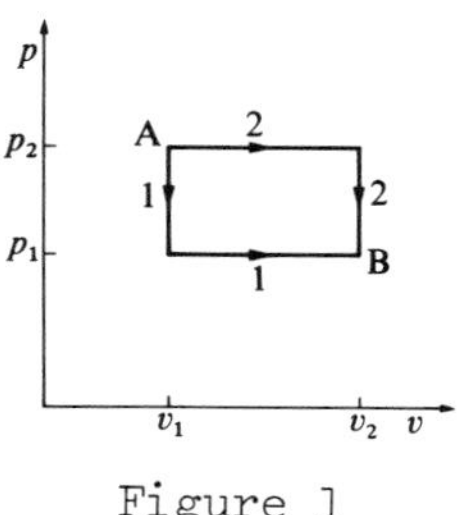

Figure 1

These are basically simple ideas (see Appendices A and B). Engineers, who always have plenty of mathematics courses, as well as chemists and of course physicists, must be exposed to them as part of their thermodynamics training. If these mathematical matters are not in a school syllabus, then thermodynamics is not suitable as a school subject ([11]).

If quantities like d'Q and d'W cannot be exact, what is the next best thing they can be, so as to have pleasing properties? Why, of course, they could have integrating factors. In other words d'Q multiplied by some appropriate thermodynamic function could be exact. In this way one arrives at the entropy and its properties for quasistatic changes. Its properties for non-static changes are also very interesting, for now one has to consider changes which are not represented wholly in a phase space for equilibrium states. One has to step outside this space to view the world of "Irreversible processes", or of non-static changes, i.e. of changes which involves sequence of states some of which are not equilibrium states, but I will not deal with this aspect here.

The mathematical content of this work can be kept quite simple and can be confined to a few lectures, provided the material is carefully tailored to the thermodynamics student, be he engineer, physicist or philosopher. One must not allow oneself to be tempted to develop the whole of the theory of partial differential equations!

FLUIDS

One may define three classes of ideal gases. The terminology is sometimes inaccurate or vague and precise results are given therefore in Table 1 ([12]). The equivalence of the two definitions in A follows from appendix equation (C.8). The equivalence of the two definitions in B follows from appendix statement (C.6).

TABLE 1

The "Ideal" Gases

		Usual name	Better name	Proposed simplified name
A	Either (i) $pv = cT$ (ii) C_v independent of T	perfect gas or ideal gas	ideal classical gas of constant heat capacities	ideal classical gas
	or (i) $pv = cT$ (ii) $pv = gU$			
B	Either $pv = cT$ or (i) $pv = ct$ (ii) $(\partial U/\partial v)_t = 0$	perfect gas or ideal gas	ideal classical gas	dilute gas
C	$pv = gU$	no name	ideal quantum gas	ideal (quantum gas)

There are two disadvantages of an excessive use of the class A gas: (a) It cannot exist near the absolute zero since C_v and C_p should both tend to zero whereas $C_p - C_v = c$ for class A gases, where c is a constant. (b) The heat increment d'Q has an integrating factor without any appeal to the second law being required (see Appendix D). This illustrates the point that the second law comes in as a new result, in the sense that it yields an integrating factor for d'Q, only if the number of independent variables is at least three. For two independent variables an integrating factor always exists ([2,13]). For class C objection (a) no longer applies since in the limit $T \to 0$

$$\left(\frac{\partial S}{\partial v}\right)_T = \left(\frac{\partial p}{\partial T}\right)_v = \frac{g}{v} C_v \to 0$$

is possible. Thus one can design examples of ideal quantum gases which satisfy the third law of thermodynamics.

A crucial advantage of class C over classes A and B is in fact that simple quantum statistics leads precisely to the class C equation of state rather than to class A or class B ([2,13]). Hence I favour using class C as basic equation of state and then approach class A as an important special case of class C.

The ideal fluids and their properties are also of importance for all thermodynamics courses since they enable students to work out simple thermodynamic problems and, at the same time, class A is a good approximation to dilute, real gases and so must remain an important ingredient of any thermodynamics course. Class C applied to black-body radiation ($g = 1/3$) and an electron gas in metals ($g = 2/3$). Logically, the common part of definitions B and C is definition A. That is the main use I would make of definition B.

FLUXES

Equilibrium is an idealisation and it cannot be realised. There is no point trying to keep this a secret from our students as if we were some modern Pythagoreans. Gravitational motion anywhere affects our system in principle if not in practice. The idea of the flux of a quantity, namely the flow of it across unit area in unit time is therefore another important concept which introduces non-equilibrium properties. One can study it by reference to radiation which is of importance to engineers (for whom it is a form of heat transfer which must be in many syllabi), to chemists and biologists (who study photochemical effects and photosynthesis), and of course to physicists (for whom quantum theory arrived via the properties of black-body radiation). A small example is given in Appendix E ([14]). It shows that there is no need to immerse the student immediately in all the details of Onsager-type irreversible thermodynamics if one wants to teach non-equilibrium properties.

CONCLUSION

As we live in one world there should be a thermodynamics course of use and importance to all who study it and I have tried to outline what it could be. What shall one include from the statistical side? My own inclination is to finish classical thermodynamics and to venture beyond it only in response to questions. Thereafter I would explain the crucial rôle of the formula

$$S' = - k \sum p_i \ \ell n \ p_i$$

and show how its maximisation leads to Maxwell's velocity distribution and the normal distribution in statistics ([15]). All students need statistics and they will hear something of information theory and these are therefore sensible examples. The striking properties of S' then suggest that this is an important quantity and a little further work shows that it has a close relationship with thermodynamic entropy.

APPENDIX A

When do integrals along a closed loop vanish?
Mechanics as a preparation for thermodynamics

Suppose a force acts so as to take a particle from a point A(0,0) to a point B(1, 2) in the (x, y)-plane. It can take two paths: along the curve C_1: $y = 2x$ or along the curve C_2: $y = 2x^2$. Note that A and B lie on both curves. What work is done if the force is $\underset{\sim}{F}_1 = y^2 \underset{\sim}{i} + x^2 \underset{\sim}{j}$?

The answer is

$$W_1 \equiv \int \underset{\sim}{F}_1 . d\underset{\sim}{r} = \int (y^2 \underset{\sim}{i} + x^2 \underset{\sim}{j}) . (\underset{\sim}{i} \ dx + \underset{\sim}{j} \ dy) = \int (y^2 dx + x^2 dy) . \quad (A.1)$$

This formula yields for C_1 and C_2 respectively

$$C_1 : \int_0^1 (4x^2 dx + x^2 . 2dx) = \int_0^1 6x^2 dx = 2x^3 \Big|_0^1 = 2,$$

$$C_2: \int_0^1 (4x^4dx+x^2.4xdx) = 4\int_0^1 x^4dx+4\int_0^1 x^3dx = \tfrac{4}{5} + 1 = \tfrac{9}{5} .$$

The work is path-dependent, so the force is non-conservative and $d'W_1 \equiv y^2dx + x^2dy$ is an inexact differential. The difference in W between any two points is therefore not defined.

Try, alternatively, the force $\underset{\sim}{F}_2 \equiv x^2\underset{\sim}{i} + y^2\underset{\sim}{j}$, whence

$$W_2 = \int (x^2dx+y^2dy), \tag{A.2}$$

$$C_1 : W_2 = \int_0^1 (x^2dx+4x^2.2dx) = \int_0^1 9x^2dx = 3 ,$$

$$C_2 : W_2 = \int_0^1 (x^2dx+4x^4.4xdx) = \left[\frac{1}{3}x^3+\frac{16}{6}x^5\right]_0^1 = \frac{1}{3} + \frac{8}{3} = 3.$$

Thus $\underset{\sim}{F}_2$ could be <u>conservative</u>. In fact if $V \equiv \frac{1}{3}(x^3+y^3)$

$$-\text{ grad } V = \underset{\sim}{x}^2 i + \underset{\sim}{y}^2 j = \underset{\sim}{F}_2 .$$

Thus $\underset{\sim}{F}_2$ possesses a potential and <u>is</u> conservative. It follows that

$$\int_A^B \underset{\sim}{F}_2.d\underset{\sim}{r} = -\int_A^B (\text{grad } V.d\underset{\sim}{r}) = -\int_A^B dV = V(A) - V(B)$$

is independent of the path, and the difference in W between any two points is defined. The "phase space" of this force is the (x, y)-plane. The work done in a closed loop in phase space is zero. Put alternatively, $dW_2 = d\left(\frac{1}{3}x^3+y^3)\right)$ is an exact differential.

Similarly the differentials of some thermodynamic functions are normally inexact (e.g. the work $d'W = pdv$ done by a fluid) and the integral along a closed loop in <u>their</u> phase space is not normally zero.

APPENDIX B
Conditions for Exact Differentials and the Existence of Integrating Factors

Consider the one-form or Pfaffian form or differential, call it what you wish,

$$d'Q = \sum_{j=1}^{n} Y_j(x_1,\ldots,x_n)dx_j = \underset{\sim}{S}.d\underset{\sim}{r} \qquad (B.1)$$

where $\underset{\sim}{S} \equiv (Y_1,Y_2,\ldots,Y_n)$, $d\underset{\sim}{r} = (dx_1,dx_2,\ldots,dx_n)$. (B.2)

Then necessary and sufficient conditions for exactness of d'Q are

$$\frac{\partial Y_j}{\partial x_k} = \frac{\partial Y_k}{\partial x_j} \quad (j,k=1,2,\ldots,n). \qquad (B.3)$$

Applying this condition to (A.1) and (A.2) with $j = 1$, $k = 2$,

$$(A.1): \frac{\partial y^2}{\partial y} \overset{?}{=} \frac{\partial x^2}{\partial x} \qquad \text{No}$$

$$(A.2): \frac{\partial x^2}{\partial y} \overset{?}{=} \frac{\partial y^2}{\partial x} \qquad \text{Yes}$$

Thus a more general theory is now available in the form of (B.3) which enables one to avoid the special study of each system. For $n = 3$ condition (B.3) becomes

$$\text{curl } \underset{\sim}{S} = 0 \qquad (B.4)$$

so that $\underset{\sim}{S}$ must be expressible as a gradient of some scalar function in the domain of interest.

A weakened form of (B.4) is

$$\underset{\sim}{S}.\text{curl } \underset{\sim}{S} = 0 \qquad (B.5)$$

and indeed it implies something less stringent about d'Q: It need no longer be exact, but it has an integrating factor.

The existence of an integrating factor is not automatic. A single example is enough to establish this point. Perhaps the simplest is one for which $\underset{\sim}{S}$ in (B.2) has the form

$$\underset{\sim}{S} = (0, x_1, k) \tag{B.6}$$

where k is a non-zero constant. Then

$$\text{curl } \underset{\sim}{S} = (0,0,1) \ , \quad \underset{\sim}{S}.\text{curl } \underset{\sim}{S} = k \neq 0 \tag{B.7}$$

One can now study the implications of a heat form or of a work form which is of the type

$$d'Q = x_1 dx_2 + k \ dx_3 \tag{B.8}$$

to see what distinguishes it from forms which have an integrating factor.

APPENDIX C
Definitions of Ideal Gases

To discuss ideal fluids, assume that they consist of a fixed number, N, of identical particles, and let t and T be respectively an empirical and the absolute temperature scale. Consider a van der Waals gas (a,b,c are constants):

$$\left(p + \frac{a}{v^2}\right) (v-b) = ct. \tag{C.1}$$

The thermodynamic relation (in a usual notation)

$$\left(\frac{\partial U}{\partial v}\right)_t = T \left(\frac{\partial p}{\partial T}\right)_v - p \quad , \tag{C.2}$$

then yields

$$\left(\frac{\partial U}{\partial v}\right)_t = \frac{ct}{v-b} \left(\frac{T}{t} \frac{dt}{dT} - 1\right) + \frac{a}{v^2} \ . \tag{C.3}$$

Integrating, with f(t) as the constant of integration,

$$U = c(T \frac{dt}{dT} - t) \ell n(v-b) - \frac{a}{v} + f(t). \tag{C.4}$$

Hence

$$C_v = c \ T \frac{d^2t}{dT^2} \ \frac{dT}{dt} \ \ell n(v-b) + \frac{df}{dt}$$

and

$$\left(\frac{\partial C_v}{\partial v}\right)_t = \frac{c}{v-b} T \ \frac{dT}{dt} \ \frac{d^2t}{dT^2} \ . \tag{C.5}$$

These straightforward operations tell us

(a) Joule's law $(\partial U/\partial v)_t = 0$ is found for a system $pv = cT$ by choosing $a = b = 0$, $t = T$ in C.3. It is not found for a system $pv = ct$, unless it is stipulated separately. (C.6)

(b) For the van der Waals gas (and therefore for the system $pv = ct$) one finds from (C.5)

$$\left(\frac{\partial C_v}{\partial v}\right)_t = 0 \quad \text{only if } t \propto T, \tag{C.7}$$

i.e. if the absolute temperature scale is adopted. If this is done in the definition of the ideal system, then C_v is a function of temperature only, by virtue of (C.7).

A functional dependence of C_v on T is not implied by (C.1). It is, however, easy to see that if one wants

$$pv = g\,U$$

to be also valid for the system $pv = NkT$, then

$$U(T) = \int_0^T C_v \, dT = NkT/g$$

so that by differentiation

$$C_v = Nk/g = \text{independent of } T.$$

Thus, summarising

(c) $pv = NkT = gU$ implies

$$C_v = Nk/g, \quad C_p = \frac{g+1}{g} Nk. \tag{C.8}$$

APPENDIX D

For an Ideal Classical Gas of Constant Heat Capacities an Entropy Function Exists Without Any Need for the Second Law

Consider an ideal classical gas $pv = NkT$ whose internal energy is given by $U = C_v T$. Then $C_p = C_v + Nk$, where C_v, C_p are the usual constant heat capacities. For quasi-static changes with $\gamma \equiv C_p/C_v$

$$d'Q = dU + pdv = C_v\left[\left(\frac{\partial T}{\partial p}\right)_v dp + \left(\frac{\partial T}{\partial v}\right)_p dv\right] + pdv$$

$$= C_v\left[\frac{T}{p}\,dp + \frac{p}{Nk}\,dv\right] + pdv$$

$$= U\,d\ln p + \gamma U\,d\ln v$$

$$= U\,d\ln(pv^\gamma).$$

Thus $d'Q$ has U^{-1} as an integrating factor, <u>there being no need to appeal to the second law</u>. Thus

$$\frac{d'Q}{T} = C_v\,d\ln pv^\gamma.$$

and

$$S = C_v \ln pv^\gamma + \text{a constant}.$$

APPENDIX E

Energy Conversion as Providing Examples for the Use of Fluxes and of the Availability Concept

For black-body radiation at temperature T note the standard results for internal energy and entropy (σ = Stefan's constant)

$$p = \frac{1}{3}\sigma T^4, \quad U = \sigma T^4 v, \quad S = \frac{4}{3}\sigma T^3 v, \quad U/S = \frac{3}{4}T. \tag{E.1}$$

We shall assume that the fluxes of energy and entropy (for example due to the emission or absorption of this radiation), ϕ, ψ say, satisfy

$$\frac{\phi}{\psi} = \frac{U}{S} = \frac{3}{4} T. \tag{E.2}$$

For a derivation see for example ([14]).

Let ϕ_p, ψ_p be energy and entropy fluxes provided by a pump to a converter and let ϕ_c, ψ_c be corresponding fluxes emitted by the converter. Let the heat energy and entropy fluxes reaching the surroundings (the "sink") be

$$\phi_s = \dot{Q}, \quad \psi_s = \dot{Q}/T_s$$

where T_s is the sink temperature. Then the balance equations are

$$\phi_c = \phi_p - \dot{W} - \dot{Q}, \qquad \psi_c = \psi_p - \dot{Q}/T_s + \dot{S}_g$$

where $\dot{W}$ is the rate of producing work per unit area and $\dot{S}_g$ is the entropy generation per unit area of the converter. It follows that in the steady state of the converter ($\phi_c = \psi_c = 0$)

$$\dot{W} = \phi_p - T_s \psi_p - (\phi_c - T_s \psi_c) - T_s \dot{S}_g.$$

The efficiency of the conversion process is

$$\eta \equiv \frac{\dot{W}}{\phi_p} = 1 - T_s \frac{\psi_p}{\phi_p} \left(1 - T_s \frac{\psi_c}{\phi_c}\right) \frac{\phi_c}{\phi_p} - \frac{T_s \dot{S}_g}{\phi_p}.$$

Taking the converter to act as a black-body essentially at the sink temperature, using (E.2), and writing an inequality by omitting the last term,

$$\eta \leq \eta^{**} \equiv 1 - \frac{4}{3} \frac{T_s}{T} + \frac{1}{3} \left(\frac{T_s}{T_p}\right)^4 \tag{E.3}$$

η^{**} is a maximum efficiency which lies for all $0 < T_s/T_p \leq 1$ below the Carnot efficiency $1 - T_s/T_p$. It is therefore a sharper inequality than would be expected on simple thermodynamic grounds.

One can obtain it by a remarkably simple argument using the exergy or availability concept. The availability or potential to produce work due to a medium ("pump") at internal energy U_p, volume v_p, entropy S_p, particle numbers N_{pi} for species i is

$$A_p \equiv U_p + p_s v_p - T_s S_p - \sum_i \mu_{is} N_{pi} .$$

Here the surroundings are specified by $(p_s, T_s, \mu_{1s}, \mu_{2s} \ldots)$. Thus the available energy of the pump leads to the efficiency $\eta' \equiv A_p/U_p$ which is

$$\eta' = 1 + p_s \frac{v_p}{U_p} - T_s \frac{S_p}{U_p} - \sum_i \mu_{is} \frac{N_{pi}}{U_p} .$$

Substituting from (E.1) and noting that $\mu = 0$ for photons in equilibrium,

$$\eta' = 1 + \frac{1}{3} \left(\frac{T_s}{T_p}\right)^4 - \frac{4}{3} \frac{T_s}{T_p} .$$

This is (E.3), except that the inequality has here been used in discussing available energy. Hence $\eta' = \eta^{**}$.

REFERENCES

1. P. T. Landsberg, "The Foundations of Thermodynamics", Rev. Mod. Phys. 28, 363-392 (1956).

2. P. T. Landsberg, "Thermodynamics with Quantum Statistical Illustrations", (New York: Interscience), 1961.

3. P. G. Wright, "Introduction of Set Theory into Chemical Thermodynamics", J. Chem. Ed. 48, 293-295 (1971).

4. P. G. Wright, "Chemical Thermodynamics in Landsberg's Formulation", Proc. Roy. Soc. A317, 477-510 (1970).

5. E. A. Guggenheim, "Modern Thermodynamics by the Method of Willard Gibbs", (London: Methuen), 1933, p. 4.

6. E. A. Guggenheim, "Thermodynamics", Amsterdam, North Holland, Third Edition, 1957, p.15.

7. K. Denbigh, "The Principles of Chemical Equilibrium", Cambridge, University Press, 1955, p.13.

8. M. Planck, "Einfübrung in die Theorie der Wärme", Leipzig, Hirzel, 1930, p.44.

9. A. B. Pippard, "Elements of Classical Thermodynamics", Cambridge, University Press, 1957, p.22.

10. See, for example, "Physics Education", 2, 225-226 (1967) in which reversibility and other matters are discussed following the article J. W. Warren, "Thermodynamics", Physics Education 2, 29-34 (1967).

11. P. G. Wright, "Against the Teaching of Thermodyanmics in School", Education in Chemistry, 11, 9-10 (1974).

12. P. T. Landsberg, "Definition of the Perfect Gas", Am. J. Phys. 29, 695-698 (1961).

13. P. T. Landsberg, "Thermodynamics and Statistical Mechanics", Oxford, University Press, 1978.

14. P. T. Landsberg, "Some Maximal Thermodynamic Efficiencies for the Conversion of Blackbody Radiation", J. App. Phys. 54, 2841-2843 (1983).

15. P. T. Landsberg, "Thermodynamics and Statistical Mechanics", Oxford, University Press, p. 148-151, 1978.

TEACHING THERMO-FLUID MECHANICS AT THE OPEN UNIVERSITY

W. K. Kennedy

Faculty of Technology
The Open University

The developments in the teaching of thermo-fluid mechanics resulting from the preparation of the Open University, 2nd Level Course: T233 'Thermo-fluid Mechanics and Energy' are outlined. The particular problems associated with teaching the subject at a distance required steps to be taken at an early stage to dispel students' misconceptions of various concepts; to introduce an alternative means of presentation of the second law; and to develop various problem-solving procedures as an aid to a logical approach to the subject material.

INTRODUCTION

The Open University Course 'Thermo-fluid Mechanics and Energy' comprises a teaching package which includes: 16 unit texts, 8 T.V. programmes, 4 audio-cassettes and a home experimental kit. The various components are closely integrated, the students being required to cover certain material before viewing the T.V. broadcast and with certain parts of the course covered solely by means of audio-cassette and/or experimental work.

The course topics covered include the following: introduction to energy and thermodynamics, modelling, the first and second laws for a control mass, engine cycles, property diagrams, available energy and entropy, modelling fluids, steady-state mass balance, similarity analysis and dimensionless groups, the first and second laws for flow processes, control volume procedure, Bernoulli's equation, force-momentum procedure, waterwheels and turbines, heat transfer analysis and vapour power cycles.

DEVELOPMENTS

An outline of some of the innovative aspects of the course follows:

Introduction to Energy - the concept of energy is introduced with reference to several examples from solid mechanics. Measures are taken to dispel misconceptions which the student may already have on, e.g., energy, work and heat; working and heating as energy transfer processes are emphasised. At this early stage of the course, the second law is introduced in terms of imposing a 'directional limitation to energy transfer by heating', a 'restriction on heating-to-working transformations' and 'energy degradation'. The introduction is completed with an audio-visual presentation (via audio-cassette) on a simple energy analysis procedure and the consideration of 'perpetual motion' machines.

The second law is introduced by considering irreversibilities and the problem of heating-to-working transformations. No attempt is made to teach the various corollaries of the second law; the emphasis is to instill a practical grasp of the implications of the law. The concept of available and unavailable energy are also introduced at this stage of the course. This aspect of the course is covered in another thermodynamics workshop paper (1).

The control-volume procedure. The material dealing with the first and second laws of thermodynamics for flow processes develops a control volume procedure as follows: (a) define the control volume, with properties specified; (b) specify/calculate the energy terms; (c) carry out a steady-state energy balance; (d) specify/calculate the entropy terms; (e) carry out a steady-state entropy balance: $\dot{Q}_{cv}/T_r + \Sigma_{in}\dot{m}s - \Sigma_{out}\dot{m}s + \dot{S}_I = 0$ where $\dot{Q}_{cv}$ is the steady heat input rate from a constant temperature reservoir at T_r, $\dot{m}$ is the mass flowrate, s is the specific entropy and $\dot{S}_I$ is the entropy creation rate.

A force-momentum control volume procedure along the same lines as the above is then introduced: (a) specify the control volume; (b) specify the co-ordinate axes direction and the relevant flow stream properties at the flow stations; (c) decide on the type of forces to be considered and evaluate the various terms in the force-momentum balance; (d) apply the force-momentum balance and calculate the required unknown:

$F_r + \Sigma_{in}\dot{M} - \Sigma_{out}\dot{M} = 0$ where F_r is the resultant force acting on the fluid in the control volume and $\dot{M}$ is the momentum flowrate.

The course concludes by summarising the control volume procedure in terms of the 'balances' already introduced: the steady-state mass balance; the steady-state energy balance; the steady-state entropy balance and the steady-state force-momentum balance.

REFERENCES

1. Armson, R., "Teaching the Second Law", (This workshop).

CHEMICAL THERMODYNAMICS FOR CHEMICAL ENGINEERS

J. C. R. Turner

Department of Chemical Engineering
University of Exeter

INTRODUCTION

Chemical engineering undergraduates will have encountered "thermodynamics" before entering University, but to varying extents. For example, not all have taken Physics A-Level, and the treatment of thermodynamics in the different Chemistry syllabuses is not standard. Irrespective of their previous experience, their present grasp of the subject is, for most students and most practical purposes, virtually zero.

They will receive chemistry courses at University, usually taught in Chemistry Departments. Such courses may, in fortunate cases, give students a good ground in chemical thermodynamics. This requires extension in four main areas, for chemical engineers.

i) Compressors, turbines, power generation and refrigeration,
ii) Enthalpy balances on process plant, often with chemical reaction,
iii) Chemical equilibrium calculations,
iv) Phase equilibrium calculations.

These topics are not sequential, but interlock like revolving gear wheels; they come round again and again. I shall not discuss i), nor the wider problem of 'energy management', which is of major importance in chemical plant complexes. The other three topics I shall treat from the point of view of teaching the early undergraduate years. Sadly, it is the case that many students fail to comprehend the topic at this level. Later sophistications go over their heads, and many will not even try to follow them.

ENTHALPY BALANCES ON CHEMICAL PLANT

In "Process Principles" classes, students are taught to cope with systems in which the stoichiometry is not simple. At school they can handle (methane + oxygen) → (water + carbon dioxide), and do the enthalpy balance, too. But they do not find the stoichiometry of the reactions leading to (naphtha + air) → (ammonia synthesis gas) simple, nor the enthalpy balances over such a plant. But they have to grasp, and early, the cost and design implications, of such material and thermal balances.

CHEMICAL EQUILIBRIUM CALCULATIONS

Students have probably met equilibrium constants at school, and may even recognise the Van't Hoff isotherm and isochore. It is a rare student who arrives at University fluent in using these concepts to obtain numbers, to put a bound on the maximum performance possible for a reactor (and at various temperatures).

Following on from the previous section, the effect of varying the temperature and pressure in design calculations for an ammonia synthesis reactor has to be grasped. This ties in thermodynamics with reactor design, where we may start with "irreversible" reactions for simplicity.

PHASE EQUILIBRIUM CALCULATIONS

Chemical engineers soon acquire some competence in liquid-vapour equilibrium calculations in connection with distillation design. Liquid-liquid equilibrium does not have the 'ideal' model to fall back on for a simple start; students do not therefore connect thermodynamics to liquid extraction so readily, and they find solid/fluid equilibrium more difficult still.

The concept of chemical potential as a tool (I leave the 'philosophy' of thermodynamics to others) is one which most students find hard to grasp. Few of them develop any real facility with handling phase-equilibrium concepts, and this is reflected in widespread inability to get started on thermodynamics examination questions. Non-ideality (fugacity coefficients and activity coefficients) raises further problems in the student mind.

I must enter a caveat here about phase-equilibrium computer programs (e.g. distillation programs). Of course these are necessary, particularly in multicomponent systems. But most chemical engineers who use these programs have little grasp of how they were written and what the limitations are of the thermodynamic models incorporated therein. I have myself seen examples in industry of nonsenses such as the use of low-pressure programs in super-critical systems!

CONCLUSION

I have tried to outline very briefly some of the main areas in which I would hope to develop some confidence in my students when teaching thermodynamics. Many years of experience has forced me to realise that the subject is difficult to teach and for the students to grasp. Many students give up; the consolation lies in the few who do finally see some of the beauty and utility of the subject. Even at a fairly lowly level it is of crucial importance in chemical engineering design.

DISCUSSION IN SESSION 6

"SYLLABUS DEVELOPMENT"

Edited by D. C. Anderson

University of Lancaster

On the paper by R. S. Silver, "Thermodynamics and Irreversibility"

Professor Silver first of all pointed out that his textbook, published by C.U.P. in 1971 is not included in the survey submitted to the Workshop by Button and Minton. A book by R. F. Silver and J. E. Nydral, published by West (1980) is included in the survey and some participants may have mistakenly assumed him, as at least one did, to be the co-author.

Professor Landsberg asked, "If you link two states will ΔW_f depend on the path taken or, to put it another way, is ΔW_f exact or inexact?"

In reply, Professor Silver said that he did not like thinking about ΔW_f in these terms. It was an important point which merited an expanded written reply as follows.

"In principle ΔW_f, and of course also ΔQ and ΔW are certainly not perfect differentials, but it is misleading merely to classify them contrarily as "imperfect differentials". The mathematically inter-related functions which we call properties in thermodynamics are properties of a particular reference mass. Energy transmissions to or from that reference mass cannot be similar functions, since they involve other bodies, although the transmitted amounts physically produce property changes of equal magnitude in the reference mass. That is why I use the symbol Δ in ΔW, ΔQ, and ΔW_f, to denote elements of energy transmission.

Consider my fundamental thermal equation

$$\Delta Q + \Delta W_f = dU + pdV \tag{1}$$

It states that the sum of two process elements on the L.H.S. produces the combined property changes on the R.H.S. The symbol d is appropriate for the elements of the perfect differential property changes dU and dV. Now in my text-book the important proof is given that even under irreversible conditions, i.e. when ΔW_f is not zero, the combined property change dU + pdV has the integrating factor 1/T, i.e. that

$$(dU + pdV)/T \equiv S \tag{2}$$

It follows then that it is also true to write

$$(\Delta Q + \Delta W_f)/T = dS \tag{3}$$

Thus formally it might seem legitimate to say that the sum $\Delta Q + \Delta W_f$ "has an integrating factor", and so that sum might be said to be an "imperfect differential". Certainly it has been the common practice in reversible thermodynamics teaching, where ΔW_f does not appear, to refer to ΔQ as an imperfect differential having the integrating factor 1/T. But that practice blurs the distinction between process and property. Thus even when I consider the ideal reversible case, where $\Delta W_f = 0$ I make the teaching points that

$$(dU + pdV)/T \equiv dS$$

and hence $\Delta Q/T = dS$

i.e. emphasise the distinction between the identity, which is a mathematical relationship between properties, and the equation, which is a physical result.

I hope that will enable participants to appreciate the use of the full irreversible equation. The important points all follow from the facts that with appropriate temperature difference, ΔQ can be made positive or negative, but ΔW_f is always positive. Hence the entropy of a reference mass can only be diminished by transmitting energy from it by the heat process, and that means you cannot reduce the entropy of a reference mass unless there is some other entity available at a lower temperature. Other obvious consequences are that the same positive entropy change value dS can be obtained either by a low ΔW_f and high positive ΔQ, or by a high ΔW_f and low positive ΔQ. Thus severe dissipation reduces the amount of energy which can be supplied by the heat process for a specified entropy increase. Conversely, severe dissipation increases the amount of energy which must be extracted by the heat process for a specified entropy decrease. It follows that the property change dS of a reference mass can be precisely reversed in a real process, but the physical interactions which occur are not precisely reversed. Less energy can be supplied by heating to increase dS than must be rejected by heating to restore the original value. The property path can be precisely reversed. The physical interactions which produce the reversed path must differ from those which produce the direct path. In short, property paths are always reversible and therefore we should not speak about "reversible" and "irreversible" paths. The terms should be used only as adjectives for processes, and all real processes are irreversible, "reversible" being an unreal idealisation.

Please note that, while I have used T and dS in the above discussion to deal with the question of nature of differentials and integrating factors, my text-book treatment discusses the path and process reversibility and irreversibility of real processes included."

Professor Landsberg communicated his thanks for Professor Silver's courtesy in sending him the above contribution for comment after the Workshop Conference. This prompted him to further consideration. As there are no inequalities in (1) and (3) one would imagine that the normal Clausius inequality

$$\Delta S > \Delta Q/T$$

is reduced to the equality (3) by the definition $\Delta W_f = T\Delta S - \Delta Q$ in Silver's thermodynamics. Here Δ is used as indicating the result of an integration over a "small" range of the relevant parameters. His way of doing things and the normal way are then - and there are now two ways of putting it - (i) identical and (ii) not in disagreement, as far as this point is concerned.

Professor Silver concurred with these remarks and said that his treatment was of course consonant with the "normal" view. The equality

$$TdS = \Delta Q + \Delta W_f$$

is the Clausius inequality, since ΔW_f is always positive. The equality is preferable because it comprehends irreversibility.

Dr. Mayhew, on Professor Silver's suggestion that if his method of irreversible thermodynamics were adopted there would be no need to define systems and their boundaries, wondered how one would cope with, say, friction in a horizontal plane. He recalled that P. W. Bridgman addressed himself to this problem in "The Nature of Thermodynamics" (Harvard U.P., 1941). Dr. Mayhew continued by asking, "If work is done to overcome friction, and some energy must flow across the surface, some up and some down, how are the energy quantities apportioned, irrespective of whether you call that plate a boundary or something else?"

Professor Silver: "If you talk about "systems", as is conventional in thermodynamics, you certainly have to define what you are talking about. I did not say there would be no need to define what you are talking about. My point is that the word "system" is redundant, because its definition means nothing more than a reference mass, and that is already defined in Newtonian mechanics.

For the case of friction in a horizontal plane, the reference mass A is made to slide along its common boundary with a stationary reference mass B. The energy is obtained from a third reference mass C which exerts on A the force necessary to cause the slide. Thus output work from C supplies the frictional dissipation to the combined reference mass (A + B). Then, on the assumptions that there are no other transmissions to or from the surroundings, that volumes are constant, and that potential and kinetic energy changes are zero, my equations (1.1) and (1.3) used formally lead directly to the obvious result that the work output from C goes to increase of the internal energy of the combined reference mass (A + B). How the internal energy change is distributed between the individual reference masses A and B cannot be known without specifying characteristics of these masses, and in general, becomes a rate problem.

The famous Joule experiment is a systematisation of that horizontal plane case, contrived so that the reference mass C is a falling weight, the reference mass A is the outer cone, and the reference mass B is the inner cone plus water content,

Concerning equating (1.3) viz.,

$$\Delta Q = dU + pdV - \Delta W_f$$

Dr. Frost asked whether a symbol convention was necessary to exclude the sign of ΔW_f being exchanged in contrast to all the other terms.

Professor Silver: "I appreciate the point and I have already emphasised the distinction between the fact that dU, dV and ΔQ are each able to be positive or negative, while ΔW_f can never be negative. In any real situation it is always positive and can only be zero in an idealised case. I would suggest that no other symbol convention is necessary however because surely the suffix "f" which defines ΔW_f as work dissipated against frictional forces is already a sufficient symbol indication that the quantity is always positive."

With reference to the discussion so far, Professor Silver went on to state positively that the results of his approach to thermodynamics are entirely consistent with those of the usual approach. He made the point that he believed it resulted in a better overall presentation more suitable for a unified thermodynamics, including irreversibilities of real processes.

Dr. Hay revealed that irreversible thermodynamics used to be included in the third year syllabus at Nottingham but was phased out because it was considered to be of limited use for mechanical engineers. Apart from thermoelectric effects they could not find suitable quantitative applications. It would therefore be helpful if some indication could be given of engineering situations where the approach can be used for quantitative analysis.

Professor Silver: "I am interested to learn that Dr. Hay stopped teaching "irreversible thermodynamics" because of limited applications for mechanical engineers, apart from the thermoelectric effects. Similar experience was one of the motives which impelled me to develop my approach. Since all mechanical, and other, engineering thermodynamic experience is of irreversible phenomena, there just has to be a better treatment than the Onsager-based procedure.

Consider the flow of a fluid in a heat-exchanger tube. It could be described in the following way. With a steady mass flow rate along the tube we can maintain a steady longitudinal distribution of temperature by supplying or extracting heat. Suppose the temperature distribution is such that the temperature falls from left to right. If the fluid flows from left to right we find we have to extract heat to maintain the temperature distribution. If now we make the water flow from right to left, but want to keep the same temperature distribution falling from left to right, we find that instead of extracting heat we have to supply heat. Thus we conclude that this effect is reversible.

You will say that is a very clumsy and roundabout way to talk about cooling or heating fluid in a tube. But that is exactly the language conventionally used for describing the Thomson Effect in a wire along which electric current is flowing. The two phenomena are essentially the same but the different language normally used in the two cases obscures their common nature.

Dr. Hay will find that my 1980 paper in the Journal of Physics gives a full account, and proves, inter alia, that the so-called "thermo-couple" relations are completely general for any loop circuit, whatever the nature of the mass flowing in it, and irrespective of load, and irrespective of extent of departure from equilibrium. That proof alone extends the utility very widely. Some other examples are given in the paper. It is obvious also that if the introductory teaching of general thermodynamics is based on my procedure which included irreversibility from the start, the rest follows in a natural continuity."

On the general question on the nature of a perfect gas and on the relation between the perfect gas temperature scale and thermodynamic temperature, Professor Silver pointed out that in his text-book he gives a new, and rigorous argument identifying the thermodynamic temperature with other temperature scales.

Professor Silver: "First I prove that, with complete generality

$$T = \left[1 + \frac{1}{p}\left(\frac{\partial U}{\partial V}\right)_T\right] \Big/ \frac{1}{p}\left(\frac{\partial p}{\partial T}\right)_V \qquad (4)$$

Hence, if we can find a substance for which U is a function of temperature only, we shall have, for that substance

$$\frac{1}{T} = \frac{1}{p}\left(\frac{\partial p}{\partial T}\right)_V \tag{5}$$

Thus by experiments on such a substance at constant volume, measuring pressure and pressure increments for temperature intervals on any arbitrary scale, we can determine the absolute value of T as a number of these arbitrary units.

I then prove that a substance for which the product pV is a function of temperature only, without restricting the form of that function, also has U a function of temperature only. Hence such a substance can be used in the experiments for equation (5). Experimentally such a substance is available in any gas over a range for which pV is constant at constant temperature.

That argument is of course more complicated than the conventional identification argument, but it is rigorous, precise, and does not require the usual assumtpion of constant specific thermal capacity. Incidentally it is also proved that, although the pV = f(T) assumption is made in general, it turns out that the only possible function is that of direct proportionality, so that the gas constant R arises naturally.

Thus my definition of a perfect gas is merely pV = f(T) and the rest follows. I refer to a gas with the additional idealisation of constant thermal capacity as a "pluperfect" gas!"

The difficulty of teaching irreversible thermodynamics at undergraduate level was referred to earlier by Dr. Hay. Some general discussion, particularly between Professor Gurney and Professor Silver, concerning the interpretation to be given to the term ΔW_f, showed that this difficulty is related to the problems of establishing a new, logical and convincing approach leading to an agreed structure of irreversible thermodynamics suitable for the undergraduate syllabus.

R. A. Haywood warmly commended the approach Professor Silver had used to introduce realistic representations of irreversibility. With reference to equation (4), he pointed out that dS was also related to ΔQ through the equation

$$dS = \frac{\Delta Q}{T} + dS_c,$$

where dS_c was the entropy creation due to irreversibility.

Thus equation (4) was an alternative way of writing

$$\Delta Q = TdS - TdS_c .$$

Comparing this with (4), one saw that

$$\Delta W_f = TdS_c ,$$

namely, that ΔW_f was simply the local lost work due to irreversibility.

When expressed per unit mass of fluid, in incompressible flow W_f/g was in fact the hydraulic engineer's "loss of total head due to friction", and W_f could be described as "the mechanical energy dissipated by friction".

Professor Silver welcomed communication with any young academic who might wish to carry.the development of irreversible thermodynamics further.

The discussion on this paper concludes with a contribution from Professor Patterson which relates to some of these difficulties.

Professor Patterson: "The various methods proposed for presenting aspects of thermodynamics to engineers appeared at times as a bewildering variety of differing concepts and subjective interpretations of the basic structure, which in my view could be avoided by adopting the following perspective in initial studies on fluids.

1. The primary law of thermodynamics, the sole fundamental law of experience peculiar to thermodynamics with a status equal to the conservation laws of mass, momentum and energy, is that expressed by the equation of state for a gas (or its equivalent bulk modulus relation for other media). There are no separate or additional fundamental equations in thermodynamics.

2. The 'first law' is a convenient title given to the conservation of energy requirement expressed in the Lagrangian reference system for a specific fluid element (rather than the alternative Eulerian system which selects spatial Cartesian co-ordinates). In the Lagrangian system during displacement, whether expansive or compressive, a small positive net amount of energy

("reheat") is always expended between molecules within the fluid element ("viscosity").

3. The first law therefore, when set down quantitatively must include a specific term with its own individual identiy (say dq_r, after Bannister) which is always positive and which must be, either deducted from the displacement work term (pdV), or added to the heat addition term (dq).

4. When the fundamental energy and state equations expressed in the most general terms are combined, it is implicit in the mathematics (for 'exact' differential equations) that the group $\frac{dq + dq_r}{T}$ has the status of, and must be treated as a genuine thermodynamic variable - i.e. one which satisfies the fundamental statement of the equation of state that the selection of any two genuine thermodynamic variable determines all others. The variable for convenience is referred to as 'entropy' and is represented symbolically by the single increment dS, which simply stands for, and has no separate existence from $\frac{dq + dq_r}{T}$. Thus, the identity $dS \equiv \frac{dq + dq_r}{T}$ completely describes the entropy variable and constitutes a statement of the "second law of thermodynamics", which is not a fundamental equation but a consequence of, completely satisfied by, and included within the fundamental equations of state and energy.

In my experience the above explanation of fundamental concepts avoids misunderstandings and mistakes in later work. The perfectly acceptable alternative mathematical approach of treating the second law as fundamental and relegating the state equation to being a derived result is not favoured by students in their initial studies, presumably because of their previous understanding of the importance of the 'gas laws' in physics."

On the paper by M. D. Dampier, "Can Thermodynamics be Made More Simple?"

Professor Gurney: "Engineers should note Dr. Dampier's remark at the beginning of § 2, that his phase space has no geometrical structure. What he calls vectors in this paper, have no general cosine dependence, and would be better called (ordered) pairs and triples. His subject is matrix algebra, not vector algebra.

This misuse of the word vector (from veho- I carry) has recently become common among mathematicians. Products between non-vectorial single column matrices P, Q, should be called dot or inner (P^TQ); cross or outer $-(PQ^T - QP^T)$ and open (PQ^T)."

Dr. Dampier: "We do use the word 'vector' in wider contexts than that of Euclidean space with its 'cosine dependence'. That does not seem to me any less legitimate than using the word 'space' in more general ways. The theory of vectors gives us an ideal calculus for talking about direction and so naturally appears wherever partial derivatives are found."

Dr. Rao: "Was the use of Jacobians considered by the author? I found the exposition on the use of Jacobians by Tribus (Thermostatics and Thermodynamics, Ch. 9, Van Nostrand, 1961) very useful and a powerful technique for developing property relations."

Dr. Dampier: "No, I did not specifically consider Jacobians; but, since the Jacobian is a determinant, the close connection between vector products and determinants make it reasonable that Jacobians could be used in a way that parallels the vector product method of my paper. One would choose the approach that best matched the students' mathematical knowledge."

Dr. Lewins: Dr. Dampier gives us a concise notation for thermodynamics of two or three independent variables in the internal state space, using the vector notation that will be familiar to students from say Maxwell's equations or the as stream/potential theory of fluid-flow. Professor Gurney has objected to the use of vector concepts in the discussion of the internal thermodynamic state space. It is true that ordered numbers are not necessarily vectors in the sense that vector addition may not follow, e.g. the ordered numbers describing a finite rotation in three dimensions can fail this test. But to deny the generalisation of vectors from three-dimensional Cartesian geometry to the state space-vector concept would be to ignore a major technique of mathematical invention. All the vector concepts, including a defintion of the cosine via a scalar or dot product, can be justified.

The proper problem, however, to be faced is that a generalisation of Maxwell's relations to a four or greater dimensional state space cannot be accomplished in the vector calculus notation. Gibbs himself, on introducing the 'del' notation, acknowledges the special structure of a three-dimensional space; the two-dimensional case we most commonly use in thermodynamics (two independent variables) is covered by the trick of a pseudo-symmetry analogous to the treatment of axial symmetry in fluid mechanics. If we want a compact notation for n-dimensions, we should turn to the Goursat theory of external differential forms and, in particular, the Poincaré integrability lemma. A readable introduction is "Differential Forms" by Harley Flanders, Academic, New York 1963, from which the following has been adapted. But before continuing, it has to be acknowledged that an extension of Maxwell relations to more than three dimensions is hardly the stuff of undergraduate thermodynamics; one hopes their teachers might be interested!

The theory uses an algebra where multiplication is defined appropriate to integration in an n-dimensional space. Thus an element of area conventionally written dxdy should be treated as antisymmetric since the signed area dxdy=-dydx. Then we write $dx \wedge dy$ with $\wedge$ a multiplication having such an antisymmetric property: $dx \wedge dy = - dy \wedge dx$. Then it follows that $dx \wedge dx = 0$. If x_i; i=1,2,..n, are the independent vectors of the state space, we may choose the dx_i as the unit vectors for expansions. Differentiation of a form in this space is _defined_ as the d operator such that

$$d(\eta + \omega) = d\eta + d\omega; \; d(\eta\wedge\omega) = d\eta \quad + (-1)^{\deg\eta} d\omega$$

$$d(d\eta) \equiv d^2\eta = 0 \text{ (Poincaré lemma) and } df = \frac{\partial f}{\partial x_i} dx_i$$
repetition summation

Now we note that each $d(x_i)$ _is_ dx_i.

In application to thermodynamics, the following scheme works for any $n \geq 2$ but for exposition, clarity and brevity we chose the conventional case n=2, starting from the fundamental equation dU = TdS - PdV.

(i) Suppose we take S and V as the independent variables so that dS and dV are the unit vectors dx_1 and dx_2. From Poincaré, d(TdS) = d(PdV) or

$$d(TdS) = \frac{\partial(TdS)}{\partial S} dS + \frac{\partial(TdS)}{\partial S} dV = \underset{0}{\cancel{T\frac{\partial dS}{\partial S}}} \wedge dS + \underset{0}{\cancel{dS\wedge\frac{\partial T}{\partial S} dS}} + \underset{0}{\cancel{T\frac{\partial dS}{\partial V}}}$$

$$+ \, dS\wedge\frac{\partial T}{\partial V} dV$$

and

$$d(PdV) = P\frac{\partial dV}{\partial S}\wedge dS + dV\wedge\frac{\partial P}{\partial S}dS + P\frac{\partial dV}{\partial V}\wedge dV + dV\wedge\frac{\partial P}{\partial V}dV$$

(the first, third and fourth terms struck through and marked 0)

and therefore $\frac{\partial T}{\partial V} = -\frac{\partial P}{\partial S}$, the first Maxwell relation.

(ii) For a more general case, take say P and T as independent variables so that $dS = \frac{\partial S}{\partial P}dp + \frac{\partial S}{\partial T}dT$. In the expansion of

d(TdS) terms in $\frac{\partial dP}{\partial P}dP$, $\frac{\partial dT}{\partial p}dP$ are zero as are $dP\wedge dP$ and $dT\wedge dT$ as before. We now obtain

$$d(TdS) = \frac{\partial S}{\partial P}\,dP\wedge dT + T\frac{\partial^2 S}{\partial T\partial P}\,dP\wedge dT + T\frac{\partial^2 S}{\partial P\partial T}\,dT\wedge dP = \frac{\partial S}{\partial P}\,dp\wedge dT$$

and

d(PdV) similarly becomes $\frac{\partial V}{\partial T}\,dT\wedge dP$ giving the usual fourth Maxwell relation, $\frac{\partial S}{\partial P} = -\frac{\partial V}{\partial T}$.

For n=3, consider $dU = TdS-PdV-Adc$ and $d^2U=0$. Taking S,V,C as independent gives

$$d(TdS) = dS\wedge\frac{\partial T}{\partial V}dV + dS\wedge\frac{\partial T}{\partial C}dC;\quad d(PdV) = dV\wedge\frac{\partial P}{\partial S}dS + dV\wedge\frac{\partial P}{\partial C}dC;$$

$$d(AdC) = dC\wedge\frac{\partial A}{\partial S}dS + dC\wedge\frac{\partial A}{\partial V}dV$$

so that $\left[\frac{\partial T}{\partial V} + \frac{\partial P}{\partial S}\right](dS\wedge dV) + \left[\frac{\partial T}{\partial C} - \frac{\partial A}{\partial S}\right](dS\wedge dC) + \left[\frac{\partial P}{\partial C} - \frac{\partial A}{\partial V}\right](dC\wedge dV) = 0$

Operating with $dS\wedge$, $dV\wedge$ and $dC\wedge$ in turn shows each bracket to be independent and yields the appropriate extended Maxwell Relation. Similar results comes with picking other base triples and one can go on to the higher order Maxwell relations.

Flanders makes a good case for engineers to be taught external differential forms when vector analysis is introduced. Until that is done, of course, there will be little advantage in introducing the above in the teaching of thermodynamics. The alternative of a matrix notation, however, would seem far too clumsy. It remains my view that Dr. Dampier's use of vector calculus has a genuine pedagogic advantage although the conventional derivation supplemented by Born's mnemonic diagram and the use of Jacobians have merit, the latter because it too can be extended to the higher dimensional case.

Dr. Dampier: "I do not think that there is much in my paper that cannot immediately be generalised to n-dimensions ($\underset{\sim}{a} \wedge \underset{\sim}{b}$ will not be a vector, but it is already not a vector in the 2-dimensional case). Such generalization, of course, could be regarded as the beginning of the theory of exterior differential forms, but it is certainly worth noting how little of that theory is used here.

<u>On the paper by W. K. Kennedy</u>, "<u>Teaching Thermo-Fluid Mechanics at the Open University</u>"

Dr. Kotas: "How do you introduce the Second Law by considering irreversibilities and the problem of heating-to-working transformations."

Dr. Kennedy answered by first illustrating the structure of Unit T233 of the Open University course as shown below.

Course Unit T233 - THERMOFLUID MECHANICS AND ENERGY

<u>Unit titles</u>

1. Introduction to Energy and Thermodynamics
2. The First Law
3. The Second Law
4. Available Energy and Entropy
5. Modelling Fluids
6. Similarity Analysis and Dimensionless Groups

7/8. The First and Second Laws: Flow Processes

9/10. Fluid Mechanics: Energy and Momentum

11/12. Water Wheels and Turbines

13/14. Heat Transfer Analysis

15. Vapour Power Cycles (Power Station Cycles)
16. Revision

The 'finale' of the course unit is the bringing together of the control volume procedure with considerations of mass, energy, entropy, force and momentum balances.

Concerning the question of Dr. Kotas, he pointed out that Unit 3, introducing the Second Law, tackles the problem of irreversibility with the help of an energy analysis of a pile driver. Heating-to-working transformations relate heating processes and temperature to the concept of energy usefulness and degradation. All of this is a preamble to a more rigourous consideration of the Second Law which is done without getting "bogged-down" in the various corollaries of that Law. Unit 4 of the course deals with available energy which is introduced before considering what is entropy.

Dr. Himsworth: "When presenting his paper, Dr. Kennedy said that he was not sure whether the equation

$$\dot{Q}_{cv}/T_R + \sum_{in}\dot{m}s - \sum_{out}\dot{m}s + \dot{S}_I = 0$$

which he referred to as the "steady-state entropy balance", could be properly called by this name.

The earliest reference to this equation of which I am aware occurs in 'The Separation of Gases' (M. Ruhemann, Oxford 1st Ed., 1940, p.85) where it is called 'the balance of entropy'. I recently asked Dr. Ruhemann (private communication, 7th July 1981) who deserved the credit for devising this extremely useful equation. He replied, with characteristic diffidence, that as far as he knew, he had introduced the idea of an entropy balance, but may well have read it somewhere and forgotten the source."

In further discussion Dr. Kennedy and Dr. Himsworth agreed that the use of the therm 'entropy balance' was perfectly proper (like a mass or energy balance) and this in turn required the idea of 'entropy creation' in order to achieve that balance. Dr. Kennedy commented that any problem, if there indeed is a problem, is more likely to be confined to mechanical engineers, rather than chemical engineers who routinely use these terms and concepts.

The discussion ended with Dr. Lewins simply noting that the Open University course uses the 'positive work' in convention for the First Law, i.e.

$$Q + W = \Delta E$$

Dr. Kennedy said that this had not caused any problems with their students and that he would support a wider adoption of this convention.

On the paper by J. C. R. Turner, "Chemical Thermodynamics for Chemical Engineers"

There was no recorded discussion on this paper but Professor Turner presented the following contribution as an extension to his paper.

Professor Turner: "I shall not speak to my paper, which though short needs no elaboration, but I wish to make a few comments which this workshop has prompted in my mind.

Firstly, chemical engineering does not see the same difficulties in thermodynamics which seem to be rife in mechanical engineering. This may be due to its comparatively small size, and world-wide uniformity. In this country chemical engineering is a virtually 100% graduate profession. There are only four Polytechnic graduate courses, and they join with the Universities in the Institution - there is no binary divide.

It has been stated here that thermodynamics is not easy, but, as Sir Brian Pippard says, that applies to other subjects, too. Thermodynamics is not unique in this respect. Scientists are still taught to think, and sometimes I fear that engineering courses, with the pressure to contain facts, are not teaching their students to think. "To know is to understand" is not true - the other way round, yes.

I look back on my own development in thermodynamics, and hope therefrom to get some guidance on how to teach the subject - I am sure we all do that. I was fortunate enough to have a physics teacher at school who liked to talk about thermodynamics. He was a dear man, if a bit short on sense of humour. I then proceeded to Professor Norrish's lectures in Physical Chemistry here (obliquely alluded to by Sir Brian in his Dinner Speech). Norrish was a great man, in whose Department I was privileged to do research, but his thermodynamic lectures were sheer chaos. So I had to think, and ask my supervisor, and subsequently myself earned for many years an honest pittance by supervising students similarly confused; and learning the subject myself while doing it.

I was a well-behaved boy, so naturally accepted that thermodynamics was right, since my elders and betters went on at such length about it. But I came to believe in it when I saw the numerical equivalence between the "statistical mechanical" entropy, and "calorimetric" entropy of gases. That two such completely different sets of experiments should give the same answer was a revelation. On the one hand spectroscopy and quantum theory, on the other "simple" calorimetry and latent heat measurements - using the third Law (Nernst's statement) which if I am not mistaken has not been mentioned so far at this workshop!

In my studies I have been much helped by the books of Denbigh, Pippard, Rowlinson, Prausnitz, etc., not one of which is to be found in the book list here!

Finally, it is in the Design Project, as John Mecklenburgh (Session 5) has described, that thermodynamics comes, together with rate processes and much else, to produce the end product of our courses. And rates are different from thermodynamics. To me, the only thermodynamics in heat transfer is that heat runs downhill. But I repeat, for chemical engineers the Design Project should encapsulate all - including thermodynamics.

On the paper by D. R. Croft, P. W. Foss and M. Denman, "Thermodynamics: A New Teaching Approach"

In the presentation of this paper, Dr. Foss showed a table of "power densities" for various devices.

Dr. Hay referred to the value of "several thousand W/m^2" given under the heading of "Wave Power". On the presumption that the reference area (m^2) was that occupied by the power extraction device, he remarked that the power density would be a lot less than "several thousand W/m^2" for wave power.

Dr. Denman replied that wave power density is quoted as several thousand W/m^2 over the first several metres of depth of water below the surface. Unlike wind power, there is no flow rate and the motion induced in the water by the passage of a wave is related to this depth of water. Usually, power/metre length of device is quoted on the basis of a typical Atlantic wave having a capacity of 80 kW/m length. The value of "several thousands W/m^2" is based on an area given by a unit length of wave times the frontal "depth" of the wave. Dr. Denman agreed with Dr. Hay that if the total wetted area of the device is used as a reference area, then this figure will be considerably reduced.

On the paper by J. D. Lewins, "Thermodynamics Sign Convention and Nomenclature"

The main contribution to the discussion of this paper consisted of the presentation given by Y. R. Mayhew (Department of Mechanical Engineering, University of Belfast) and J. W. Rose (Department of Mechanical Engineering, Queen Mary College, University of London), given in full earlier in this session.

The discussion continued with Dr. Kennedy commenting on points raised by Dr. Lewins which relate to the Open University course T233.

Dr. Kennedy: "We do use 'specific heat capacity' although we point out that it is better to realise that it should be 'specific heat capacity for internal energy increase'. Also, as I pointed out in my presentation on our course, we refer to 'energy transfer by heating' rather than 'heat' in the introductory units of the course. After students have become familiar with what they are dealing with they can shorten it to 'heat' or 'heat transfer' for convenience."

A general problem was then illustrated by a contribution from Professor M. J. French.

Professor French: "A problem arises in writing equations where some of the algebraic quantities have been given numerical values. Consider the equation

$$M = \rho v \text{ (mass = density x volume)}$$

It is sometimes done, deplore it as we may, and it may even be convenient, to substitute for V a particular volume, say 8 cubic metres.

It seems to me logical to write

$$M = 8\rho \text{ m}^3$$

Strange as it may appear, it will be right if we substitute the value of ρ in kg/m^3, and it will still be right, if strange, if we use say, the value of ρ in lbm/ft^3 and obtain a result lbm m^3/ft^3."

Dr. Lewins: "Indeed we can see the logic but also how clumsy the result is. Such clumsiness is the source of the common advice, "do not substitute numbers until the end so that you can continue to check the dimensions". If numbers <u>are</u> substituted earlier, surely we must recognise that '8' is now not non-dimensional. An equation can have its dimensions/units added in brackets, e.g.,

$$M[kg] = \rho[kg/m^3]v[m^3]$$

so that we can write, for $V = 8\ m^3$

$$M[kg] = 8[m^3]\rho[kg/m^3]$$

implying the final result

$$M = x.kg.$$

I would prefer this to writing $M = 8\rho\ m^3$

Dr. Mayhew also responded to the contribution by Professor French.

Dr. Mayhew: "I am glad that Professor French raises this problem because it is one which troubles many and can lead to breaches of the convention that

$$\text{physical quantity} = \text{numerical value} \times \text{unit} \qquad (1)$$

It is necessary to make up one's mind whether one is ready to accept the convention consistently or whether one jumps, at the drop of a hat and without warning as many engineers are in the habit of doing, from the above convention to one where a symbol can stand for the numerical value alone. Reference (7) of the paper by Mayhew and Rose only lays down the ground rules. The method which I favour and which is fully consistent with these rules, is described in my paper "Physical and Numerical Equations" (Int. J. Mech. Engg. Ed. 1975, Vol. 3, No. 1, pp. 85-7). In that paper I propose a routine procedure at arriving at a numerical working equation which satisfies the conventions and which should satisfy Professor French also.

Taking the example given by Professor French, viz. the physical equation

$$M = V\rho \qquad (2)$$

let us divide both sides of the equation by the unit in which we wish to obtain the value of M and, to be awkward, let this be the lb_m. This gives

$$\frac{M}{lb_m} = V\rho\frac{1}{lb_m} \tag{3}$$

Let us now substitute the known fixed values, here only $V = 8\ m^3$, and multiply and divide the variable quantities by the units intended for them, e.g. kg/m^3 for ρ. Hence we obtain

$$\frac{M}{lb_m} = (8\ m^3)\left(\frac{\rho\ kg/m^3}{kg/m^3}\right)\frac{1}{lb_m} = 8\left(\frac{\rho}{kg/m^3}\right)\frac{kg}{lb_m} \tag{4}$$

If we remember that $1\ lb_m = 0.453\ 592\ 37$ kg, and hence $kg/lb_m = 1/0.453\ 592\ 37$, the numerical equation follows as

$$\left(\frac{M}{lb_m}\right) = \left(\frac{8}{0.453\ 592\ 37}\right)\left(\frac{\rho}{kg/m^3}\right) \tag{5}$$

an equation in which each bracketed quantity is a pure number. We can rewrite this as

$$M = 17.637\left(\frac{\rho}{kg/m^3}\right)\ lb_m \tag{6}$$

Equation (5) is preferable to the equation

$$M = 8\rho\ m^3 \tag{7}$$

given by Professor French which contains a mixture of numerical and physical quantities, and which leaves it still wide open which units are to be used in substituting the variables. Equation (5) leaves one in no doubt: e.g. only if the value of ρ is substituted in kg/m^3 units, would the denominator cancel.

Those who argue that equation (5) is too cumbersome would normally write the equation as

$$M = 17.637\rho \tag{8}$$

This equation is not only convention-breaching and dimensionally inconsistent, but needs to be followed, for completeness, by the statement "in which M is in lb_m and ρ in kg/m^3", and this is certainly more cumbersome than equation (5).

A further advantage of the way equation (5) is written is that this way is consistent with the recommendations for labelling coordinates of graphs and headings of table columns contained in Refs. (1) and (4) of the paper by Mayhew and Rose, and in the Report "SI units, signs, symbols, and abbreviations for use in school science" of the Association of Science Education."

(Editors note: The paper by Mayhew and Rose referred to is that presented earlier in this session).

CLOSING DISCUSSION

The Workshop has drawn together a variety of thermodynamic principles and teaching approaches. There was general agreement that it was difficult to instil the principles of thermodynamics in one pass to undergraduate students. It was clear to participants that there was scope for variety in approach and that new applications requiring developments in thermodynamic theory were maintaining the life of the subject.

Many of the authors presenting papers to the meeting had done valuable service in collecting concepts and references. A particularly useful overview was the Button and Munton paper reviewing mechanical engineering texts on thermodynamics in print. This is given as an annexure to the Workshop papers together with supplementary remarks by Turner on chemical engineering texts. The reader might particularly note the suggestion that students should be engaged formally in reviewing textural sources.

The following remarks by Professor French are pertinent to an overall view.

GENERAL COMMENT

"The workshop as a whole was very interesting, with a wealth of thought-provoking papers and many useful ideas on material and methods for teaching. However, it seemed to me that the questions with which it began were not specifically addressed in conclusions, and it is on this I wish to comment.

My conclusion is that many of the difficulties students have with thermodynamics arise from the so-called 'classical' approach usually adopted, which differs from that adopted in most subjects, in that physical understanding is not attempt-ted. That this understanding would be imperfect is inevitable, for all human understanding is, but it is none the less desirable. We can explain thermodynamics well in physical terms with which the student is already familiar and which he can readily grasp, for the subject is not really difficult. Moreover, the student so taught might be able to cope with those cases where the 'classical' approach is difficult.

For example, a printed example paper which had been used for years in a very large and distinguished engineering school contained an error in principle about which there was no dispute when it was pointed out. In the meantime, however, hundreds of students had dutifully produced the wrong answer as required, and no member of teaching staff had realised it was wrong. However, a little coarse and unrefined considera-tion of the actual physical process revealed the error. This was a first-year problem involving only the first law, and I believe the reason none of these students, some of them very bright indeed, had spotted the mistake was that they were overawed by the paraphernalia and mystique of the classical approach as deployed at that place and no longer trusted their own thinking.

Among the perversities usual in the classical exposition is the distinction made between heat and intrinsic energy without it being made plain this is purely a matter of _accounting_, there being no more difference in reality than there is between a pound of income and a pound of capital. Indeed, the muddle into which some exponents have got them-selves is extreme - there can be heat flow in a body, but no heat, and so on. A particular weakness is that 'heat' is all of one kind, divorced from the idea that it has a temperature associated with it, which involves further semantic acrobatics when radiation is encountered.

The idea of a system in stable equilibrium, if we are to be rigorous, and that is the claim often made by the 'classi-cist', presents severe difficulties. To avoid more than one equilibrium state being possible, we must think of a steel pressure vessel as 'unsteady', while a bubble is everlasting, which is a difficult idea to justify unless we use our know-ledge of physics, which would defeat the object of the exer-cise. In short, the whole classical approach, far from being rigorous, depends on special pleading derived from a consideration of mechanics. We disallow the stress sustained by a solid wall but not that sustained by a liquid film because one conflicts with the development of the argument and the other does not, notwithstanding the fact that the 'unstable' equilibrium may endure for centuries and the 'stable' one for only milliseconds.

However, manifest as are the weaknesses of the classical treatment, a great deal of thought has gone into its development and any alternative will also require a substantial study. Many proposals have been made in the past and some in this workshop. As one element, I find Professor Silver's approach very promising, although perhaps some term such as 'Joulean' instead of 'frictional' might cause less trouble: the description of I^2R work as 'frictional' jars somewhat.

I believe it is possible to give a good, insightful and adequate treatment of basic engineering thermodynamics, demystified and lucid, which will satisfy students and enable them to obtain the right answers to simple questions, without any pretence of 'rigour' (which is illusory anyway) using statistical mechanics in a descriptive role only (or perhaps with just a few very simple analytical results) and based on the recognition of the Second Law as a statement about probabilities in large collections of bodies, subject to some conditions on the actions between them, e.g., microscopic reversibility. (It is, of course, a very difficult problem to determine what are sufficient conditions, but one into which it is quite unnecessary to go).

Is it possible that some group of thermodynamicists, younger than me, should put in the very considerable amount of work necessary to devise and test such a treatment, with all the thoroughness necessary, and launch it on the world at a sufficient scale to gain a bridgehead of acceptance? Earlier attempts have failed, either because they were insufficiently well worked out, or fell on stony ground, or tried to retain too much of the old, or went too deeply into the statistical mechanics, or introduced too many complexities, or for some other reason. But once a sufficient bridgehead was made, the common-sensical and more satisfactory nature of the new approach should assure its eventual success, and students would no longer find special difficulty with thermodynamics."

The discussion identified three areas of consideration worth further action. To this end, three working groups were established. The Chairman of each is identified below and welcomes contact from those concerned. It is envisaged that reports of further Working Parties may be published in due course and their views made known through, for example, the U.K. Conference of Engineering Professors.

1. Nomenclature, Symbols and Units. Chairman: Dr. Yon Mayhew, Department of Nuclear Engineering, University of Bristol.

The major difficulty was in the use of names associated with caloric concepts of heat. While the meeting had generally recognised a number of unsatisfactory terms, there was no clear agreement on what might be used instead or indeed of the desirability of change. Working Party 'A' would reflect on the points made at the meeting and take up the problem of establishing a concensus.

2. Philosophy and Teaching Structure of Thermodynamics. Chairman: Dr. W. Kennedy, The Open University, Walton Hall, Milton Keynes.

The meeting had received reports not only on conventional lecturing but on a variety of integrated teaching approaches, many of which had been pioneered in the Polytechnics. The developments in microcomputers and new audio-visual techniques suggested that the philosophy and practice of education were as important as the principles of thermodynamics in the discipline. Working Party 'B' would explore the opportunities for promulgating new teaching techniques.

3. The Exergy Method of Analysis. Chairman: T. J. Kotas, Department of Mechanical Engineering, Queen Mary College, London.

Several papers had pursued the developments of available energy flows or exergy. These approaches promised to provide a rational analysis of engineering systems and a route to design and synthesis. Working Party 'C' would convene to review the implications for syllabus inherent in these trends.

The Workshop closed with expressions of thanks to the organisers and our hosts at Emmanuel College and the Department of Engineering, University of Cambridge.

A BRIEF SURVEY OF BOOKS ON THERMODYNAMICS FOR ENGINEERING STUDENTS

B. L. Button and R. J. Munton

Department of Mechanical Engineering
Trent Polytechnic, Nothingham

SYNOPSIS

The paper covers about sixty books which are in print. A short description of each book is included in which the aims or intentions and readership, as stated by the author or authors, are given. The books are those which are published in English and which are suitable at different levels for engineers and for engineering students on certificate, diploma, degree and first year graduate courses. Not included are books which deal essentially with statistical and chemical thermodynamics.

A proposal is made for the extension of this survey with the aim of eventually providing all forms of information, on demand, from video discs.

INTRODUCTION

We are inundated with many forms of information every day. This is in addition to the information which we seek out. In the case of thermodynamics, as with any other subject, we need to apply and use information science (the collection, classification, storage, retrieval and distribution of knowledge) so that we can easily establish and use the information that is available. We recognise that libraries will contain a large number of books on thermodynamics which have been published over the years. If students are to build up their own reference library, they need to be aware of the books which are currently available to them. This paper, which is the record of a brief survey of the books in print which we have reviewed so far, represents a first, primitive step in these endeavours.

We believe that, as part of their course, our students should be involved in helping to prepare learning material for the students who will follow them. To this end, we plan to require some of our students to complete this survey and to start and complete a further survey on books which are out of print. We will then require each of our students to review a different book, having regard to its relevance to their course and engineering applications. At a later stage, some students, as part of their coursework, will be asked to obtain the permission of the publishers and to store the indexes of all these books on a computer, using a data search and retrieval package. Next should follow, with a similar approach, the collection and storage about other forms of information, e.g. films, video tapes and tape/slide programmes. It is conceivable that all the forms of information for subjects like thermodynamics will eventually be available in their entirety, on demand, from video discs.

BOOK SURVEY

ABBOTT, M. M. and VAN NESS, H. C., Thermodynamics, SI (metric) edition, McGraw-Hill, 1976.
The fundamental principles of classical thermodynamics are presented and illustrated by numerous worked examples and problems with many of their applications in science and engineering. For undergraduate and first year graduate level course.

ADKINS, C. J., Equilibrium Thermodynamics, Cambridge University Press, 1983.
Intended as a thorough, but concise course on the fundamentals of classical thermodynamics. A clear and stimulating exposition and one that is easy to learn from. Primarily for undergraduate physicists, but also suitable for use in materials sciences, engineering and chemistry.

BACON, D. H., Engineering Thermodynamics, Butterworths and Company (Publishers) Limited, 1972.
A short text outlining the thermodynamic material for a mechanical engineering degree course. It could be used to replace lecture notes so that students would be able to concentrate on understanding a lecture rather than on note-taking.

BACON, D. H., BASIC Thermodynamics and Heat Transfer, Butterworths and Company (Publishers) Limited, 1983.
Not a comprehensive treatise on BASIC, thermodynamics or heat transfer. Instead, it could help readers to become proficient at BASIC programming, by using it to solve problems in thermodynamics and heat transfer.

BACON, D. H. and STEPHENS, R. C., Thermodynamics for Technicians: Level 3-4, Butterworths and Company (Publishers) Limited, 1982.
Primarily intended to cover levels III and IV units in thermodynamics for higher technician certificate and higher national diploma courses.

BALZHISER, R. E. and SAMUELS, M. R., Engineering Thermodynamics, Prentice Hall Incorporated, 1977.
For students from all engineering disciplines. An introductory course including a treatment of chemical as well as classical thermodynamics.

BEATTIE, J. A. and OPPENHEIM, I., Principles of Thermodynamics, Elsevier Scientific Publishing Company, 1979.
Presents a rigorous and logical discussion of the fundamentals of thermodynamics.

BENSON, R. S., Advanced Engineering Thermodynamics, Pergamon Press Limited, second edition, 1977.
Directed to the undergraduate final-year honours course in engineering.

BOXER, G., Engineering Thermodynamics: Theory, Worked Examples and Problems, The Macmillan Press, Limited, 1976.
Offered as a reasonably self-contained basis of fundamental work. For first year mechanical engineering undergraduates.

BURGHART, M. D., Engineering Thermodynamics with Applications, Harper and Row Publishers, Incorporated, second edition, 1982.
To help undergraduate students learn thermodynamics. It develops the students' ability to undertake thermodynamic analyses and to apply these skills.

CALLEN, H. B., Thermodynamics, John Wiley and Sons, 1960.
An introduction to the physical theories of equilibrium thermostatics and irreversible thermodynamics. For advanced undergraduate and first year graduate students in the departments of physics, chemistry and engineering.

CHUE, S. H., Thermodynamics, John Wiley and Sons, 1977.
A rigorous, postulatory approach which should be particularly beneficial to the physicist at undergraduate level.

CRAVALHO, E. G. and SMITH, J. L., Engineering Thermodynamics, Pitman Books Limited, 1981.
Developed to be an integral part of the modern undergraduate engineering curriculum, with much use made of the modelling techniques of system dynamics. Seven distinguishing objectives of this new method of presentation are given.

Intended for a one semester first course in thermodynamics for students studying mechanical, electrical, oceanic, aerospace, nuclear and civil engineering, but not those studying chemical engineering or materials science.

DIXON, J. R., Thermodynamics: An Introduction to Energy, Prentice Hall International Incorporated, 1975.
Attempts to: develop laws of thermodynamics inductively; to introduce second law through the concept of degradation of energy and to develop a problem solving approach. For engineering students.

DOOLITTLE, J. S. and HALE, F. J., Thermodynamics for Engineers, John Wiley and Sons Incorporated, 1984.
A general approach where all engineering students should gain an understanding of energy and energy transformations as formulated in the first and second laws of thermodynamics.

EASTOP, T. D. and McCONKEY, A., Applied Thermodynamics for Engineering Technologists, Longmans Group Limited, third edition, 1978.
A text designed for undergraduate courses in engineering.

ELWELL, P. and POINTON, A. J., Classical Thermodynamics, Penguin Books Limited, 1972.
Presents the basic ideas of classical thermodynamics in a concise and systematic form complementary to a normal lecture course. For students of physics.

FABRY, C., The Elements of Thermodynamics, J. G. Miller Limited, 1953.
Provides those starting the study of thermodynamics for the first time with an exposition divested of all needless complication.

FAIRES, V. M., SIMMANG, C. M. and BREWER, A. V., Problems in Thermodynamics, Collier-Macmillan Ltd., fifth edition, 1970.
Intended for use in the first course in thermodynamics taken by students in engineering.

FERMI, E., Notes on Thermodynamics and Statistics, University of Chicago Press, 1966.
Not intended for publication or as a book on the subject, but could be useful to students and staff alike.

GOODGER, E. M., Principles of Engineering Thermodynamics, The Macmillan Company, 1974.
Being concerned directly with principles, the book lends itself to most syllabuses on the subject, up to and including first degree level.

GRANET, I., Thermodynamics and Heat Power, Reston Publishing Company Incorporated, second edition, 1980.
A general approach to the subject giving a self-contained coverage for general engineering students and serving as a first text for those students studying the subject to greater depth.

GROSSMAN, L. M., Thermodynamics and Statistical Thermodynamics, McGraw-Hill Incorporated, 1969.
A self-contained text on equilibrium thermodynamics and statistical mechanics for use in intermediate or advanced courses or for self study by graduate students in engineering and the sciences. A first course in classical thermodynamics and some previous exposure to elementary statistical physics are required.

HAYWOOD, R. W., Analysis of Engineering Cycles, Pergamon Press, third edition in SI units, 1980.
A comprehensive text dealing with the analysis of the overall performance under design conditions of power producing and power absorbing plants. For students studying the subject to final honours degree level.

HAYWOOD, R. W., Equilibrium Thermodynamics for Engineers and Scientists, John Wiley and Sons, Ltd., 1980.
A comprehensive text dealing with the basic concepts of the first and second laws of thermodynamics. Aimed to be readily digestible by reasonably competent undergraduates.

HOLMAN, J. P., Thermodynamics, McGraw-Hill International Book Company, third edition, 1980.
Intended for use in the first course in thermodynamics taken by students in engineering.

HOYLE, R. and CLARKE, P. H., Thermodynamic cycles and processes, Longman Group Limited, 1973.
A general approach and coverage of the subject matter suitable for undergraduate engineering students.

HSIEH, J. S., Principles of Thermodynamics, Scripta Book Company, 1975.
Primarily for a second course in classical thermodynamics at first year graduate level for mechanical and chemical engineering students.

HUANG, F. F., Engineering Thermodynamics: Fundamentals and Applications, Collier Macmillan Publishers, 1976.
A text for the first course in thermodynamics taken by students in all branches of engineering. Its aims to introduce the student to design and decision making.

JOEL, R., Basic Engineering Thermodynamics in SI Units, Longmans Group, Limited, third edition, 1971.
General coverage, directed primarily at the technician engineer. It will prove useful in the early stages of study for an engineering degree or the Higher National Diploma.

JOEL, R., Thermodynamics: Level 3, Longman Group Limited, 1984.
Written primarily to cover the syllabus aims for the BTEC level III Unit: thermodynamics.

JOHN, J. E. A. and HABERMAN, W. L., Engineering Thermodynamics, Allyn and Bacon, 1980.
A general approach and coverage of fundamental concepts and applications.

JOHNSTON, R. M., BROCKETT, W. A., BOCK, A. E. and KEATING, E. L., Elements of Applied Thermodynamics, Naval Institute Press, fourth edition, 1978.
Written to present to the undergraduate the fundamental concepts of thermodynamics.

KARLEKAR, B. V., Thermodynamics for Engineers, Prentice Hall Incorporated, 1983.
This text on classical thermodynamics is student-oriented and is designed for a first level undergraduate engineering course in thermodynamics.

KESTIN, J., A Course in Thermodynamics, McGraw Hill Book Company, 1979.
Introduces the subject of thermodynamics to the novice, using the first and second laws.

KESTIN, J., The Second Law of Thermodynamics, Dowden, Hutchinson and Ross Incorporated, 1976.
Fifteen papers have been carefully selected to portray the development of the second law and, to a large extent, the development of classical thermodynamics in general.

LOOK, D. C. and SAUER, H. J., Thermodynamics, Brooks/Cole, Engineering Division, 1982.
Provides a basic introduction to the art and science of engineering thermodynamics for general engineering students.

MARK, M., Concepts of Thermodynamics, West, 1980.
Uses the 'auto-instructional' technique and may be used to supplement a text in a conventional course.

MODELL, M. and REID, R. C., Thermodynamics and its Applications, Prentice Hall, Inc., second edition, 1983.
A text with a rigorous theoretical and conceptual basis, interspersed with a relatively large number of examples and solutions. Intended as a learning text, rather than a teaching text. Offered as a one-semester, graduate-level course.

NAG, P. K., Engineering Thermodynamics, Tata McGraw Hill Publishing Company, Ltd., 1981.
Provides a general coverage of the basic principles and applications of classical thermodynamics.

OPEN UNIVERSITY. Thermofluid Mechanics and Energy, The Open University Press, 1982.
Provides a good introduction to the laws of thermodynamics and their applications.

REDLICH, O., Thermodynamics: Fundamental, Applications. Elsevier Scientific Publishing Company, 1976.
Designed to present the essential tools for independent application of thermodynamics, not only to the chemist and chemical engineer, but also to the physicist, to the electrical and mechanical engineer and to the biochemist.

REYNOLDS, W. C., Thermodynamics, McGraw-Hill Kogakosha Limited, second edition, 1968.
Provides a fundamental first course in thermodynamics for engineers.

REYNOLDS, W. C. and PERKINS, H. C., Engineering Thermodynamics, McGraw Hill Inc., second edition, 1977.
Developed for a fundamental first course in thermodynamics for engineers.

ROGERS, G. F. C. and MAYHEW, Y. R., Engineering Thermodynamics Heat and Work Transfer, Longman Group Limited, third edition, 1980.
Intended for engineering students and covers the fundamentals of applied thermodynamics up to honours degree standard.

SCHMIDT, F. W., HENDERSON, R. E. and WOLGEMUTH, C. H., Introduction to Thermal Sciences, Thermodynamics, Fluid Mechanics and Heat Transfer, John Wiley and Sons Incorporated, 1984.
Written to introduce engineering undergraduates, who are not majoring in mechanical engineering, to the thermal sciences.

SILVER, H. F. and NYDAHL, J. E., An Introduction to Engineering Thermodynamics, West, 1980.
Presents the basic concepts of thermodynamics to undergraduate engineering students in all fields.

SONNTAG, R. E. and VAN WYLEN, G. J., Introduction to Thermodynamics: Classical and Statistical, John Wiley and Sons Incorporated, 1982.
Presents a comprehensive and rigorous treatment of thermodynamics while retaining an engineering perspective.

SPALDING, D. B. and COLE, E. H., Engineering Thermodynamics: in SI Units, Edward Arnold Publishers Limited, 1973.

Designed to present the fundamentals of classical thermodynamics to students of all branches of engineering.

SUSSMAN, M. V., Elementary General Thermodynamics, Addison-Wesley Publishing Company, 1972.
Presents a broad introduction to thermodynamics thought and methodology and applications to many branches of engineering and science. Intended for a variety of undergraduate students.

TODD, J. P. and ELLIS, H. B., An Introduction to Thermodynamics for Engineering Technologists, John Wiley and Sons, 1981.
Written to present clearly to the engineering technology student the basic principles and equations of thermodynamics relevant to current industrial and scientific applications and to provide practising technologists and engineers with a convenient and useful reference book.

TYLDESLEY, J. R., An Introduction to Applied Thermodynamics and Energy Conversion, Longman Group Limited, 1977.
Applies basic thermodynamics in conjunction with basic economics to practical systems. A one-year course for undergraduates from a number of different fields of study.

TYLER, F., Heat and Thermodynamics: SI Units, Edward Arnold (Publishers) Limited, 1973.
An aid to students coping with 'A' Level courses in these aspects of Physics.

WARK, K., Thermodynamics, McGraw Hill Book Company, 1983.
Introductory textbook designed for undergraduate students in the field of engineering.

WILKS, K., The Third Law of Thermodynamics, Oxford University Press, 1961.
Aims to cover all significant aspects of the third law in a manner intelligible to an Honours undergraduate.

WILL, R. K., Thermodynamics for Engineering Technologists, Marcel Dekker Ltd., 1979.
Presents the basic thermodynamic principles followed at the earliest possible time by applications to real devices. For students from many disciplines.

WOOD, B. D., Applications of Thermodynamics, Addison-Wesley Publishing Company, 1982.
Builds on the foundation of a prerequisite first course in the fundamentals of thermodynamics in order to explain a large number of engineering systems.

ZEMANSKY, M. W., ABBOTT, M. M. and VAN NESS, H. Z., Basic Engineering Thermodynamics, McGraw Hill Kogakuska Limited, second edition, 1975.
An initial course in thermodynamics for engineers of any discipline.

ZEMANSKY, M. W. and DITTMAN, R. H., Heat and Thermodynamics, McGraw-Hill Book Company, sixth edition, 1981.
Provides a fundamental foundation of thermodynamics which is well within the abilities of undergraduate students.

CONCLUSIONS

Very few authors have expressed the intention of writing their book for students to learn from. Similarly, very few have given the prerequisites for the use of their book. While many of the books (and this is increasingly the case) are concerned with applying fundamental thermodynamics to up-to-date engineering applications, there are only isolated examples where economics are used. If we are to implement in thermodynamics the recommendations of the Committee of Inquiry into the Engineering Profession (HMSO, 1980) concerned with costs, manpower and manufacture, then there is a need for the existing books to be amended or for others to be written.

There are many styles and strategies for approaching thermodynamics, just as there are styles and strategies for learning by students (Entwistle, 1981). Each individual will have his own approach and, if each student is to be able to discover the approach which is best for him (or her), there is a need for the essence of each approach to be prepared.

ACKNOWLEDGEMENTS

The authors would like to express their thanks to Mrs. J. Dawson, Librarian, Coventry (Lanchester) Polytechnic, and Mr. J. Corlett, Librarian, Trent Polytechnic, for their assistance in compiling the list.

REFERENCES

1. H.M.S.O., "Engineering our Future: Report of the Committee of Inquiry into the Engineering Profession", H.M.S.O., London, 1980.

2. Entwistle, N., "Styles of Learning and Teaching", J. Wiley, 1981.

DISCUSSION OF PAPER

To complement the mechanical engineering books, Professor Turner proposed to list the following.

Thermodynamics books which I have found helpful:

1. Denbigh, K. G., "The Principles of Chemical Equilibrium", Cambridge University Press, 4th edition, 1981.

2. Kyle, B. G., "Chemical and Process Thermodynamics", Prentice-Hall, Inc., 1984.

3. Pippard, A. B., "Classical Thermodynamics, Cambridge University Press, 1957 and reprinted in paperback 1964.

4. Prausnitz, J. M., "Molecular Thermodynamics of Fluid-Phase Equilibria", Prentice-Hall Inc., 1969.

5. Rowlinson, J. S., "The Perfect Gas", Pergamon, 1963.

6. Rowlinson, J. S. and Swinton, F. L., "Liquids and Liquid Mixtures", 3rd edition, Butterworths, 1982.

Others referred to the availability of classic treatise in the series of Dover Reprints, particularly perhaps Carnot's monograph.

A historical background, for <u>teachers</u>, is given in the two treatise:

7. Cardwell, D. S. L., "From Watt to Clausius. The Role of Thermodynamics in the Early Industrial Age", Heinemann, 1971.

8. Brush, S. G., "The Kind of Motion we call Heat." A History of the Kinetic Theory of Gases in the 14th Century. Vols. I and II, North-Holland, 1976.

STEERING COMMITTEE

D.C.Anderson	(U Lancaster)
F.J.Bayley	(U Sussex)
G.Boxer	(U Aston)
P.Brazier	(Imperial C)
B.L.Button	(Coventry P)
R.I.Crane	(Imperial C)
M.J.French	(U Lancaster)
B.N.Furber	(NNC)
N.Hay	(U Nottingham)
T.Hinton	(U Surrey)
G.S.Holister	(O U)
R.G.Huggett	(I Mech E)
J.D.Lewins	(U Cambridge)
G.D.S.MacLellan	(U Leicester)
Y.R.Mayhew	(U Bristol)
K.W.Ramsden	(Cranfield I)
V.Snairey	(I Nuc E)
V.Walker	(U Bradford)
D.Walton	(U Manchester)
B.R.Wakeford	(U Heriot-Watt)
H.K.Zienkiewicz	(U Exeter)

THERMODYNAMIC WORKSHOP - RECORD OF ATTENDANCE

Alder	Dr E M	U Edinburgh	Dept Mech Engr	King's Building	Edinburgh EH9 3JL
Anderson	Mr D C	U Lancaster	Dept Engr	U Lancaster Bailrigg	Lancaster
Andrew	Dr S P S	ICI	ICI Agricultural Div	PO Box 1 Billingham	Cleveland TS23 1LB
Andrews	Prof G E	U Leeds	Dept Fuel & Energy	Univiversity of Leeds	Leeds L32 9JT
Armson	Miss R	Open University	Open University	Walton Hall	Milton Keynes
Bacon	Mr D H	Plymouth Poly	Dept Mech Engr	Drake Circus	Plymouth PL4 8AA
Bajhtar	Dr	U Birmingham	U Birmingham		Birmingham
Bowen	Mr R J	U Southampton	Dept Mech Engr	U Southampton	Highfield Southampton
Brazier	Mr P H	Imperial College	Mech Engr Dept	Exhibition Road	London SW7 2BX
Buckingham	Dr D J	U Exeter	Dept Engr Science	North Park Road	Exeter EX4 4QF
Burnside	Dr B M	Heriot-Watt U	Heriot-Watt University	Riccarton	Edinburgh EH14 4AS
Burnside	Mrs K	Guest			
Button	Prof B L	Trent Polytechnic	Trent Polytechnic	Nottingham	
Button	Mrs	Guest			
Campbell	Dr A M	U Cambridge	Engr Laboratories	Trumpington St	Cambridge CB2 1PZ
Collings	Dr N	U Cambridge	Engr Laboratories	Trumpington Street	Cambridge CB2 1PZ
Cottrell	Sir Alan	Jesus College	The Master	Jesus College	Cambridge
Cowell	Dr T A	Brighton Poly	Brighton Polytechnic	Moulsecoomb	Brighton
Cowley	Dr M D	U Cambridge	Engr Laboratories	Trumpington St	Cambridge CB2 1PZ
Crane	Dr R I	Imperial College	Dept Mech Engr	Exhibition Road	London SW7 2BX
Dampier	Dr M D	U Leicester	Dept Mathematics	U Leicester	Leicester LE1 7RH
Daneshyar	Dr H	U Cambridge	Engr Laboratories	Trumpington St	Cambridge CB2 1PZ
Dobbins	Mr B N	Coventry Poly	Dept Mech Engr	Cox St	Coventry CV1 5FB
Dobbins	Mrs H M	Guest			
Exell	Prof R H B	Asian Inst Tech	PO Box 2754	Bangkok 10501	Thailand
Foss	Dr P W	Sheffield City P	Dept Mech/Prod Engr	Pond St	Sheffield
French	Prof M J	U Lancaster	Dept Engr U Lncaster	Bailrigg	Lancaster LA1 5B
Frost	Dr T H	U Newcastle-U-Tyne	Dept Mech Engr	Stephenson Building	U Newcastle-U-Tyne
Frost	Mrs S	Guest			
Furber	Dr B N	N N C	Nat Nucl Corp	Warrington Rd Risley	Cheshire WA3 6BZ
Goodger	Dr E M	Cranfield	School of Mech Engr	Cranfield Inst Tech	Bedford MK43 OAL
Gregory	Mr R D	Hatfield Poly	Hatfield Polytechnic	Hatfield	Herts
Gurney	Prof C	U Hong Kong	Brooklyn Bow Rd	Harpertonporel	Devon
Hardisty	Dr H	U Bath	School of Engr	Claverton Down	Bath BA2 7AY
Hay	Dr N	U Nottingham	Dept Mech Engr	University Nottingham	Nottingham NG7 2RD
Hay	Mrs Y	Guest			
Haynes	Mr T R	U Adelaide	Audio Visual Dept	U Cambridge	Cambridge

Haywood	Mr R W	U Cambridge	Flat F Craig-y-Don	Belle Vue Road	Swanage BH19 2HP
Heikal	Dr M	Brighton Poly	Brighton Polytechnic	Moulsecoomb	Brighton
Himsworth	Dr J R	UMIST	UMIST	Sackville St	Manchester M60 1QD
Hinton	Dr T	U Surrey	University Surrey	Guildford	Surrey GU2 5XH
Jervis	Mr J M	N Staffs Poly	N Staffs Polytechnic	Beaconside	Stafford
Johnston	Mr J A	U Cambridge	Engr Laboratories	Trumpington St	Cambridge CB2 1PZ
Jones	Dr D R H	U Cambridge	Engr Laboratories	Trumpington St	Cambridge CB2 1PZ
Kennedy	Dr W	Open University	Open University	Walton Hall	Milton Keynes
Kotas	Dr T J	Queen Mary C London	Dept Mech Engr	Mile End Road	London E1 4NS
Landsberg	Prof P T	U Southampton	Univ Southampton	Southampton	SO9 5NH
Lewins	Dr J D	U Cambridge	Engr Laboratories	Trumpington St	Cambridge CB2 1PZ
Lewins	Mrs S	Guest			
Linnhoff	Professor	UMIST	Dept Chem Engr	UMIST PO BOx 88	Manchester M60 1QD
MacLellan	Prof G D S	U Leicester	Dept Engr	The University	Leicester LE1 7RH
Mayhew	Dr Y R	U Bristol	Dept Mech Engr	Queen's Building	Bristol BS8 1TR
Mayhew	Mrs Y	Guest			
Mecklenburgh	Dr J C	U Nottingham	Dept Chem Engr	University Park	Nottingham NG7 2RD
Mukundan	Dr P N S	Newcastle U Tyne P	Ellison Building	Ellison Place	Newcastle U T NE1 8ST
Munton	Dr R J	Trent Polytechnic	Dept Mech Engr	Burton St	Nottingham NG1 4BU
Muskett	Mr W J	Liverpool P	Liverpool Polytechnic		Liverpool
Organ	Dr A J	U Cambridge	Engr Laboratories	Trumpington St	Cambridge CB2 1PZ
Packer	Dr J P	UMIST	UMIST	Sackville St PO Box 88	Manchester M60 1QD
Packer	Mrs R T F	Guest			
Patterson	Prof J	Huddersfield P	Huddersfield Polytech		Huddersfield
Pilkington	Mr D W	Manchester Poly	Dept Mech Engr	Chestol St	Manchester M1 5GD
Pippard	Sir Brian	U Cambridge	30 Porson Road	Cambridge	CB2 2EU
Ramsden	Mr K W	Cranfield	School Mech Engr	Cranfield Inst Tech	Cranfield MK43 0AL
Rao	Dr H V	Huddersfield Poly	Dept Mech Engr	Polytechnic Queensgate	Huddersfield HD1 3DH
Rose	Dr J W	Queen Mary C London	Queen Mary College	Mile End Road	London E1 4NS
Saunders	Mr R J	U Sheffield	Mech Engr Dept	Mappin St	Sheffield S1 3JD
Sharma	Prof C S	Birkbeck C London	Birkbeck College	Malet St	London WC1 7HX
Silver	Prof R S	U Glasgow	Oakbank	Tobermory	Isle of Mull
Slowley	Mr J	U Bath	School of Engr	Claverton Down	Bath BA2 7AY
SO'Brien	Dr S	U Trinidad	Port of Spain	Trinidad	West Indies
SO'Brien	Mr D	Guest			
Stone	Dr C R	Brunel U	Brunel University	Uxbridge	Middx

(continued)

THERMODYNAMIC WORKSHOP -RECORD OF ATTENDANCE (CONTINUED)

Szymanski	Mr R W	Middlesex P	Sch Mech Engr	Bounds Green Rd	London N11 2NQ
Tinson	Mr P A	ESPE Queen Mary C	ESPE Dept Mech Engr	Queen Mary College	Mile End Rd E1 4NS
Tucker	Mr P	Preston Polytechnic	Preston Polytechnic	Corporation St	Preston Lancs PR1 2TQ
Turner	Prof J C	RU Exeter	Dept Chem Engr	North Park Road	Exter EX4 4QF
Wakeford	Mr B R	Heriot-Watt U	Dept Mech Engr	Riccarton Campus	Edinburgh EH14 4AS
Wakeford	Mrs B R	Guest			
Walton	Mr D G	U Manchester	Simon Engr Labs	Oxford Rd	Manchester M13 4PL
Woods	Prof W A	Queen Mary C London	Dept Mech Engr	Mile End Road	London E1 4NS
Wright	Mr M C J	BBC TV	O U Production Centre	Walton Hall	Milton Keynes MK7 6BH
Young	Dr J B	U Cambridge	Engr Laboratories	Trumpington St	Cambridge CB2 1PZ
Zienkiewicz	Dr H K	U Exeter	Dept Engr Science	North Park Road	Exeter EX4 4QF

INDEX